APPLIED MATRIX
ALGEBRA

Lawrence Harvill

Emeritus Professor of Engineering and Applied Mathematics
and
Director, John Stauffer Center for Science, Mathematics and
Environmental Studies, retired
University of Redlands
Redlands, CA 92374

To order additional copies of this book, contact:
Xlibris Corporation
1-888-795-4274
www.Xlibris.com
Orders@Xlibris.com
98048

DEDICATION

This book is dedicated to two individuals who had a great impact and significance in my career and life. The first is Dr. Louis A. Pipes who was my professor in many applied mathematics courses at UCLA and my PhD thesis advisor. Louis was an amazing lecturer who could present complex material so that it was clear and understandable. He posessed an amazing sense of humor and a photographic memory from which he could cite chapter and verse of source material from obscure references.

The second individual was my, Dr. Robert Engel, best friend, mentor, and colleague for thirty-five years. We met as undergraduate engineering students at UCLA, continued through graduate school to our PhD's, and became colleagues as professors at the University of Redlands in Redlands, California. He was a brilliant man of endless patience with a delightful humor and, above all, an outstanding teacher. Bob taught me the unforgettable lesson that, "life is too short to waste doing things one does not enjoy!"

LH, Redlands, CA 2011

CONTENTS

PREFACE

This text has been written to satisfy the needs of a one-semester introductory course in linear or matrix algebra. Theoretical aspects of linear algebra have been minimized, and particular emphasis has been placed on applications to aid in stimulating the reader's interest. Every effort has been made to provide a range of applications to appeal to students of diverse interests. The main audience is assumed to be students from the physical and biological sciences, mathematics, and engineering. It is hoped that students from the social and behavioral sciences including economics and business will also find the text of use as several examples and problems from those areas have been included as well.

The first two chapters are fairly traditional comprising the topics of "Fundamentals of Matrix Algebra" and "Linear Algebraic Equations and Vectors." Chapter III, "Application of Linear Algebraic Equations," provides a broad introduction to a number of practical areas. These are least squares, steady state linear electrical circuits, input-output economic analysis, the solution of potential problems, difference equations (including biological growth, random walk, and gambling problems), and the analysis of plane trusses. The fourth chapter is "Determinants and the Inverse Matrix."

Chapter V provides a thorough introduction to the eigenvalue problem, which includes a geometric as well as an algebraic presentation of the concept of eigenvalues and eigenvectors. In addition to covering the characteristic equation and orthogonal transformations, this chapter presents quadratic forms and circulant matrices. Chapter VI is devoted to the aspects of the calculus of matrices and matrix functions, which includes the Cayley-Hamilton and Sylvester's theorems. Special attention is paid to similarity transformations as a means for the evaluation matrix functions as well as the situation of multiple eigenvalues and the Jordan canonical form. The focus of chapter VII is on the topic of "Linear Difference and Differential Equations." Stability of solutions to finite difference equations is introduced along with eigenvector expansions and the adjoint method for nonhomogeneous systems. Applications of finite difference and differential equations are the subject of chapter VIII with topics including four terminal networks, electrical circuits, translational and rotational mechanical systems, and flow or process problems.

The final chapter of the text is devoted to a presentation of the prominent aspects of "Numerical Methods" as related to matrix computations. Topics included are Gaussian elimination, LU factorization, Cholesky's method, Jacobi iteration, Gauss-Seidel iteration, ill-conditioned problems, Householder transformations, and QR factorization as well as the shifted inverse power method among.

The evolution of computers and the rise of the personal computer over the past several decades have had a marked impact on matrix algebra as a tool for solving applied problems. Sophisticated, efficient algorithms for matrix manipulations have been developed by extensive, painstaking efforts and are available for use at the click of a mouse. The upshot of these events is that the content and approach of introductory courses are changing to reflect these impacts. Many teachers are beginning to agree that a first course should no longer be a theoretical linear algebra course with mostly pencil-and-paper assignments and some programming of algorithms. With powerful algorithms readily available through a number of matrix computational programs, first courses should have students working with them hands on as soon as possible to solve a variety of extensive problems.

This text has been developed in response to these changes with the assumption that detailed theory will be left to follow-on courses. Every effort should be made to encourage that students use one of several matrix computational programs currently available as early as possible in the course. In particular, this text has not been written with any particular software program in mind. The focus is on matrix algebra and its applications rather than being an instructional manual for a particular software package. Such packages are to be utilized as tools for solving matrix problems. It is my desire that students will have to use such programs for at least part of every assignment. Many of the exercises in the text are by design of such size and complexity that they can only be solved in a reasonable amount of time through the use of such programs. These exercises as well as those for which solution by means of a matrix computational program would be advantageous have been designated with an asterisk (*).

As indicated, there are no direct references to the various software packages in the text. However, as illustration of their use, several of the example problems have been solved with EXCEL, MATLAB, and Mathematica to name a few. There also exist today a number of web-based sites providing access to a wide variety of standard solution methodologies.

Another aspect of using a calculator or computer, which students must become familiar with, is that of computational accuracy and other sources of error in solving problems. This topic is discussed in the introduction and illustrated at various other points in the text to serve as a reminder to students of its importance. Another item to be mentioned is that the author does not believe that students should ever be given assignments to write computer programs for any matrix computational algorithms. My experience has been that these almost always become exercises in programming technique and style rather than the particular mathematical aspect being presented. Finally, solutions to the exercises is available from the author as a PDF file.

In preparing this text, I want to extend my appreciation and thanks to my institution, the University of Redlands, whose support and encouragement throughout my thirty-five years of tenure there provided an invaluable experience. My appreciation also goes to my former colleagues at the University of Redlands, Drs. Dave Bragg and Don Winter, for their comments and suggestions. Special recognition goes to my former colleague Dr. Janet Beery, who tirelessly checked the solutions for all the exercise problems and provided invaluable suggestions as to content and approach. I particularly would also like to recognize the students who were enrolled in my linear algebra course during the spring of 1991 when a draft of this text was used on a trial basis. They all helped greatly in spotting various errors, omissions, and confusing language. Special thanks go to Mr. Darren Rose and Dr. Keith Schubert (now professor of computer science at CSUSB) of that class who undertook the special project of a detailed review from a student's perspective. They provided several important suggestions for improvements and clarifications. Finally, special thanks go to my wife, Evelyn Ifft, for her continuous support, encouragement, and patience.

INTRODUCTION

The term *matrix* in mathematics seems to have been used for the first time by the mathematician James Sylvester. In the year 1850, Sylvester applied the name "matrix" to a rectangular array of numbers "out of which determinants can be formed." The modern concept of a matrix as a hypercomplex number is due to William Rowland Hamilton. In his paper, various properties of linear transformations use arrays of quantities without giving these arrays an explicit name. However, it was not until 1857 that the eminent English mathematician Cayley in his celebrated paper[1] "A Memoir on the Theory of Matrices," presented the modern basis of the theory of matrices. It is of historical interest to note that during this same period of time, Charles Babbage was developing the first mechanical computer based on the concept of a stored program. Additionally, the Countess of Lovelace (Ada) was writing the first numerical algorithms (programs) for Babbage's machine.

Matrices, as shown by Cayley, provided a compact and flexible notation particularly useful in studying linear transformation and presented an organized method for the solution of systems of linear algebraic and linear differential equations. For a long time, matrices and matrix algebra (also known as linear algebra) were subjects studied almost exclusively by pure mathematicians, and most physicists and engineers were unaware of the existence of matrix algebra. It was not until 1925 that the physicists Heisenberg, Born, and Jordan discovered the utility of matrices in the study of problems in quantum mechanics. The subject of matrices and matrix algebra was introduced to the engineering profession by two English aeronautical engineers, W. J. Duncan and A. R. Collar. In 1934, Duncan and Collar published the paper[2] "A Method for the Solution of Oscillation Problems by Matrices." This paper, which is of fundamental importance in the theory of vibrations, greatly stimulated the interest in the use of matrices by engineers and physicists.

Since Cayley's original memoir, a vast literature dealing with the applications of matrices to many branches of mathematics, physics, engineering, economics, and the social sciences has developed. This literature reflects the fact that matrix algebra has been found to be extremely useful in the solution of many problems that arise in these and other applied fields.

The development of modern digital computing machines and in particular the small desktop personal computer has had a great impact on the evolution of matrix algebra from essentially a convenient, useful notation to a powerful

---

[1]   *Philosophical Transactions*, vol. 148, London, pp. 17–37, 1857.
[2]   *Philosophical Magazine*, series 7, vol. 17, p. 865, 1934.

computational tool. Complex problems that could only be discussed in general terms a few years ago may now be analyzed in extremely fine detail. For example, by the techniques of modern numerical analysis, partial differential equations are transformed into linear algebraic equations that can be studied by the use of matrices. This approach is currently embodied in the topic of finite element analysis, which is a general tool based on matrix analysis and may be used to solve a great variety of engineering problems such as laminar and turbulent fluid flow, heat transfer, and stress analysis of parts with complex geometries. Other areas of application today include linear programming, graph theory, game theory, Markov chain problems, economic and biological modeling, to name just a few.

Spurred by the development of small powerful computing machines has been the development of a number of sophisticated mathematical computational programs. Some of these software programs are devoted expressly to manipulating matrices, and others are applicable to a wider range of mathematical topics. Further yet, in recent years, there has been the development of a number of "online" or web-based linear algebra programs that provide a more available source of access to a wide variety of solving problems without the need for a stand-alone application package. Some include symbolic processors that provide functional or algebraic solutions in addition to standard number-crunching power. These systems allow users to focus attention on the specific aspects of a problem without becoming bogged down in cumbersome algebraic computations. They also permit one to deal with very large and complex problems with relative ease.

It is the purpose of this book to develop an understanding of the fundamentals of matrix algebra as well as the differential and integral calculus of matrices that are fundamental for the analysis of a wide range of applied problems. When used in conjunction with a matrix computational program, those readers who make their way through the end should be in a position to readily analyze sophisticated and complex applied problems. Completion of the text should also prepare the reader for moving on to much more theoretical and advanced topics in linear algebra. Such additional study will enable one to more fully understand not only the mathematical complexities of the subject but to gain a greater insight also to the intricate details of the computational algorithms.

A Comment on Numbers and Accuracy

In using calculators or computers, it must always be remembered that in almost all cases, the numerical values used and computed have some degree of uncertainty or inaccuracy. These errors arise from three general sources: the input values themselves, their representation within the computer and round off, and other computational errors.

At the very beginning of the analysis of a problem, the values used are frequently determined from experimental measurements. Instrumentation accuracy typically varies from ± 0.01 % to ± 2.0 %. In many situations, this is the largest source of error.

Two other sources of error result from the fact that computers work in the binary or base two system, and storage media retain only a finite number of significant digits. These two aspects combine to generate some additional inaccuracies. For example, the fraction 1/2 would be accurately stored as 0.500 . . . to as many significant digits as the machine can handle. However, the fraction 2/3 would be stored as 0.666 . . . The last digit is truncated at the maximum capacity of a computer's memory for a specific number. It is not rounded because computers have no way of "knowing" what the next digit is. If this internal value is displayed on a screen or printed out, one might see the value 0.6667 because the output software usually rounds the internal value whenever fewer digits are displayed than stored. This is generally the case as most modern calculators and computers work on the basis of a large number of significant digits, and results are rarely requested at the maximum level.

In addition to truncation error as a source of inaccuracy, in some situations, binary representation of digital values also produces uncertainties. This is illustrated by the example of representing the fraction 1/3 in binary form. Binary fractions are powers of 1/2 or 1/2, 1/4, 1/8, 1/16, etc. The fraction 1/3 can only be represented by an infinite series of binary fractions, and with only a fixed number of places available, the result produces another source of error.

Another aspect to be aware of is that numerical values may be stored in either integer or floating-point form. Integer form means that the numbers dealt with are whole or integer values. If integer arithmetic is performed on such numbers, the result is always an integer. This is a situation where if you divide one by three, you get an answer of zero! Floating-point arithmetic is where numerical values are stored in a scientific notation form such as $\pm 0.ddddd \, E \pm xx$. The d's represent the significant digits and the x's the power of ten, which is indicated by the E. This type of arithmetic is used in almost all calculations, with some exceptions such as accounting where amounts must be accurate to the last penny. In this case, you would use integer arithmetic and the amounts in cents.

Floating-point arithmetic allows you to maintain a constant precision (number of digits of accuracy) but generates other computational problems. If, for example, you were working with a machine that maintained four-digit accuracy, adding 7.263 to 0.0009926 completely loses the second number. As you can see, adding numbers of greatly different magnitudes is a source of error. A somewhat more sinister situation occurs when subtracting numbers of nearly equal magnitude. Given our four-digit machine, the result of subtracting $0.7252 \, E \, 04$

from 0.7265 *E* 04 is 0.0013 *E* 04 or 0.1300 *E* 02. We have moved from four-digit accuracy to two in a single operation.

The many matrix computational programs available utilize almost as many different programming languages and numerous variations in algorithms. Because of this, the algorithms themselves can introduce errors for a variety of numerical reasons. In any situation, a user must be sure to carefully note the particular aspects of the system and programs they are using.

When Zero Is Not Really Zero

When using any computer one must be aware of the presence of round-off errors as discussed above. This probably occurs most graphically when results that should be zero are computed and displayed. Depending on the system being used, one might see results like 0 or 0.0000 or −0.0000. In the floating-point arithmetic, the appearance of 0 means true zero. In integer arithmetic, it might not. The 0.0000 is clearly floating-point, but it means only that the value is zero to four places. This is emphasized when one sees −0.0000 for which the internal value could be something like −0.0000437 . . .

A Few Final Suggestions

In matrix computations, well-designed algorithms are developed to minimize these sources of error. Unfortunately, they cannot be eliminated due to the nature of computers. In addition, there are some applied problems for which even the best algorithms fail because the exact solution is extremely sensitive to very small variations in values of various coefficients. Because of all this, one must never accept any generated solution on blind faith. First, answers should always be checked against approximate answers that can be produced out of one's knowledge of the problem and some reasonable simplifications in the analysis. Second, answers should always be checked by substitution into the original equations to be sure they satisfy them up to an expected level of accuracy. At various points in the following chapters, a few lines will be taken to point out the presence of errors or show their existence and magnitude by testing computed answers against the stated problem.

I

FUNDAMENTALS OF MATRIX ALGEBRA

1. INTRODUCTION

Matrices and matrix notation enable one to write equations encountered in a wide variety of applied fields in more compact and efficient forms. The use of matrices introduces a well-designed notation that clarifies the underlying essence of the mathematics and the relationships being represented. Matrix notation expresses linear transformations in a concise and lucid manner. It is only natural that they are employed in the formulation and the solution of linear problems.

2. DEFINITION OF A MATRIX

The most common use of matrix notation is to represent systems of linear algebraic equations. For illustration purposes, consider the following system of three equations in the three unknowns x, y, and z.

$$4x \quad -3y \quad +3z = 0$$
$$-2x \quad -5y \quad +2z = 1$$
$$3x \quad +2y \quad +6z = 4$$

Focusing our attention on the coefficients of the unknowns, we see a square array of numbers.

$$
\begin{array}{rrr}
4 & -3 & 3 \\
-2 & -5 & 2 \\
3 & 2 & 6
\end{array}
$$

This array of numbers is the coefficient matrix for the given system of equations. The matrix notation for this array is

$$
A = \begin{bmatrix}
4 & -3 & 3 \\
-2 & -5 & 2 \\
3 & 2 & 6
\end{bmatrix}.
$$

Brackets delineate the array, and the capital letter A is a convenient symbol or name representing the matrix. Additionally, we note that this matrix contains three separate rows and three separate columns of values.

A matrix can have different numbers of rows and columns. In general, a matrix could have m rows and n columns.

DEFINITION

Matrix

A matrix is a rectangular array of elements arranged in m rows and n columns.

A matrix is represented in the following general manner:

(2.1)
$$A = \begin{bmatrix} a_{11} & a_{12} & a_{13} & \cdots & a_{1n} \\ a_{21} & a_{22} & a_{23} & \cdots & a_{2n} \\ \cdot & \cdot & \cdot & \cdots & \cdot \\ a_{m1} & a_{m2} & a_{m3} & \cdots & a_{mn} \end{bmatrix}$$

The elements a_{ij} of the matrix A may be real or complex numbers, general functions, or even other matrices. A matrix, such as the one given above, is of order m by n in that it has m rows and n columns. The notation for the order is $m \times n$. From equation (2.1), the element a_{23} lies in the second row and the third column. In referring to a general element, the term a_{ij} is used. Since the matrix of (2.1) has m rows and n columns, we note that the subscript i ranges over the values from 1 to m, and the subscript j ranges over the values from 1 to n.

EXAMPLE 1

Identify the order of the following matrix and the value of the element a_{12}.

(2.2)
$$A = \begin{bmatrix} 1 & -2 & 0 \\ -3 & 2 & 4 \end{bmatrix}$$

SOLUTION

Since there are two rows and three columns, this matrix is of order 2×3. For this matrix, the value of the element a_{12} is -2.

***

In the introductory illustration of three linear algebraic equations in three unknowns, the matrix of coefficients has three rows and columns. Its order is 3 × 3. In such situations, the matrix is referred to as square and its order is 3. Square matrices occur quite commonly in many applications.

DEFINITION

Square Matrix

When the number of rows equals the number of columns, then the matrix is a square matrix of order $m \times m$ or a matrix of the mth order.

EXAMPLE 2

Another example of a third order square matrix is the following.

(2.3)
$$M = \begin{bmatrix} 1 & -3 & 6 \\ 0 & -2 & 4 \\ 5 & 2 & 3 \end{bmatrix}$$

***

Besides referring to the rows and columns of matrices, it is also useful to refer to elements lying along diagonal lines. In particular, the elements 1, −2, 3 of the matrix M above from the upper left to lower right corners comprise the main diagonal of the matrix. For the opening example system of three linear algebraic equations and three unknowns, the coefficient matrix is

$$A = \begin{bmatrix} 4 & -3 & 3 \\ -2 & -5 & 2 \\ 3 & 2 & 6 \end{bmatrix}.$$

For this matrix, the main diagonal elements are 4, −5, 6.

DEFINITION

Main Diagonal

For any matrix, the set of elements a_{11}, a_{22}, a_{33}, ..., a_{mm} is the principal or main diagonal.

Even if the matrix is not square, the diagonal starting at the upper left corner is the principal or main diagonal.

EXAMPLE 3

For the matrix given in (2.2) above, identify the elements on the main diagonal.

SOLUTION

The matrix given in (2.2) above is

$$A = \begin{bmatrix} 1 & -2 & 0 \\ -3 & 2 & 4 \end{bmatrix}.$$

The main diagonal elements are 1 and 2.

***

We now introduce another common operation that is useful in a variety of basic operations. Consider the following matrix:

$$B = \begin{bmatrix} 1 & -3 \\ -2 & 2 \\ 0 & 4 \end{bmatrix}$$

Comparing this with the matrix A of (2.2) above, we see that the first column of B is the first row of A and that the second column of B is the second row of A. The matrix B is the transpose of A and noted as $B = A^T$.

DEFINITION

The Transpose of a Matrix

The transpose A^T of a matrix A is the matrix whose columns are the rows of A taken in order.

Suppose the elements of a matrix A are a_{ij}. That is, $A = [a_{ij}]$. If a new matrix is formed such that the rows of the new matrix are the respective columns of the original, the new matrix is defined as the transpose of the original matrix. The ith row of A is identical with the ith column of A^T.

EXAMPLE 4

For the matrix M given in (2.3) above, find its transpose.

SOLUTION

In (2.3) above, the M matrix is

$$M = \begin{bmatrix} 1 & -3 & 2 \\ 0 & -2 & 4 \\ -1 & 4 & 3 \end{bmatrix}.$$

The transpose is formed by taking the first row of M and making it the first column of M^T. The second row of M becomes the second column of M^T, etc.

$$M^T = \begin{bmatrix} 1 & 0 & -1 \\ -3 & -2 & 1 \\ 2 & 4 & 3 \end{bmatrix}$$

***

The process of transposition may be graphically visualized as a reflection about the main diagonal.

3. PRINCIPAL TYPES OF MATRICES

Let us now return to the introductory illustration of a system of three linear algebraic equations and three unknowns, namely

$$
\begin{aligned}
4x &\quad -3y &\quad +3z &= 0 \\
-2x &\quad -5y &\quad +2z &= 1 \\
3x &\quad +2y &\quad +6z &= 4
\end{aligned}
$$

For this system, we have defined the coefficient matrix as A, where

$$
A = \begin{bmatrix} 4 & -3 & 3 \\ -2 & -5 & 2 \\ 3 & 2 & 6 \end{bmatrix}.
$$

Looking at the right-hand sides of the equations, we see that the numbers there form a matrix with a single column. This matrix would be

$$
B = \begin{bmatrix} 0 \\ 1 \\ 4 \end{bmatrix}.
$$

This is a matrix of order 3×1. This type of matrix occurs with great frequency, and it is called a column matrix or column vector.

DEFINITION

Column Vector

A set of m elements arranged in a column is a matrix of order $m \times 1$ and is called an mth order column matrix or an mth order column vector.

Equation (3.1) below illustrates the notation to be used in this text unless stated otherwise.

$$(3.1) \qquad X = \begin{bmatrix} x_{11} \\ x_{21} \\ \vdots \\ x_{m1} \end{bmatrix} = \begin{bmatrix} x_1 \\ x_2 \\ \vdots \\ x_m \end{bmatrix}$$

As the column subscript $_1$ is redundant, it is dropped, and the second form shown above is the standard notation. When referring to column vectors, the following notation is also convenient: $[x_1 \, x_2 \ldots x_m]^T$.

In a similar manner, a matrix with a single row of elements is defined as a row matrix or row vector. For example, the elements of the first row of the above equations, 4, −3, 3, could be considered as the following row vector.

$$R = \begin{bmatrix} 4 & -3 & 3 \end{bmatrix}$$

DEFINITION

Row Vector

A set of n elements arranged in a row is a matrix of order $1 \times n$. Such a matrix is an nth-order row matrix or an nth-order row vector .

For convenience, we usually will designate a row vector as the transpose of a column vector. For example, if Y is a row vector, it will be represented as

$$Y = \begin{bmatrix} y_1 & y_2 & \cdots & y_n \end{bmatrix} = \begin{bmatrix} y_1 \\ y_2 \\ \vdots \\ y_n \end{bmatrix}^T .$$

EXAMPLE 5

A specific example of a fourth order row vector is

$$Y = \begin{bmatrix} -2 & 1 & 0 & 3 \end{bmatrix} = \begin{bmatrix} -2 \\ 1 \\ 0 \\ 3 \end{bmatrix}^T .$$

***

It is sometimes useful to note that a matrix may be considered as an ordered collection of row or column vectors. For example, if we define the following row vectors

$$R_1 = \begin{bmatrix} 4 & -3 & 3 \end{bmatrix}$$
$$R_2 = \begin{bmatrix} -2 & -5 & 2 \end{bmatrix}$$
$$R_3 = \begin{bmatrix} 3 & 2 & 6 \end{bmatrix} ,$$

it can be seen that the coefficient matrix for our example system of linear algebraic equations may be expressed as

$$A = \begin{bmatrix} R_1 \\ R_2 \\ R_3 \end{bmatrix} = \begin{bmatrix} 4 & -3 & 3 \\ -2 & -5 & 2 \\ 3 & 2 & 6 \end{bmatrix} .$$

This example shows that the elements of a matrix can even be matrices. This method of grouping elements of a matrix as smaller or submatrices is called partitioning. This concept will be developed in more detail in section 9. Proceeding in the other direction, we can ask if there is such a thing as a matrix of order 1×1 or a matrix with only one row and one column or simply a single element? The answer is yes. Each single coefficient in our system of linear equations can be regarded as a matrix of order 1.

DEFINITION

Scalar

A scalar is a matrix of order 1 × 1.

A matrix of order 1 is a single element. A single number or quantity is called a scalar to distinguish it from an array of numbers or a matrix. In our notation, lowercase letters refer to scalar quantities such as the elements of a matrix.

Now let us consider a very simple system of three linear algebraic equations and three unknowns, namely

$$
\begin{aligned}
4x & & = -1 \\
-5y & & = 1 \\
6z & = 3
\end{aligned}
$$

The missing unknowns may be included in each equation by using a zero coefficient as follows.

$$
\begin{aligned}
4x \ +0y \ +0z &= -1 \\
0x \ -5y \ +0z &= 1 \\
0x \ +0y \ \ 6z &= 3
\end{aligned}
$$

In this form we see that the matrix of coefficients would be the following.

$$
C = \begin{bmatrix} 4 & 0 & 0 \\ 0 & -5 & 0 \\ 0 & 0 & 6 \end{bmatrix}
$$

This matrix is somewhat special in that all the elements except those on the main diagonal are zero. This is termed a diagonal matrix.

DEFINITION

Diagonal Matrix

A square matrix is a diagonal matrix if all the elements except those along the main or principal diagonal are identically zero. Some, but not all, of the main diagonal elements may be zero.

EXAMPLE 6

An example of a fourth-order diagonal matrix is

$$A = \begin{bmatrix} 2 & 0 & 0 & 0 \\ 0 & 0 & 0 & 0 \\ 0 & 0 & -3 & 0 \\ 0 & 0 & 0 & 1 \end{bmatrix}.$$

***

For convenience, the following notation is employed to designate diagonal matrices in this text.

$$A = \text{Diag}\,(2, 0, -3, 1)$$

In any of our later considerations, we will find use for a special type of diagonal matrix, namely one in which the diagonal elements are all equal to 1.

DEFINITION

The Unit or Identity Matrix

The identity matrix of the nth order is the nth-order diagonal matrix whose diagonal elements are all equal to one. An identity matrix of the nth order is denoted by I_n.

For example, the third-order identity matrix is

$$I_3 = \begin{bmatrix} 1 & 0 & 0 \\ 0 & 1 & 0 \\ 0 & 0 & 1 \end{bmatrix}.$$

Another matrix that is useful in dealing with matrix equations and the partitioning of large matrices is a matrix in which all the elements are zero. This is the so-called null or zero matrix.

DEFINITION

The Null or Zero Matrix

A matrix of order $m \times n$ that has all of its elements equal to zero is a zero or null matrix. As a special notation, we shall denote a square null matrix as 0_n and a null column as $0_{n,1}$.

In applied problems, one frequently encounters systems of linear algebraic equations with the following form.

$$
\begin{array}{rrrcr}
x & -2y & +3z & = & 2 \\
-2x & +5y & +4z & = & -1 \\
3x & +4y & +0z & = & 1
\end{array}
$$

In this case, the coefficient matrix is the following.

$$
C = \begin{bmatrix}
1 & -2 & 3 \\
-2 & 5 & 4 \\
3 & 4 & 0
\end{bmatrix}
$$

After careful inspection, the reader will note the following relation between the elements, namely $a_{12} = a_{21}$, $a_{13} = a_{31}$, and $a_{23} = a_{32}$. This situation is called symmetry, and the matrix is termed a symmetric matrix.

A square matrix for which the elements satisfy the relation $a_{ij} = a_{ji}$ for all i and j is called a symmetric matrix. In other words, for all i, the ith row is identical to the ith column.

DEFINITION

Symmetric Matrix

A symmetric matrix is a matrix that has the property $A^T = A$.

EXAMPLE 7

Which of the following matrices is symmetric?

$$A = \begin{bmatrix} 1 & -5 & 2 \\ 5 & 0 & 4 \\ -2 & -4 & 3 \end{bmatrix}, \quad B = \begin{bmatrix} 8 & 0 & -5 \\ 0 & -2 & 3 \\ -5 & 3 & 7 \end{bmatrix}, \quad C = \begin{bmatrix} -4 & 0 & 5 \\ 3 & -1 & 8 \\ 2 & 0 & 3 \end{bmatrix}$$

SOLUTION

Only matrix B is symmetric.

***

It is helpful for the reader to note that for symmetric matrices, the part below the main diagonal is a reflection of the part above the main diagonal.

Another type of matrix closely allied with symmetric matrices is a skew symmetric matrix. These matrices are defined as follows.

DEFINITION

Skew Symmetric Matrix

If a square matrix A has the property $a_{ij} = -a_{ji}$ or $A^T = -A$, then A is said to be skew symmetric.

By this definition, all diagonal elements of a skew symmetric matrix must be identically zero, that is, $a_{11} = a_{22} = \ldots = a_{nn} = 0$.

The following is an example of a skew symmetric matrix.

$$A = \begin{bmatrix} 0 & -2 & 3 \\ 2 & 0 & 4 \\ -3 & -4 & 0 \end{bmatrix} = -A^T$$

EXAMPLE 8

Which of the following matrices is skew symmetric?

$$A = \begin{bmatrix} 1 & -5 & 2 \\ 5 & 0 & 4 \\ -2 & -4 & 3 \end{bmatrix}, \quad B = \begin{bmatrix} 8 & 0 & -5 \\ 0 & -2 & 3 \\ -5 & 3 & 7 \end{bmatrix}, \quad C = \begin{bmatrix} -4 & 0 & 5 \\ 3 & -1 & 8 \\ 2 & 0 & 3 \end{bmatrix}$$

SOLUTION

None of these are skew symmetric even though A appears to satisfy the condition that $a_{ij} = -a_{ji}$. The matrix A is not skew symmetric since the diagonal elements are not all zero. The only way a_{ii} may equal $-a_{ii}$ is for a_{ii} to be zero.

$$***$$

Now let us return to the very first set of linear algebraic equations we considered at the beginning of this chapter.

$$\begin{array}{rrrcl} 4x & -3y & +3z & = & 0 \\ -2x & -5y & +2y & = & 1 \\ 3x & +2y & +6z & = & 4 \end{array}$$

Suppose we wish to attempt to solve this system by successively eliminating the unknowns. There are many ways of developing this approach. We are going to develop one method in particular that is the basis of an important technique known as Gaussian elimination. This method will be developed in detail in chapter III.

To begin, let us to eliminate the x term from the second equation. This is accomplished by multiplying the first equation by 1/2 and adding the result to the second equation. Multiplying the first equation by 1/2 gives

$$2x - (3/2)y + (3/2)z = 0$$

Adding this to the second equation generates a new second equation which is

$$0x - (13/2)y + (7/2)z = 1.$$

For convenience, let us clear the fractions by multiplying through by 2 to obtain

$$-13y + 7z = 2.$$

Before continuing, we must make it clear that operating on the equations algebraically to eliminate unknowns does not change the solution.

To continue, we can also remove the x term from the third equation by multiplying the first equation by $-3/4$ and adding the result to the third equation. Now $-3/4$ times the first equation gives

$$-3x + (9/4)y - (9/4)z = 0.$$

We now generate a new third equation by adding this result to the original third equation to obtain

$$0x + (17/4)y + (15/4)z = 4.$$

Again, the fractional coefficients are cleared by multiplying through by 4. The result is the following.

$$17y + 15z = 16$$

By means of this algebraic manipulation, the original three equations have been converted to the following.

$$4x - 3y + 3z = 0$$
$$-13y + 7z = 2$$
$$17y + 15z = 16$$

In a similar manner, we may combine the second and third equations to remove the y term from the last equation. This requires multiplying the second equation by the fraction $17/13$ and adding the result to the third equation. Performing this operation gives

$$0y + (314/13)z = 242/13.$$

Simplifying this gives as a new third equation

$$157y = 121.$$

Now the original system of three equations has been converted to the following important form.

$$
\begin{array}{rrrcr}
4x & -3y & +3z & = & 0 \\
 & -13y & +7z & = & 0 \\
 & & 157z & = & 121
\end{array}
$$

In this case, the matrix of coefficients is

$$C = \begin{bmatrix} 4 & -3 & 3 \\ 0 & -13 & 7 \\ 0 & 0 & 157 \end{bmatrix}$$

This form of a matrix is called triangular, or more particularly, upper triangular. This terminology results from the fact that all the elements below the main diagonal are zero.

It is important to again remind the reader that any of the above algebraic operations on the original set or system of three linear algebraic equations changed the solution. In other words, the same values of x, y, and z that satisfy the original equations also satisfy the triangularized system above and vice versa. As we will develop in a later chapter, the process of converting a system of linear algebraic equations to triangular form provides a very efficient way of finding the solution.

Before we continue, however, the reader is likely interested in knowing how to complete the solution to the above equations as well as knowing what the solution is. The reason that the system of equations has been reduced to upper triangular form is that the process of finding the unknowns x, y, and z may now be accomplished by means of a procedure called back substitution. Although the details of this method will be developed later, we will employ it now to find the solution to this example system of equations.

The reader should easily see, from the upper triangular form, the last equation may be directly solved for the last unknown z. The solution is $z = 121/157$. If we now replace z in the second equation with this value, it in turn may be solved for y. The result is $y = 533/2{,}041$. Finally, knowing the solution for y and z, these values may be substituted into the remaining (first) equation that is then solved for x. The solution is $x = -780/2{,}041$. If we wished to express the solution as a column matrix, it would appear as follows.

$$X = \begin{bmatrix} x \\ y \\ z \end{bmatrix} = \begin{bmatrix} -780/2041 \\ 533/2041 \\ 121/157 \end{bmatrix} = \left(\frac{1}{2041}\right) \begin{bmatrix} -780 \\ 533 \\ 1573 \end{bmatrix}$$

In this expression, we have used a capital X to designate the solution vector, and in the last term, the scalar of $1/2041$ has been factored out.

In the above, we have found the exact solution by carrying out the computations in terms of fractions. This could be done by hand or using a calculator that handles fractional notation. In most cases, however, one would use a calculator or matrix computational program with decimal notation. To illustrate the consequence of using decimal notation, suppose we are using a calculator that carries only three places following the decimal. If we repeated the above procedure using this calculator, the following solution vector is obtained.

$$X = \begin{bmatrix} x \\ y \\ z \end{bmatrix} = \begin{bmatrix} -0.382 \\ 0.261 \\ 0.771 \end{bmatrix}$$

To check the accuracy of these answers, these values are substituted into the original equations to see how closely they come to satisfying them. Taking each equation in order, the results are

$$4(-0.382) \quad -3(0.261) \quad +3(0.771) \quad = \quad 0.002$$
$$-13(0.261) \quad +7(0.771) \quad = \quad 2.004$$
$$157(0.771) \quad = \quad 121.047$$

The discrepancies between these values and the one that should result are 0.002, 0.004, and 0.047 respectively. Even though we used a calculator that was accurate to three decimal places, the solution did not satisfy the equations to the same degree of accuracy. The point of this illustration is to emphasize that the use of a calculator or computer to determine solutions will only produce results that are, at best, accurate to the number of digits the machine will carry. This is also assuming that one knows all the input values, the coefficients, and the constants perfectly. In real applications, this is not the usual case. This means that any solution obtained can even be more inaccurate. As we have already discussed in the introduction to this text, if the number of equations is large, then additional sources of error or inaccuracy may arise from the computation algorithms themselves. These effects contribute further to the inaccuracy of any calculated result.

One valuable point to remember in performing numerical solutions to algebraic equations is to always check your answers by substitution into the original equations whenever possible. This will provide a means of evaluating the amount of error that has accrued in the answers obtained.

Now let us return to the presentation at hand and state the next definition we were developing.

DEFINITION

Triangular Matrix (Upper, Lower)

A triangular matrix is one in which either all the elements lying below the main diagonal are zero (upper) or all the elements above the main diagonal are zero (lower).

By this definition, an upper triangular matrix has zero elements below the main diagonal, and all the elements on and above the main diagonal may take any value but not all zero. A lower triangular matrix is one for which all the elements lying above the main diagonal are zero but those on and below the main diagonal have any value but not all zero. The terms *upper* or *lower* refer to the triangular regions in which not all the elements are zero.

Now consider the following matrix.

$$B = \begin{bmatrix} 1 & -2 & 3 & 0 & -1 \\ 0 & 1 & 4 & -3 & 0 \\ 0 & 0 & 1 & 2 & 1 \\ 0 & 0 & 0 & 1 & -2 \end{bmatrix}$$

This matrix is upper triangular and illustrates that a triangular matrix does not have to be square. However, some authors refer to nonsquare matrices in this form as upper or lower trapezoidal matrices. Because all the elements on the main diagonal of this matrix are ones, it is also referred to as a unit upper triangular matrix.

EXERCISES

Classify each of the following matrices as one or more of the following types: row, column, diagonal, identity, null, symmetric, skew symmetric, upper triangular, lower triangular.

1. $\begin{bmatrix} 2 & -1 & 3 \\ -1 & 0 & 4 \\ 3 & 4 & 1 \end{bmatrix}$ 2. $\begin{bmatrix} 0 & 0 & 0 \\ 0 & 0 & 0 \\ 0 & 0 & 0 \end{bmatrix}$ 3. $\begin{bmatrix} -5 & 0 & 0 \\ 1 & 0 & 0 \\ 2 & 7 & -4 \end{bmatrix}$

4. $\begin{bmatrix} 1 & 0 & 0 \\ 0 & 1 & 0 \\ 0 & 0 & 1 \end{bmatrix}$ 5. $\begin{bmatrix} 0 & 7 & -4 \\ -7 & 0 & -12 \\ 4 & 12 & 0 \end{bmatrix}$ 6. $\begin{bmatrix} 1 & -2 & 3 \\ 0 & 1 & 2 \\ 0 & 0 & 1 \end{bmatrix}$

7. $\begin{bmatrix} 2 & 0 & 0 \\ 0 & 0 & 0 \\ 0 & 0 & -1 \end{bmatrix}$ 8. $\begin{bmatrix} 3 & -1 & 2 \end{bmatrix}$ 9. $\begin{bmatrix} -5 \\ 1 \\ 2 \end{bmatrix}$

4. EQUALITY, ADDITION, AND SUBTRACTION OF MATRICES

From the above definitions, it should be clear that a matrix is a mathematical entity that is merely a square or rectangular array of numbers, functions, or even other matrices. Now the algebraic rules for manipulating these arrays are presented. In most cases, the rules for algebraic operations with matrices are the same as those of scalar algebra, with a few modifications.

Now consider the following three matrices.

$$C = AB = \begin{bmatrix} c_{ij} \end{bmatrix}$$

$$c_{ij} = a_{i1}b_{1j} + a_{i2}b_{2j} + a_{i3}b_{3j} + \ldots + a_{ip}b_{pj}$$

$$= \sum_{k=1}^{k=p} a_{ik}b_{kj}$$

Of these matrices, A and C are said to be equal. The concept of equality is of fundamental importance in ordinary algebra and is similarly so in matrix algebra.

DEFINITION

Equality of Matrices

Two matrices $A = [a_{ij}]$ and $C = [c_{ij}]$ are equal if and only if they are of the same order and their corresponding elements are identical.

That is, by definition, we have $A = C$ provided that they have the same number of rows and columns and that $a_{ij} = c_{ij}$ for all i and j.

Now let us examine how matrices are added by means of the following example.

EXAMPLE 9

Find the sum of the two second-order matrices

$$A = \begin{bmatrix} 1 & 4 \\ 0 & -2 \end{bmatrix} \text{ and } B = \begin{bmatrix} 2 & 3 \\ -1 & 0 \end{bmatrix}.$$

SOLUTION

The sum is

$$C = A + B = \begin{bmatrix} 1 & 4 \\ 0 & -2 \end{bmatrix} + \begin{bmatrix} 2 & 3 \\ -1 & 0 \end{bmatrix}$$

$$= \begin{bmatrix} 1+2 & 4+3 \\ 0+(-1) & -2+0 \end{bmatrix} = \begin{bmatrix} 3 & 7 \\ -1 & -2 \end{bmatrix}.$$

***

Here we see that the addition of matrices is accomplished by adding the respective elements. From this it should be clear that to add two matrices, they must be of the same order.

DEFINITION

Addition and Subtraction

If A and B are matrices of the same order, then the sum or difference, $A \pm B$ is a matrix C whose general element c_{ij} is given by the relation

$$c_{ij} = a_{ij} \pm b_{ij},$$

and we write

$$C = A \pm B.$$

From the above definitions and the basic rules of scalar algebra it can be shown that the following operations are valid:

$$A + B = B + A, \text{ the commutative law for matrices}$$

$$A + (B + C) = (A + B) + C, \text{ the associative law for matrices}$$

EXERCISES

Note: If you have access to a mathematical computational program that performs matrix operations, you are encouraged to use it to solve the applicable problems below or to check your work. You should also use such programs to investigate the principles involved in the exercises by experimenting with variations in the example or exercise problem values. This practice should be continued for the exercises throughout this text. Exercises marked with an asterisk are recommended for, or may require, the use of a matrix computational software package in determining the solution.

For the matrices given below, evaluate the indicated expressions. If they do not exist, then indicate the reason.

$$A = \begin{bmatrix} 5 & -1 & 3 \\ 2 & 0 & -4 \end{bmatrix} \qquad B = \begin{bmatrix} -2 & 1 & 1 \\ 3 & -3 & 2 \end{bmatrix}$$

$$C = \begin{bmatrix} -2 & 1 \\ 2 & -3 \\ 3 & 1 \end{bmatrix} \qquad D = \begin{bmatrix} 1 & -2 \\ 0 & 1 \\ -1 & 2 \end{bmatrix}$$

$$E = \begin{bmatrix} -2 & 1 & 3 \\ 0 & 1 & -5 \\ 2 & -3 & 3 \end{bmatrix} \qquad F = \begin{bmatrix} 7 & -1 & 3 \\ 2 & 0 & -2 \\ 1 & 4 & -3 \end{bmatrix}$$

1. $A + B$ 2. $C - D$ 3. $A - B$
4. $C + B$ 5. $A + D^T$ 6. $C - B^T$
7. $C^T + B$ 7. $D + E$ 8. $E + F^T$
9. $(A + B)^T$ 10. $(B - D^T)^T$

11. For the matrix E above, show that the sum $E + E^T$ is a symmetric matrix. Can you develop a proof of this for a general matrix A in terms of its elements a_{ij}?

12. For the matrix E above, show that the difference $E - E^T$ is a skew symmetric matrix. Can you develop a proof of this for a general matrix A in terms of its elements a_{ij}?

13. It is a fact that any real square matrix may be expressed as the sum of a symmetric and a skew symmetric matrix. For a general n^{th}-order matrix, derive expressions by which the symmetric and skew symmetric components may be calculated.

***

5. MATRIX MULTIPLICATION

Consider the following matrix.

$$A = \begin{bmatrix} 3 & 7 \\ -1 & -2 \end{bmatrix}$$

If we add this matrix to itself, we find that

$$A+A=2A=\begin{bmatrix} 6 & 14 \\ -2 & -4 \end{bmatrix}.$$

By comparing the result with the given A matrix, we see that all the elements have a common factor of two. This implies that we may write

$$\begin{bmatrix} 6 & 14 \\ -2 & -4 \end{bmatrix}=\begin{bmatrix} 2(3) & 2(7) \\ 2(-1) & 2(-2) \end{bmatrix}=2\begin{bmatrix} 3 & 7 \\ -1 & -2 \end{bmatrix}$$

Here the scalar factor 2 that is common to all the elements is factored out as shown. This illustrates what we define as scalar multiplication since the factor 2 is a scalar.

DEFINITION

Multiplication of a Matrix by a Scalar

The multiplication of a matrix A by an ordinary number or scalar k is accomplished by multiplying each element of A by the scalar k and obtaining a new matrix whose elements are ka_{ij}. That is, by definition,

$$B = kA, \text{ where } b_{ij} = ka_{ij}.$$

EXAMPLE 10

Evaluate 1.5 times the matrix A.

$$A=\begin{bmatrix} 1 & 4 \\ 3 & -2 \end{bmatrix}$$

SOLUTION

$$(1.5)A=\begin{bmatrix} 1.5(1) & 1.5(4) \\ 1.5(3) & 1.5(-2) \end{bmatrix}=\begin{bmatrix} 1.5 & 6 \\ 4.5 & -3 \end{bmatrix}$$

***

The rules for the multiplication of a matrix by another matrix have been established to facilitate representation of linear algebraic equations and linear transformations by matrix algebra. As a specific example, let us return to the set of three linear algebraic equations in three unknowns that were introduced as an example at the beginning of section 2 of this chapter.

$$4x - 3y + 3z = 0$$
$$-2x - 5y + 2z = 1$$
$$3x + 2y + 6z = 4$$

Our goal is to show how this system may be written in matrix form as $AX = B$, where A is the matrix of coefficients, X the column matrix of unknowns and B the column matrix of right-hand constants. Before we establish the general rules for multiplying matrices, let us develop the product of a row matrix times a column matrix. From the equations above form the row vector R_1, which contains the coefficients of the first equation.

$$R_1 = [\,4 \; -3 \; 3\,]$$

Now let the unknowns x, y, and z be placed in order as the elements of a third-order column vector that will be called X.

$$X = \begin{bmatrix} x \\ y \\ z \end{bmatrix}$$

We now want the product of the row vector R_1 times the column vector X to represent the left-hand side of the first algebraic equation. That is,

$$R_1 X = \begin{bmatrix} 4 & -3 & -3 \end{bmatrix} \begin{bmatrix} x \\ y \\ z \end{bmatrix} = 4x - 3y + 3z$$

From this we can see the pattern of multiplying a row times a column vector, namely, we sum the product of the first element of the row times the first element of the column with the product of the second element of the row times the second element of the column, etc. In this illustration we can see that the number of elements in the row must be the same as in the column. The reader should also note that we have formed the product of a matrix of order 1×3 times a matrix of order 3×1. The resulting algebraic equation is a scalar or a matrix of order 1×1.

If we now let the zero on the right-hand side of the first equation be represented by b_1, then the first equation above may be expressed as follows in terms of the matrices we have defined.

$$R_1 X = b_1$$

Let us define R_2 and R_3 as row matrices whose elements are the coefficients of the second and third equations above respectively. If b_2 and b_3 are assigned the values 1 and 4, then the second and third equations may be written as follows.

$$R_2 X = b_2$$

$$R_3 X = b_3$$

Now assemble the right-hand side constants b_1, b_2, and b_3 into a third-order column vector B.

$$B = \begin{bmatrix} b_1 \\ b_2 \\ b_3 \end{bmatrix}$$

Replacing each element with its matrix equivalent gives

$$\begin{bmatrix} b_1 \\ b_2 \\ b_3 \end{bmatrix} = \begin{bmatrix} R_1 X \\ R_2 X \\ R_3 X \end{bmatrix} = \begin{bmatrix} R_1 \\ R_2 \\ R_3 \end{bmatrix} X .$$

Finally, let us substitute the individual values for the row matrices and the unknown column as well as the constants b_1, b_2, and b_3.

$$\begin{bmatrix} R_1 \\ R_2 \\ R_3 \end{bmatrix} X = \begin{bmatrix} 4 & -3 & 3 \\ -2 & 5 & 2 \\ 3 & 2 & 6 \end{bmatrix} \begin{bmatrix} x \\ y \\ z \end{bmatrix} = \begin{bmatrix} 0 \\ 1 \\ 4 \end{bmatrix}$$

Defining the matrix of coefficients as A, we can see that the system of linear algebraic equations is represented in the following matrix form where B is the column vector at the right.

(5.1) $$AX = B$$

The first equation is given by the product of the first row of the coefficient matrix times the column matrix of unknowns and equated to the first element of the B matrix.

In this scheme, we see that the multiplication of one matrix times another involves the multiplication of the rows of the first times the columns of the second. In our illustration above, the second matrix contains only a single column. The only restriction we have noted thus far is that the number of elements in the rows of the first matrix must be the same as the number of elements of the columns in the second matrix.

EXAMPLE 11

Write the following set of two linear algebraic equations in two unknowns in matrix form.

$$3x + 2y = 5$$
$$2x - 4y = 7$$

SOLUTION

We first define the column vector of unknowns by letting $x = x_1$ and $y = x_2$ so that

$$X = \begin{bmatrix} x_1 \\ x_2 \end{bmatrix} = \begin{bmatrix} x \\ y \end{bmatrix}.$$

Now the coefficient matrix A and the constant vector B are

$$A = \begin{bmatrix} 3 & 2 \\ 2 & -4 \end{bmatrix}, B = \begin{bmatrix} 5 \\ 7 \end{bmatrix}.$$

According to the above statement, the two equations are given by

$$AX = B$$

$$\begin{bmatrix} 3 & 2 \\ 2 & -4 \end{bmatrix} \begin{bmatrix} x \\ y \end{bmatrix} = \begin{bmatrix} 3(x) & 2(y) \\ 2(x) & -4(y) \end{bmatrix} = \begin{bmatrix} 5 \\ 7 \end{bmatrix}.$$

EXAMPLE 12

Based on the previous example, compute the product of the following two matrices.

$$A = \begin{bmatrix} 3 & 2 \\ 2 & -4 \end{bmatrix}, B = \begin{bmatrix} 2 \\ -1 \end{bmatrix}$$

SOLUTION

To form the product AB, we treat the B column in the same manner as the X column in the foregoing example.

$$AB = \begin{bmatrix} 1 & 2 \\ 3 & 4 \end{bmatrix} \begin{bmatrix} 2 \\ -1 \end{bmatrix} = \begin{bmatrix} (1)(2)+(2)(-1) \\ (3)(2)+(4)(-1) \end{bmatrix} = \begin{bmatrix} 0 \\ 2 \end{bmatrix}$$

***

From these examples, we see that the first element in the resulting column, 0, is the sum of the products of the elements of the first row of the first matrix, 1 and 2, times the elements of the column of the second matrix, 2 and −1, taken in order. To generalize a bit more, we note that the 1,1 element of the result is the sum of the products of elements in the first row of A times the elements of the first and only column of B. Similarly, the 2,1 element in the answer is the sum of the products of the elements in the second row of A times the elements in the first column of B.

From the foregoing discussion, the following general rules may be stated. Two matrices can be multiplied together only when the number of columns of the first is equal to the number of rows of the second. That is, if A is a matrix of order $m \times n$ and B is a matrix of order $p \times q$, the product AB is not defined unless $n = p$. When $n = p$, the matrices A and B are said to be conformable. In other words, this means that the number of elements in the each row of the first matrix equals the number of elements in each column of the second. Also, a product only exists if the matrices are conformable.

DEFINITION

Matrix Multiplication

The product of a matrix A of order $m \times p$ by a matrix B of order $p \times q$ is a matrix C whose elements are given by

$$C = AB = \left[c_{ij} \right],$$

where

$$c_{ij} = a_{i1}b_{1j} + a_{i2}b_{2j} + a_{i3}b_{3j} + \dots + a_{ip}b_{pj}$$

(5.3)

$$= \sum_{k=1}^{k=p} a_{ik}b_{kj} \ .$$

The element c_{ij} is the product of the ith row of the first matrix A and the jth column of the second matrix B.

From the above, it is seen that the product matrix, C, will have as many rows as the first matrix, A, and as many columns as the second, B. The fact that a matrix of order $m \times p$ is multiplied by a matrix of order $p \times q$ results in a matrix of order $m \times q$ may be expressed symbolically in the following manner:

(5.4) $(m \times p)(p \times q) = m \times q$

This symbolic equation is useful in determining the order of the resulting matrix when several matrices are multiplied together.

EXAMPLE 13

Evaluate the product $C = AB$, where

$$A = \begin{bmatrix} 1 & -2 & 1 \\ 0 & 3 & 2 \\ -5 & 0 & 1 \end{bmatrix}, \quad B = \begin{bmatrix} 2 & 1 \\ -1 & 0 \\ 3 & -1 \end{bmatrix}.$$

SOLUTION

In this case we have matrices of order 3 × 3 and 3 × 2. Accordingly, by equation (5.4), the product AB must be a matrix of order 3 × 2. For this example, the elements of the product matrix $C = AB$ are given by (5.3). The upper left or c_{11} element of C is

$$c_{11} = \begin{bmatrix} 1 & -2 & 1 \end{bmatrix} \begin{bmatrix} 2 \\ -1 \\ 3 \end{bmatrix} = (1)(2) + (-2)(-1) + (1)(3) = 7$$

Similarly, the c_{12} element is

$$c_{12} = \begin{bmatrix} 1 & -2 & 1 \end{bmatrix} \begin{bmatrix} 1 \\ 0 \\ -1 \end{bmatrix} = (1)(1) + (-2)(0) + (1)(-1) = 0 \quad .$$

Calculating all the elements gives

$$C = AB = \begin{bmatrix} 1 & -2 & 1 \\ 0 & 3 & 2 \\ -5 & 0 & 1 \end{bmatrix} \begin{bmatrix} 2 & 1 \\ -1 & 0 \\ 3 & -1 \end{bmatrix} = \begin{bmatrix} 7 & 0 \\ 3 & -2 \\ -7 & -6 \end{bmatrix} \quad .$$

Properties of Matrix Multiplication

Because of the special character of matrix multiplication, namely the summing of the products of the elements of the rows of the first matrix times the elements of the columns of the second, the question arises as to what happens when matrices are multiplied in reverse order. To begin our investigation of this, let us consider the next example.

EXAMPLE 14

Evaluate the products $C = AB$ and $D = BA$ where

$$A = \begin{bmatrix} 2 & -3 \\ 4 & 1 \end{bmatrix}, \ B = \begin{bmatrix} -1 & 0 \\ 2 & 3 \end{bmatrix} .$$

SOLUTION

$$C = AB = \begin{bmatrix} (2)(-1) + (-3)(2) & (2)(0) + (-3)(3) \\ (4)(-1) + (1)(2) & (4)(0) + (1)(3) \end{bmatrix} = \begin{bmatrix} -8 & -9 \\ -2 & 3 \end{bmatrix}$$

$$D = BA = \begin{bmatrix} (-1)(2) + (0)(4) & (-1)(-3) + (0)(1) \\ (2)(2) + (3)(4) & (2)(-3) + (3)(1) \end{bmatrix} = \begin{bmatrix} -2 & 3 \\ 16 & -3 \end{bmatrix}$$

***

As this example illustrates, the products AB and BA exist, but they are not the same. In general, matrix multiplication is not commutative—that is, in almost all situations,

$$AB \neq BA.$$

This inequality can occur in three ways. Assume the product AB exists, which means that the number of columns of A equals the number of rows of B. First, the product BA may not even exist because the number of columns of B may not necessarily be the same as the number of rows of A. In other words, even though the product AB is conformable, the reverse product BA may not be. This is the case for the matrices of example 4 above. If we attempt to form the product BA, the orders are 3×2 and 3×3, which means that the product in this order is not conformable even though AB is.

For the second situation, consider a case in which A is $m \times n$ with $m \neq n$ and B is $n \times m$ so that both product AB and BA are conformable. The product AB will produce a square matrix of order m, and BA will produce a square matrix of order n. Clearly, these two products cannot be equal since their orders are not equal as demonstrated by the following example.

EXAMPLE 15

Evaluate AB and BA for the following matrices:

$$A = \begin{bmatrix} 1 & -1 & 2 \\ 0 & 2 & -1 \end{bmatrix} \qquad B = \begin{bmatrix} 2 & 1 \\ -3 & 2 \\ -1 & 0 \end{bmatrix}$$

SOLUTION

$$AB = \begin{bmatrix} 1 & -1 & 2 \\ 0 & 2 & -1 \end{bmatrix} \begin{bmatrix} 2 & 1 \\ -3 & 2 \\ -1 & 0 \end{bmatrix} = \begin{bmatrix} 3 & -1 \\ -5 & 4 \end{bmatrix}$$

$$BA = \begin{bmatrix} 2 & 1 \\ -3 & 2 \\ -1 & 0 \end{bmatrix} \begin{bmatrix} 1 & -1 & 2 \\ 0 & 2 & -1 \end{bmatrix} = \begin{bmatrix} 2 & 0 & 3 \\ -3 & 7 & -8 \\ -1 & 1 & -2 \end{bmatrix}$$

***

In this example, the product AB yields a matrix of order 2 whereas the product BA has yielded a matrix of order 3.

The third case to consider is when both A and B are square and of the same order, say n. This is the situation illustrated in example 14 above. In this situation, both products AB and BA produce square matrices of order 2 as well (n in general) so that equality is possible.

At this point, the question occurs as to whether the products AB and BA can ever be equal even if they both exist. The answer is that there are a few special situations in which the equality

$$AB = BA$$

is true. When this occurs, the matrices A and B are said to commute or to be permutable. Two cases for which this occurs are when either A or B is an identity matrix I or both are diagonal matrices. A third case in which the equality holds is when one matrix is the so-called inverse of the other. The inverse matrix will be introduced a little later in this chapter. A final case occurs when the matrix B is a power of the matrix A.

What we have been stating about matrix products is that they do not commute except in a few special cases.

Because of the situation expressed by the inequality above, it is sometimes necessary in matrix algebra to identify the order of multiplication. For the product AB, we use the terminology that A premultiplies B or B postmultiplies A.

Continued Products of Matrices

All the ordinary laws of algebra apply to the multiplication of matrices as long as the order of multiplication is obeyed. Of particular importance is the associative law of continued products. This property is illustrated in the following example.

46

EXAMPLE 16

For the given matrices A, B, and C, evaluate the products $A(BC)$ and $(AB)C$. The parentheses have been used to indicate which pair of matrices are to be multiplied together first.

$$A = \begin{bmatrix} 1 & -2 \\ 0 & 3 \end{bmatrix}, B = \begin{bmatrix} 2 & 1 \\ -1 & 0 \end{bmatrix}, C = \begin{bmatrix} -1 & 1 \\ 0 & 2 \end{bmatrix}$$

SOLUTION

$$AB = \begin{bmatrix} 1 & -2 \\ 0 & 3 \end{bmatrix} \begin{bmatrix} 2 & 1 \\ -1 & 0 \end{bmatrix} = \begin{bmatrix} 4 & 1 \\ -3 & 0 \end{bmatrix}$$

$$BC = \begin{bmatrix} 2 & 1 \\ -1 & 0 \end{bmatrix} \begin{bmatrix} -1 & 1 \\ 0 & 2 \end{bmatrix} = \begin{bmatrix} -2 & 4 \\ 1 & -1 \end{bmatrix}$$

$$(AB)C = \begin{bmatrix} 4 & 1 \\ -3 & 0 \end{bmatrix} \begin{bmatrix} -1 & 1 \\ 0 & 2 \end{bmatrix} = \begin{bmatrix} -4 & 6 \\ 3 & -3 \end{bmatrix}$$

$$A(BC) = \begin{bmatrix} 1 & -2 \\ 0 & 3 \end{bmatrix} \begin{bmatrix} -2 & 4 \\ 1 & -1 \end{bmatrix} = \begin{bmatrix} -4 & 6 \\ 3 & -3 \end{bmatrix}$$

***

As may be seen by this example, the result is the same independent of the order in which the product is evaluated. The associative law of continued products is stated as

$$(AB)C = A(BC) = D.$$

which allows one to dispense with the parentheses and to write $D = ABC$ without ambiguity. This result may be verified by application of the multiplication rule expressed in equation (5.3). When applied to the product ABC to evaluate the elements of D, the result is a double summation

$$d_{ij} = \sum_k^{\text{no cols of A}} \left(\sum_r^{\text{no cols of b}} a_{ik} b_{kr} c_{rj} \right)$$

$$= \sum_r^{\text{no cols of A}} \left(\sum_k^{\text{no cols of b}} a_{ik} b_{kr} c_{rj} \right).$$

These summations may be carried out in either of the orders indicated. However, it must be noted that the product of a chain of matrices will have meaning only if the adjacent matrices of the chain are conformable.

As we will see in later parts of this text, we will need to evaluate the powers of a given matrix. One common case in which this occurs is in evaluating a matrix polynomial.

Positive Powers of a Square Matrix

If a square matrix A is multiplied by itself n times, the resultant matrix is denoted as A^n. That is,

$$A^n = AAA \ldots A \text{ to } n \text{ factors.}$$

EXERCISES

For the matrices given below, evaluate the indicated products and state their order. If the product is undefined, state the reason why. An asterisk by a problem number indicates that it is recommended for solution by use of a matrix computational software package.

$$A = \begin{bmatrix} 5 & -1 \\ 2 & 3 \end{bmatrix}, \quad B = \begin{bmatrix} -2 & 1 & 1 \\ 3 & -3 & 2 \end{bmatrix}$$

$$C = \begin{bmatrix} -2 & 1 \\ 2 & -3 \\ 3 & 1 \end{bmatrix}, \quad D = \begin{bmatrix} 1 \\ 2 \\ -1 \end{bmatrix}$$

$$E = \begin{bmatrix} -2 & 1 & 3 \\ 0 & 1 & -5 \\ 2 & -3 & 3 \end{bmatrix}, \quad F = \begin{bmatrix} 7 & -1 & 3 \\ 2 & 0 & -2 \\ 1 & 4 & -3 \end{bmatrix}$$

1.	AB	2.	$A^T B$	3.	$B^T A$
4.	AC	5.	AF	6.	CF
7.	CB	8.	BC	9.	$B^T C^T$
10.	CD	11.*	ABC	12.	$D^T F$
13.	$C^T F$	14.	CE	15.*	FCD
16.*	BED	17.*	$CABED$	18.*	$C + 2B^T$
19.*	$3E + F^T$	20.*	$I_3 + 2E - E^2$	21.*	$CC^T - 3F$

Evaluate the following products where possible. State the reasons when an expression cannot be evaluated.

$$A = \begin{bmatrix} 1 & 2 \\ -2 & 1 \\ 0 & 3 \end{bmatrix}, \quad B = \begin{bmatrix} 1 & -1 & 0 \\ 0 & 2 & 1 \\ 3 & -2 & 1 \end{bmatrix}, \quad C = \begin{bmatrix} 3 \\ -1 \\ 2 \end{bmatrix}$$

22.* AB	23.* BA	24.* $A^T B$
25.* $BC + C^T B$	26.* $C^T B^T$	27.* $A^T C$
28.* $C^T BA$	29.* $A^T B^T C$	30.* CC^T
31.* $C^T C$	32.* B^3	33.* A^2

34. Prove for any general matrix A with real elements that $B = AA^T$ is a symmetric matrix.

35. If D is an nth-order diagonal matrix, $D = \text{Diag}(d_1, d_2, \ldots, d_n)$, prove that any integral power, say k, of D is also diagonal and has the value $D^k = \text{Diag}(d_1^k, d_2^k, \ldots, d_n^k)$.

36.* A store sells a certain product that is available in three sizes: regular, giant, and supergiant. The price of each size respectively are given in the matrix $P = [1.75 \ 2.15 \ 2.99]$. During a typical week, the quantity of sales of each of these items is given by the column matrix $S = [50, 87, 62]$. Compute the product PS and determine what it means.

37.* The closing prices of an individual's investment stock portfolio, which contains four stocks, over a one-week period are given by the following matrix P.

$$P = \begin{bmatrix} 75\frac{1}{2} & 71 & 68\frac{3}{4} & 70\frac{1}{2} & 74 \\ 21 & 19 & 18\frac{1}{2} & 23 & 28 \\ 10\frac{1}{2} & 12\frac{5}{8} & 14 & 19\frac{1}{2} & 25 \\ 55\frac{1}{2} & 55 & 50\frac{3}{4} & 52\frac{1}{2} & 47 \end{bmatrix}$$

The rows represent the four individual stocks and the columns the five daily prices. Hence, each entry representing the number of shares of each stock held by the individual is given by the row vector $H = [200 \ 500 \ 400 \ 100]$. Evaluate the product HP and determine its meaning.

***

6. THE INVERSE MATRIX

In ordinary algebra, if the product of two quantities a and x is such that

$$ax = 1,$$

then x is said to be the reciprocal of a, and it is written in the following manner:

$$x = \frac{1}{a} = a^{-1}$$

In matrix algebra, the identity matrix I_n behaves in a manner similar to unity in ordinary algebra. That is, if A is an nth-order square matrix and I_n is the nth-order unit matrix, we have

$$I_n A = A I_n = A.$$

This is analogous to the relation $1 \times a = a \times 1 = a$ in ordinary algebra.

To introduce the inverse of a matrix, consider the following example.

EXAMPLE 17

For the given matrices A and B, evaluate the products AB and BA.

$$A = \begin{bmatrix} 2 & 5 \\ 1 & 3 \end{bmatrix}, \quad B = \begin{bmatrix} 3 & -5 \\ -1 & 2 \end{bmatrix}$$

SOLUTION

$$AB = \begin{bmatrix} 2 & 5 \\ 1 & 3 \end{bmatrix} \begin{bmatrix} 3 & -5 \\ -1 & 2 \end{bmatrix} = \begin{bmatrix} 1 & 0 \\ 0 & 1 \end{bmatrix}$$

$$BA = \begin{bmatrix} 3 & -5 \\ -1 & 2 \end{bmatrix} \begin{bmatrix} 2 & 5 \\ 1 & 3 \end{bmatrix} = \begin{bmatrix} 1 & 0 \\ 0 & 1 \end{bmatrix}$$

***

In this example, we see that the matrices A and B not only commute, but their product is also the identity matrix. By comparison with the previous algebraic discussion, we can see that the matrix B may be called the reciprocal or inverse of A or conversely A the inverse of B. We may now proceed with the following general formulation.

DEFINITION

The Inverse Matrix

Let A be a given square matrix of the nth order and I_n be the identity matrix of the nth order. If a square matrix B can be determined so that

$$AB = BA = I_n,$$

then B is said to be the inverse of A, and it is denoted in the form

$$B = A^{-1}.$$

EXAMPLE 18

Show that the matrix B is the inverse of A.

$$A = \begin{bmatrix} 1 & -1 & -2 \\ -1 & 0 & 3 \\ -2 & 3 & 4 \end{bmatrix}, \quad B = \begin{bmatrix} 9 & 2 & 3 \\ 2 & 0 & 1 \\ 3 & 1 & 1 \end{bmatrix}$$

SOLUTION

First, we evaluate the product AB.

$$AB = \begin{bmatrix} 1 & -1 & -2 \\ -1 & 0 & 3 \\ -2 & 3 & 4 \end{bmatrix} \begin{bmatrix} 9 & 2 & 3 \\ 2 & 0 & 1 \\ 3 & 1 & 1 \end{bmatrix} = \begin{bmatrix} 1 & 0 & 0 \\ 0 & 1 & 0 \\ 0 & 0 & 1 \end{bmatrix}$$

Although it is not necessary, we may verify that the matrix B is the inverse of A by evaluating the product BA.

$$BA = \begin{bmatrix} 9 & 2 & 3 \\ 2 & 0 & 1 \\ 3 & 1 & 1 \end{bmatrix} \begin{bmatrix} 1 & -1 & -2 \\ -1 & 0 & 3 \\ -2 & 3 & 4 \end{bmatrix} = \begin{bmatrix} 1 & 0 & 0 \\ 0 & 1 & 0 \\ 0 & 0 & 1 \end{bmatrix}$$

***

As we will see later, even if A is square, it may not have an inverse.

If we employ the symbol A^{-1} to denote the inverse of A, then the above definition of the inverse states that

$$AA^{-1} = A^{-1}A = I_n$$

This expresses the fact that a matrix and its inverse commute.

The reader should be aware of the fact that the matrices A and B do not have to be square to yield an identity matrix as a product. For example, A could be a 10×3 matrix and B a 3×10 matrix, and the product AB could yield a 10×10 identity matrix. It could additionally be true that the reverse product, BA, could give a 3×3 identity matrix. However, by the definition, B is not the inverse of A because the products AB and BA do not give identity matrices of the same order. Such considerations lead to the concepts of left, right, and pseudo inverses, which will not be developed here and are left for consideration in later more advanced courses.

The concept of matrix inversion leads to certain operations involving matrices that are similar in a formal manner to those of division in ordinary algebra. For example, if one has the equation $ab = c$ in scalar algebra, one can multiply both sides of this equation by a^{-1} and obtain $b = a^{-1}$, $c = c/a$.

In matrix algebra, if one has the product

$$AB = C$$

and if A is such that its inverse A^{-1} exists, then it is possible to solve for the matrix B by premultiplying both sides of the expression above by A^{-1} and thus obtaining

$$A^{-1}(AB) = A^{-1}C$$

$$(A^{-1}A)B = A^{-1}C$$

$$I_nB = A^{-1}C$$

or finally

$$B = A^{-1}C.$$

Another important use of the inverse matrix is to express the solution of a system of linear algebraic equations as previously shown in equation (5.1). In cases where the coefficient matrix A is square, then equation (5.1) may be premultiplied on both sides by A^{-1} to give the solution in the following form.

$$X = A^{-1}B$$

Although we will present methods by which the inverse of a matrix may be computed, the evaluation of the inverse of a coefficient matrix is not the preferred way of actually determining the solution to a system of linear algebraic equations. As we have already begun to develop earlier in this chapter, the most commonly preferred method of solution is that of Gaussian elimination.

It must be noted that there are some differences between matrix algebra and ordinary algebra. For example, in ordinary algebra, the necessary and sufficient condition that the product $ab = 0$ is that $a = 0$ or $b = 0$. However, in matrix algebra, if we have

$$AB = 0_n$$

It is sufficient that either $A = 0_n$ or $B = 0_n$, but not necessary. This is illustrated in the following example.

EXAMPLE 19

Evaluate and discuss the product AB.

$$A = \begin{bmatrix} 1 & 0 \\ 0 & 0 \\ 1 & 0 \end{bmatrix}, \quad B = \begin{bmatrix} 0 & 0 & 0 \\ 1 & 1 & 0 \end{bmatrix}$$

SOLUTION

$$AB = \begin{bmatrix} 1 & 0 \\ 0 & 0 \\ 1 & 0 \end{bmatrix} \begin{bmatrix} 0 & 0 & 0 \\ 1 & 1 & 0 \end{bmatrix} = \begin{bmatrix} 0 & 0 & 0 \\ 0 & 0 & 0 \\ 0 & 0 & 0 \end{bmatrix}$$

This illustrates that although neither matrices A and B are null (zero) matrices, their product is null, which deviates from what we normally would expect in scalar algebra. That is, if a and b are scalars and if $ab = 0$, then either a or b or both must be zero.

***

As another example of differences between scalar and matrix algebra, consider the following example problem.

EXAMPLE 20

For the given matrices A, B, and C, evaluate the products AB and AC.

$$A = \begin{bmatrix} 3 & 0 & 0 \\ 0 & -2 & 0 \end{bmatrix}, \quad B = \begin{bmatrix} 1 & 0 \\ 0 & 2 \\ 3 & 0 \end{bmatrix}, \quad C = \begin{bmatrix} 1 & 0 \\ 0 & 2 \\ 0 & 0 \end{bmatrix}$$

SOLUTION

$$AB = \begin{bmatrix} 3 & 0 & 0 \\ 0 & -2 & 0 \end{bmatrix} \begin{bmatrix} 1 & 0 \\ 0 & 2 \\ 3 & 0 \end{bmatrix} = \begin{bmatrix} 3 & 0 \\ 0 & -4 \end{bmatrix}$$

$$AC = \begin{bmatrix} 3 & 0 & 0 \\ 0 & -2 & 0 \end{bmatrix} \begin{bmatrix} 1 & 0 \\ 0 & 2 \\ 0 & 0 \end{bmatrix} = \begin{bmatrix} 3 & 0 \\ 0 & -4 \end{bmatrix}$$

***

In this example, the matrix B is not equal to the matrix C, yet the products AB and AC produce the same result. In ordinary algebra, if $ab = ad$, then $b = d$ provided that $a \neq 0$. However, as we have just seen in example 20, in matrix algebra, if

$$AB = AC \text{ and } A \neq 0_n,$$

one cannot say in general that the matrices B and C must be equal.

If, however, $AB = AC$ and the inverse of A exists, then it is possible to premultiply both members of the relation by A^{-1} and thus obtain

$$A^{-1}(AB) = A^{-1}(AC),$$

$$(A^{-1})B = (A^{-1}A)C,$$

$$I_nB = I_nC,$$

or $B = C.$

Thus, the relation $AB = AC$ implies $B = C$ when A has an inverse.

DEFINITION

Negative Powers of a Matrix

If the inverse of the matrix A exists, then negative powers of A are defined by raising the inverse matrix A^{-1} to positive powers. That is,

$$A^{-n} = (A^{-1})^n.$$

EXERCISES

1.   Given the general second-order matrix

$$A = \begin{bmatrix} a & b \\ c & d \end{bmatrix},$$

show that its inverse is given by the expression

$$A^{-1} = \frac{1}{(ad - bc)} \begin{bmatrix} d & -b \\ -c & a \end{bmatrix}$$

Under what condition will this inverse exist?

Use the result of exercise 1 to evaluate the inverse of each of the following matrices.

2. $\begin{bmatrix} 3 & -7 \\ 4 & 2 \end{bmatrix}$ 3. $\begin{bmatrix} 1 + j & 2 - j \\ 1 - j & 1 + 2j \end{bmatrix}$, where $j = \sqrt{-1}$

***

7. THE REVERSAL RULE IN TRANSPOSED AND RECIPROCAL PRODUCTS

One of the fundamental consequences of the rules relating to the commutativity of matrix multiplication is the reversal rule. This is exemplified in transposing or reciprocating a continued product of matrices.

Transposition of a Matrix Product

Let us begin our discussion by considering the following example.

EXAMPLE 21

For the given matrices, evaluate the products AB, $(AB)^T$, A^TB^T and B^TA^T.

$$A = \begin{bmatrix} 1 & 2 & -1 \\ -3 & 2 & 1 \end{bmatrix}, \quad B = \begin{bmatrix} 2 & -3 \\ 1 & 4 \\ 3 & -2 \end{bmatrix}$$

SOLUTION

$$AB = \begin{bmatrix} 1 & 2 & -1 \\ -3 & 2 & 1 \end{bmatrix} \begin{bmatrix} 2 & -3 \\ 1 & 4 \\ 3 & -2 \end{bmatrix} = \begin{bmatrix} 1 & 7 \\ -1 & 15 \end{bmatrix}$$

$$(AB)^T = \begin{bmatrix} 1 & -1 \\ 7 & 15 \end{bmatrix}$$

$$A^TB^T = \begin{bmatrix} 1 & -3 \\ 2 & 2 \\ -1 & 1 \end{bmatrix} \begin{bmatrix} 2 & 1 & 3 \\ -3 & 4 & -2 \end{bmatrix} = \begin{bmatrix} 11 & -11 & 9 \\ -2 & 10 & 2 \\ -5 & 3 & -5 \end{bmatrix}$$

$$B^TA^T = \begin{bmatrix} 2 & 1 & 3 \\ -3 & 4 & -2 \end{bmatrix} \begin{bmatrix} 1 & -3 \\ 2 & 2 \\ -1 & 1 \end{bmatrix} = \begin{bmatrix} 1 & -1 \\ 7 & 15 \end{bmatrix}$$

***

Careful examination of the results of this example indicates that $(AB)^T = B^TA^T$. This leads us to the following generalization.

Let A be a matrix of order $n \times p$, that is, one having n rows and p columns, and let B be a order $p \times m$ matrix. The matrices A and B are conformable in the order AB. Let the matrix C be the product AB so that its elements are given by

(7.1)
$$c_{ij} = \sum_{r=1}^{p} a_{ir}b_{rj}, \quad \begin{array}{l} i=1,2,\cdots,n \\ j=1,2,\cdots,m \end{array}$$

By the use of the symbolic equation (5.4), the order of C may be seen to be $n \times m$. The matrix C has n rows and m columns.

When the matrices A and B are transposed, the matrices A^T and B^T are obtained. Now A^T is a matrix of order $p \times n$, and B^T is a matrix of order $m \times p$. The matrices A^T and B^T are conformable when they are multiplied in the order $B^T A^T$. This product is seen to be an $m \times n$ matrix. It shall be seen that $B^T A^T$ is the transpose of $C = AB$ given by (7.1), so

(7.2) $$C^T = B^T A^T$$

To see this, note that the typical element of C^T, that is, c'_{ij}, is given by $c'_{ij} = c_{ji}$.

(7.3) $$c_{ji} = \sum_{k=1}^{p} a_{jk} b_{ki} = \sum_{k=1}^{p} b_{ki} a_{jk} = \sum_{k=1}^{p} b'_{ik} a'_{kj} ,$$

where b'_{ik} is the ik^{th} element of B^T and a'_{kj} is the kj^{th} element of A^T. Hence,

$$\sum_{k=1}^{p} b'_{ik} a'_{kj}$$

is the ij^{th} element of $B^T A^T$ and $C^T = B^T A^T$.

It follows from (7.1), (7.2), and (7.3) that when a product of matrices is transposed, the matrices forming the product must be transposed and multiplied in the reverse order. That is,

$$(AB)^T = B^T A^T$$

The following is another example of this rule.

EXAMPLE 22

For the matrices A and B verify that $(AB)^T = B^T A^T$.

$$A = \begin{vmatrix} 1 & -2 \\ 0 & 3 \end{vmatrix}, \quad B = \begin{vmatrix} 2 & 1 \\ -1 & 0 \end{vmatrix}$$

SOLUTION

$$AB = \begin{bmatrix} 4 & 1 \\ -3 & 0 \end{bmatrix}, \quad (AB)^T = \begin{bmatrix} 4 & -3 \\ 1 & 0 \end{bmatrix}$$

$$B^T A^T = \begin{bmatrix} 2 & -1 \\ 1 & 0 \end{bmatrix} \begin{bmatrix} 1 & 0 \\ -2 & 3 \end{bmatrix} = \begin{bmatrix} 4 & -3 \\ 1 & 0 \end{bmatrix}$$

Since $(AB)^T$ and $B^T A^T$ yield the same matrix as a result, we must have $(AB)^T = B^T A^T$.

$$***$$

Similarly, for a continued product,

$$(ABCD\ldots)^T = \ldots D^T C^T B^T A^T$$

The Inverse of a Matrix Product

Suppose that in the equation

$$BCD = A,$$

the matrices are all square, of appropriate order, and they possess inverses. Let this equation be premultiplied by A^{-1}, the inverse matrix of A. The result is

$$A^{-1}BCD = A^{-1}A = I_n,$$

Where I_n is the identity matrix of the same order as A. If we now postmultiply by D^{-1}, the following equation is obtained.

$$ABC = D^{-1}$$

This equation may now be postmultiplied by C^{-1} and the result then postmultiplied by B^{-1}. The result of these operations is

$$A^{-1} = D^{-1}C^{-1}B^{-1}$$

Now using the original statement that defines the matrix A, we have

$$(BCD)^{-1} = D^{-1}C^{-1}B^{-1}.$$

The relation can be extended to the product of any number of matrices provided they are conformable in the reverse order and their inverses exist.

EXERCISES

Evaluate the following products where possible by two ways. State the reasons when an expression cannot be evaluated.

$$A = \begin{bmatrix} 1 & 2 \\ -2 & 1 \end{bmatrix}, \quad B = \begin{bmatrix} 3 & -1 & 5 \\ -1 & 4 & 0 \\ 5 & 0 & 2 \end{bmatrix}, \quad C = \begin{bmatrix} 1 & 2 & 3 \\ 2 & 5 & -2 \\ 3 & -2 & 1 \end{bmatrix},$$

$$D = \begin{bmatrix} 1 & -2 & 1 \\ 3 & 0 & 2 \\ -1 & 2 & 3 \end{bmatrix}$$

1.* $(ABC)^T$ 2.* $(BC)^T$ 3.* $(BCD)^T$

4.   Given the matrix E below, verify that $(AE)^{-1} = E^{-1}A^{-1}$. Refer to exercise 1 of the previous section to evaluate the required inverses. The matrix A is given above.

$$E = \begin{bmatrix} 3 & -7 \\ 4 & 2 \end{bmatrix}$$

***

8. DIAGONAL MATRICES AND THEIR PROPERTIES

As previously defined, a square matrix of order n that has all its nondiagonal elements equal to zero is called a diagonal matrix. For example, a diagonal matrix A of the third order has the form

$$A = \begin{bmatrix} a_{11} & 0 & 0 \\ 0 & a_{22} & 0 \\ 0 & 0 & a_{33} \end{bmatrix}$$

and can be defined by the equation

$$A = \text{Diag}(a_{11}, a_{22}, a_{33}).$$

As a consequence of the definition of the product of two matrices given by equation (5.3), properties of a diagonal matrix can be established. For the first property, consider the following example.

EXAMPLE 23

For the matrices D and B, evaluate the product DB and discuss the result.

$$D = \begin{bmatrix} 4 & 0 & 0 \\ 0 & -2 & 0 \\ 0 & 0 & 3 \end{bmatrix}, \quad B = \begin{bmatrix} 2 & 1 \\ 0 & -3 \\ -1 & 2 \end{bmatrix}$$

SOLUTION

$$DB = \begin{bmatrix} 4 & 0 & 0 \\ 0 & -2 & 0 \\ 0 & 0 & 3 \end{bmatrix} \begin{bmatrix} 2 & 1 \\ 0 & -3 \\ -1 & 2 \end{bmatrix}$$

$$= \begin{bmatrix} (4)(2) & (4)(1) \\ (-2)(0) & (-2)(-3) \\ (3)(-1) & (3)(2) \end{bmatrix} = \begin{bmatrix} 8 & 4 \\ 0 & 6 \\ -3 & 6 \end{bmatrix}$$

Careful inspection shows that the product DB results in the rows of B being multiplied by the respective diagonal elements of D.

***

We may now state the first general property for premultiplication by a diagonal matrix.

PROPERTY 1

Premultiplication by a Diagonal Matrix

If D is a diagonal matrix of order m, $D = \text{Diag}(d_1, d_2, \ldots, d_m)$, and B is any matrix of order $m \times n$, the product DB is obtained from B by multiplying the rows of B respectively by the elements $d_{11}, d_{22}, \ldots, d_{mm}$. In particular, if X is a column vector $X = [x_1, x_2, \ldots, x_m]$, then for the general case,

$$DX = \begin{bmatrix} d_{11}x_1 \\ d_{22}x_2 \\ \vdots \\ d_{mm}x_m \end{bmatrix}$$

In a similar fashion, we may easily deduce the next property that deals with the postmultiplication of a general matrix by a diagonal matrix.

PROPERTY 2

Postmultiplication by a Diagonal Matrix

If $D = \text{Diag}(d_1, d_2, \ldots, d_m)$ is a diagonal matrix of order m and B is a general matrix of order $n \times m$, the product BD is obtained from B by multiplying the columns of B respectively by $d_{11}, d_{22}, \ldots, d_{mm}$.

EXAMPLE 24

For the matrices D and B, evaluate the product BD.

$$D = \begin{bmatrix} 4 & 0 & 0 \\ 0 & -2 & 0 \\ 0 & 0 & 3 \end{bmatrix}, \quad B = \begin{bmatrix} 2 & 7 & -1 \\ 0 & -3 & 5 \end{bmatrix}$$

SOLUTION

$$BD = \begin{bmatrix} 2 & 7 & -1 \\ 0 & -3 & 5 \end{bmatrix} \begin{bmatrix} 4 & 0 & 0 \\ 0 & -2 & 0 \\ 0 & 0 & 3 \end{bmatrix}$$

$$= \begin{bmatrix} 4(2) & -2(7) & 3(-1) \\ 4(0) & -2(-3) & 3(5) \end{bmatrix} = \begin{bmatrix} 8 & -14 & -3 \\ 0 & 6 & 15 \end{bmatrix}$$

***

In particular, if X is a column vector $X = [x_1, x_2, \ldots, x_m]$, the product $X^T D$ will be

$$X^T D = [x_1 x_2 \cdots x_n] \begin{bmatrix} d_{11} & 0 & \cdots & 0 \\ 0 & d_{22} & \cdots & 0 \\ \vdots & \vdots & & \vdots \\ 0 & 0 & \cdots & d_{nn} \end{bmatrix}$$

$$= [x_1 d_{11} \quad x_2 d_{22} \quad \cdots \quad x_n d_{nn}]$$

Now let us consider what happens when two diagonal matrices are multiplied as shown in the next example.

EXAMPLE 25

Evaluate the products AB and for the following two third-order diagonal matrices.

$$A = \text{Diag}(2, -3, 5), B = \text{Diag}(-1, 4, 3)$$

SOLUTION

For the first product, we have

$$AB = \begin{bmatrix} 2 & 0 & 0 \\ 0 & -3 & 0 \\ 0 & 0 & 5 \end{bmatrix} \begin{bmatrix} -1 & 0 & 0 \\ 0 & 4 & 0 \\ 0 & 0 & 3 \end{bmatrix} = \begin{bmatrix} -2 & 0 & 0 \\ 0 & -12 & 0 \\ 0 & 0 & 15 \end{bmatrix}$$

$$BA = \begin{bmatrix} -1 & 0 & 0 \\ 0 & 4 & 0 \\ 0 & 0 & 3 \end{bmatrix} \begin{bmatrix} 2 & 0 & 0 \\ 0 & -3 & 0 \\ 0 & 0 & 5 \end{bmatrix} = \begin{bmatrix} -2 & 0 & 0 \\ 0 & -12 & 0 \\ 0 & 0 & 15 \end{bmatrix}$$

***

This example leads us to the final property we wish to state concerning diagonal matrices.

PROPERTY 3

Products of Diagonal Matrices

If A and B are diagonal matrices of the same order, they are commutative in multiplication with each other so that $AB = BA$.

EXERCISES

Given matrices below

$$A = \begin{bmatrix} 2 & 0 & 0 \\ 0 & -3 & 0 \\ 0 & 0 & 1 \end{bmatrix} \quad B = \begin{bmatrix} 3 & 0 & 0 \\ 0 & 2 & 0 \\ 0 & 0 & 1 \end{bmatrix} \quad C = \begin{bmatrix} 1 & 0 & -2 \\ 3 & 4 & 1 \\ -2 & 1 & 5 \end{bmatrix}$$

$$D = \begin{bmatrix} 2 & -3 \\ 1 & 0 \\ -4 & 5 \end{bmatrix} \quad E = \begin{bmatrix} 6 & -3 & 5 \\ 2 & 1 & -4 \end{bmatrix}$$

,

evaluate the following:

1. $A + B$ 2. AB 3. AC
4. CA 5.* BCA 6. AD
7. EB

8. Verify property 3 above, that $AB = BA$.

***

9. ELEMENTARY OPERATIONS AND THE RANK OF A MATRIX

As the reader will discover at several places in this text, many situations arise in which it is desirable to perform various operations on the rows of a matrix. At times, column operations are also useful; however, they do not occur nearly as often as basic row operations. Because of this, the following development will be devoted to row operations, even though everything presented is equally applicable to columns.

The basic operations we wish to introduce at this point are part of a wider group called elementary operations that will be presented in more detail in the next chapter. In particular, the elementary operations to be considered here are exchanging rows, multiplying a row by a scalar, adding or subtracting one row from another, and adding a scalar multiple of one row to another. These operations are accomplished by premultiplying the given matrix by a special square matrix whose order equals the number of rows of the given matrix.

To illustrate row exchange, consider the following three by four matrix.

$$M = \begin{bmatrix} 1 & 2 & 3 & 0 \\ 0 & -1 & 2 & 1 \\ -1 & 0 & 2 & -1 \end{bmatrix}$$

Let us assume that we wish to exchange the first and third rows by premultiplying with a special matrix. We know that multiplying M by a third-order identity matrix produces no change at all. Now consider another matrix that is the third-order identity matrix with its first and third rows exchanged.

$$I_3 = \begin{bmatrix} 1 & 0 & 0 \\ 0 & 1 & 0 \\ 0 & 0 & 1 \end{bmatrix} \rightarrow \begin{array}{c} \text{Exchange } 1^{st} \\ \text{and } 3^{rd} \text{ rows} \end{array} \rightarrow P_{13} = \begin{bmatrix} 0 & 0 & 1 \\ 0 & 1 & 0 \\ 1 & 0 & 0 \end{bmatrix}$$

This matrix is called a permutation matrix, and the subscripts denote which rows have been exchanged. Premultiplying M with this matrix accomplishes the desired operation.

$$P_{13}M = \begin{bmatrix} -1 & 0 & 2 & -1 \\ 0 & -1 & 2 & 1 \\ 1 & 2 & 3 & 0 \end{bmatrix}$$

To accomplish the multiplying of one row by a scalar constant, say c, we need only to replace the one in the corresponding row of an identity matrix with the scalar and premultiply the given matrix with this modified identity matrix.

DEFINITION

Elementary Matrix

Any matrix which, when pre- or postmultiplying a given matrix, performs elementary operations on the rows or columns is called an elementary matrix.

The operation described above of multiplying a row by a scalar is demonstrated in the following example.

EXAMPLE 26

Determine the required elementary matrix that will result in the multiplication of the third row of the matrix M above by the scalar -3.

SOLUTION

The desired matrix is

$$E = \begin{bmatrix} 1 & 0 & 0 \\ 0 & 1 & 0 \\ 0 & 0 & -3 \end{bmatrix}.$$

Premultiplying the matrix M given above by E accomplishes the desired operation.

$$EM = \begin{bmatrix} 1 & 0 & 0 \\ 0 & 1 & 0 \\ 0 & 0 & -3 \end{bmatrix} \begin{bmatrix} 1 & 2 & 3 & 0 \\ 0 & -1 & 2 & 1 \\ -1 & 0 & 2 & -1 \end{bmatrix}$$

$$= \begin{bmatrix} 1 & 2 & 3 & 0 \\ 0 & -1 & 2 & 1 \\ 3 & 0 & -6 & 3 \end{bmatrix}$$

***

As stated, this type of matrix is called an elementary matrix. It should be noted that elementary matrices are always square. By the above definition, is should be clear that permutation matrices are also elementary matrices.

Now let's examine what happens when M is premultiplied by the following elementary matrix. Notice that it is an identity matrix with an additional one placed in the first element of the third row.

(9.1)
$$E = \begin{bmatrix} 1 & 0 & 0 \\ 0 & 1 & 0 \\ 1 & 0 & 1 \end{bmatrix}$$

Premultiplying M by this matrix gives the following result.

(9.2) $EM = \begin{bmatrix} 1 & 2 & 3 & 0 \\ 0 & -1 & 2 & 1 \\ 0 & 2 & 5 & -1 \end{bmatrix}$

Careful examination of this result shows that the first two rows are unchanged but the last row is now replaced by the sum of the original first and third rows. The substitution of a one for the first element of the third row of the identity matrix has produced the addition of the first and third rows. If we wished to add the second and third rows and place the result in the third row, we would use

$$E = \begin{bmatrix} 1 & 0 & 0 \\ 0 & 1 & 0 \\ 0 & 1 & 1 \end{bmatrix}.$$

At this point, we may combine these operations to develop an elementary matrix that will result in the addition of a scalar multiple of one row to another.

EXAMPLE 27

Construct the elementary matrix that will add twice the first row to the second row of the matrix M given above.

SOLUTION

From the operations described above, the following new elementary matrix may be constructed.

$$E = \begin{bmatrix} 1 & 0 & 0 \\ 2 & 1 & 0 \\ 0 & 0 & 1 \end{bmatrix}$$

The 2 in the first element of the second row should produce the desired result. Evaluating *EM* gives

$$EM = \begin{bmatrix} 1 & 0 & 0 \\ 2 & 1 & 0 \\ 0 & 0 & 1 \end{bmatrix} \begin{bmatrix} 1 & 2 & 3 & 0 \\ 0 & -1 & 2 & 1 \\ -1 & 0 & 2 & -1 \end{bmatrix}$$

$$= \begin{bmatrix} 1 & 2 & 3 & 0 \\ (2)(1)+(1)(0) & (2)(2)+(1)(-1) & (2)(3)+(1)(2) & (2)(0)+(1)(1) \\ -1 & 0 & 2 & -1 \end{bmatrix}$$

$$= \begin{bmatrix} 1 & 2 & 3 & 0 \\ 2 & 3 & 8 & 1 \\ -1 & 0 & 2 & -1 \end{bmatrix}$$

***

Now let us take the operations one step further. Assume that we wish to reduce the original matrix to upper triangular form. Examining M, we can see that nothing needs to be done to the second row as it already has a zero in the first position. The first element of the third row is -1. It can be reduced to zero by replacing the third row with the sum of the first and third rows. This was accomplished by the second elementary matrix (9.1) above, so let us designate that matrix as E_1. The result of the product E_1M is given in (9.2), which is repeated here for convenience.

(9.3)
$$E_1 M = \begin{bmatrix} 1 & 2 & 3 & 0 \\ 0 & -1 & 2 & 1 \\ 0 & 2 & 5 & -1 \end{bmatrix}$$

This is nearly upper triangular except for the 2 in the second position of the third row. If we add twice the second row to the third, we will achieve the desired upper triangular form. This operation can be accomplished by premultiplying the matrix in (9.3) by the second elementary matrix.

$$E_2 = \begin{bmatrix} 1 & 0 & 0 \\ 0 & 1 & 0 \\ 0 & 2 & 1 \end{bmatrix}$$

Performing the indicated multiplication leads us to the result.

$$(9.4) \qquad E_2 E_1 M = \begin{bmatrix} 1 & 2 & 3 & 0 \\ 0 & -1 & 2 & 1 \\ 0 & 0 & 9 & 1 \end{bmatrix}$$

Now let us consider another example.

EXAMPLE 28

Reduce the following matrix to an upper triangular form in that only zeros appear below the first nonzero element of any row as counted from the left edge.

$$A = \begin{bmatrix} 1 & -3 & 0 & -2 & 0 \\ 1 & -3 & 1 & 3 & 0 \\ -2 & 6 & -1 & -1 & 1 \\ -1 & 3 & 1 & 7 & 3 \end{bmatrix}$$

SOLUTION

The first elementary matrix is

$$E_1 = \begin{bmatrix} 1 & 0 & 0 & 0 \\ -1 & 1 & 0 & 0 \\ 2 & 0 & 1 & 0 \\ 1 & 0 & 0 & 1 \end{bmatrix}.$$

Premultiplying A by this matrix yields

$$E_1 A = A_1 = \begin{bmatrix} 1 & 0 & 0 & 0 \\ -1 & 1 & 0 & 0 \\ 2 & 0 & 1 & 0 \\ 1 & 0 & 0 & 1 \end{bmatrix} \begin{bmatrix} 1 & -3 & 0 & -2 & 0 \\ 1 & -3 & 1 & 3 & 0 \\ -2 & 6 & -1 & -1 & 1 \\ -1 & 3 & 1 & 7 & 3 \end{bmatrix}$$

$$= \begin{bmatrix} 1 & -3 & 0 & -2 & 0 \\ 0 & 0 & 1 & 5 & 0 \\ 0 & 0 & -1 & -5 & 1 \\ 0 & 0 & 1 & 5 & 3 \end{bmatrix}$$

If we were to follow the standard Gaussian elimination process to reduce this result to upper triangular form, we would construct an elementary matrix to reduce the elements below the (3, 3) element to zero. This would be the process noted above since zeros have appeared in the (2, 2) element and in all elements below it. However we will introduce a slight variation at this point because the first non zero element in the second row is in the (2, 3) position. Let us construct an elementary matrix that will reduce the elements below the (2, 3) element to zero. This elementary matrix is

$$
E_2 = \begin{bmatrix} 1 & 0 & 0 & 0 \\ 0 & 1 & 0 & 0 \\ 0 & 1 & 1 & 0 \\ 0 & -1 & 0 & 1 \end{bmatrix}.
$$

What this matrix accomplishes is the addition of the second row to the third and subtracts the second row from the fourth. Premultiplying A_1 by this matrix gives

$$
E_2 A_1 = A_2 = \begin{bmatrix} 1 & -3 & 0 & -2 & 0 \\ 0 & 0 & 1 & 5 & 0 \\ 0 & 0 & 0 & 0 & 1 \\ 0 & 0 & 0 & 0 & 3 \end{bmatrix}.
$$

The final step will be to eliminate the value 3 in the (4, 5) position by use of the last elementary matrix.

$$
E_3 = \begin{bmatrix} 1 & 0 & 0 & 0 \\ 0 & 1 & 0 & 0 \\ 0 & 0 & 1 & 0 \\ 0 & 0 & -3 & 1 \end{bmatrix}
$$

The premultiplication of A_2 by this matrix leads us to the final form.

$$
E_3 A_2 = A_3 = \begin{bmatrix} 1 & -3 & 0 & -2 & 0 \\ 0 & 0 & 1 & 5 & 0 \\ 0 & 0 & 0 & 0 & 1 \\ 0 & 0 & 0 & 0 & 0 \end{bmatrix}
$$

This form of upper triangular matrix is sometimes referred to as a reduced upper triangular or reduced row echelon form. A formal detail that is required for this terminology to be accurate is for the leading non zero element of each row to be reduced to one.

***

We are now at a place where we can introduce the concept of the rank of a matrix although its real significance cannot be presented until a later chapter.

DEFINITION

Rank of a Matrix

If a general matrix is transformed to reduced row echelon form by a series of elementary operations, its rank is defined as the number of nonnull rows.

In almost all applied problems, the direct reduction by Gaussian elimination produces a reduced row echelon form with nonzero and zero elements on the main diagonal. In most applied problems, the rank of a matrix can be evaluated by counting the number on nonzero elements lying on the main diagonal after reduction to upper triangular form. This is an alternative to the formal definition above, and its failure is demonstrated by the following special example.

EXAMPLE 29

Determine the rank of the following matrix.

$$A = \begin{bmatrix} 1 & -3 & 0 & -2 & 0 \\ 1 & -3 & 1 & 3 & 0 \\ -2 & 6 & -1 & -1 & 1 \\ -1 & 3 & 1 & 7 & 3 \end{bmatrix}$$

SOLUTION

This is the same matrix of example 25 above. Through the use of elementary operations, it was transformed into the following row echelon form. This form is also upper triangular.

$$A_3 = \begin{bmatrix} 1 & -3 & 0 & -2 & 0 \\ 0 & 0 & 1 & 5 & 0 \\ 0 & 0 & 0 & 0 & 1 \\ 0 & 0 & 0 & 0 & 0 \end{bmatrix}$$

Using the formal definition above, the rank is seen to be three because of the three nonnull rows. If one were to apply the modified definition by counting the number of nonzero elements lying on the main diagonal, the rank would appear to be one which is the incorrect answer.

EXAMPLE 30

Determine the rank of the following matrix.

$$M = \begin{bmatrix} 1 & 2 & 3 & 0 \\ 0 & -1 & 2 & 1 \\ 1 & 0 & 2 & -1 \end{bmatrix}$$

SOLUTION

This is the matrix considered in example 26 above. Although it will not be obvious to the reader, this matrix is more typical of those encountered in applied problems. We wish to make use of elementary operations to transform the M matrix into upper triangular form. Since there is already a zero in the second row of the first column, we only need to concern ourselves with the -1 in the last row of that column. The required elementary matrix to reduce this element to zero is

$$E_1 = \begin{bmatrix} 1 & 0 & 0 \\ 0 & 1 & 0 \\ 1 & 0 & 1 \end{bmatrix}$$

Premultiplying M by this matrix gives the following intermediate result.

$$E_1 M = M_1 = \begin{bmatrix} 1 & 0 & 0 \\ 0 & 1 & 0 \\ 1 & 0 & 1 \end{bmatrix} \begin{bmatrix} 1 & 2 & 3 & 0 \\ 0 & -1 & 2 & 1 \\ -1 & 0 & 2 & -1 \end{bmatrix}$$

$$= \begin{bmatrix} 1 & 2 & 3 & 0 \\ 0 & -1 & 2 & 1 \\ 0 & 2 & 5 & -1 \end{bmatrix}$$

We now only need an elementary matrix to reduce the last element of the second column to zero. This matrix is

$$E_2 = \begin{bmatrix} 1 & 0 & 0 \\ 0 & 1 & 0 \\ 0 & 2 & 1 \end{bmatrix}.$$

Now premultiplying the previous intermediate matrix, M_1, by this elementary matrix leads us to the final result.

$$E_2 M_1 = M_2 = \begin{bmatrix} 1 & 0 & 0 \\ 0 & 1 & 0 \\ 0 & 2 & 1 \end{bmatrix} \begin{bmatrix} 1 & 2 & 3 & 0 \\ 0 & -1 & 2 & 1 \\ 0 & 2 & 5 & -1 \end{bmatrix}$$

$$= \begin{bmatrix} 1 & 2 & 3 & 0 \\ 0 & -1 & 2 & 1 \\ 0 & 0 & 9 & 1 \end{bmatrix}$$

Using the modified definition above, the rank is seen to be three because of the three nonzero elements on the main diagonal. The formal definition also indicates that the rank is three.

***

It should be noted that the use of column operations is equally possible; however, they are not utilized nearly as often for reasons that will be explained when we get to the subject of linear algebraic equations. For our example matrix above, its rank is three. In general, the rank of a matrix can be determined by transforming it to upper triangular form and counting the number of nonzero elements remaining on the main diagonal. Formally, the rank of a matrix is the number of linear independent rows or columns in the matrix. The concept of linear independence will be developed in chapter II.

There is a practical point the reader should be aware of in using a matrix computational program that evaluates the rank of a matrix. Because of round-off error buildup in the computational algorithm, a diagonal element that should be identically zero may have a very small nonzero value. For example, after reduction to triangular form, a diagonal element could have a magnitude of ten to the minus twelfth ($E-12$). The user is faced with the problem of deciding whether this is purely the result of round off and should be called zero or it is actually a small but nonzero value. The decision clearly alters the rank count by one. Unfortunately, there are no clear answers to this dilemma at present. Many current matrix computational programs allow the user to specify a tolerance value below which numbers are considered to be truly zero. If this feature is available, it should be used with some caution and various tolerance values tested experimentally. The best rank-evaluating algorithm currently available is known as singular value decomposition, but it too has the problem of tolerance levels just mentioned. The theory of this algorithm is quite complex and is left for study in more advanced texts.

Another term that can be defined at this point in our development is row-echelon form. A matrix is in row-echelon form when it is upper triangular like (9.4) above with an additional condition that if any rows are entirely filled with zeros, then they are the bottom rows of the matrix.

Let us now generalize the process of reduction to upper triangular form by considering a general matrix of order $m \times n$.

$$(9.5) \qquad A = \begin{bmatrix} a_{11} & a_{12} & \cdots & a_{1n} \\ a_{21} & a_{21} & \cdots & a_{2n} \\ \cdot & \cdot & \cdots & \cdot \\ a_{m1} & a_{m1} & \cdots & a_{mn} \end{bmatrix}$$

For convenience, let us also consider that there are more columns than rows, that is, $n \geq m$. This means that we want to establish a series of elementary matrices, and permutation matrices if needed, that will perform the appropriate row operations on the given matrix A to reduce it to the following row-echelon form.

$$(9.6) \qquad B = \begin{bmatrix} a_{11} & a_{12} & a_{13} & \cdots & a_{1m} & \cdots & a_{1n} \\ 0 & b_{22} & b_{23} & \cdots & b_{2m} & \cdots & b_{2n} \\ 0 & 0 & b_{33} & \cdots & b_{3m} & \cdots & b_{3n} \\ \cdot & \cdot & \cdot & \cdots & \cdot & \cdots & \cdot \\ 0 & 0 & 0 & \cdots & b_{mm} & \cdots & b_{mn} \end{bmatrix}$$

As we noted above, some of the resulting rows could be filled with zeros that should be shifted to the bottom of B by the use of an appropriate permutation matrix. This result is not reflected in the general form of (9.6). As previously suggested, the rank of the given matrix A is the number of nonzero elements on the main diagonal.

Following the procedure described above, the following m^{th} order elementary matrix would reduce the first element of the second row of A to zero.

$$E_1 = \begin{bmatrix} 1 & 0 & 0 & \cdots & 0 \\ -\dfrac{a_{21}}{a_{11}} & 1 & 0 & \cdots & 0 \\ 0 & 0 & 1 & \cdots & 0 \\ \vdots & \vdots & \vdots & \ddots & \vdots \\ 0 & 0 & 0 & \cdots & 1 \end{bmatrix}$$

It should be noted that should a_{11} be zero, then a suitable permutation matrix would have to be generated that would exchange the first row with another whose first element were not zero.

Premultiplying A by E_1 affects only the elements of the second row. The elementary matrix required to reduce the first element of the third row of A is

$$E_2 = \begin{bmatrix} 1 & 0 & 0 & \cdots & 0 \\ 0 & 1 & 0 & \cdots & 0 \\ -\dfrac{a_{31}}{a_{11}} & 0 & 1 & \cdots & 0 \\ \vdots & \vdots & \vdots & \ddots & \vdots \\ 0 & 0 & 0 & \cdots & 1 \end{bmatrix}$$

.

In a similar fashion, elementary matrices can be generated that reduce all the nonzero elements lying below the main diagonal to zeros. The last elementary matrix for operating on the first column would be

$$
E_{m-1} = \begin{bmatrix}
1 & 0 & 0 & \cdots & 0 \\
0 & 1 & 0 & \cdots & 0 \\
0 & 0 & 1 & \cdots & 0 \\
\vdots & \vdots & \vdots & \ddots & \vdots \\
-\dfrac{a_{m1}}{a_{11}} & 0 & 0 & \cdots & 1
\end{bmatrix}
$$

At this point, the following continued product of matrices would result in a modified A matrix with all zeros below the first element.

$$
E_{m-1}E_{m-2}\ldots E_2E_1A
$$

It is left to the reader to show that this continued product of elementary matrices actually produces a single elementary matrix given below, which could be constructed form the original A matrix.

$$
E = \begin{bmatrix}
1 & 0 & 0 & \cdots & 0 \\
-\dfrac{a_{21}}{a_{11}} & 1 & 0 & \cdots & 0 \\
-\dfrac{a_{31}}{a_{11}} & 0 & 1 & \cdots & 0 \\
\vdots & \vdots & \vdots & \ddots & \vdots \\
-\dfrac{a_{m1}}{a_{11}} & 0 & 0 & \cdots & 1
\end{bmatrix}
$$

This procedure can now be continued by generating another elementary matrix that would reduce all the elements lying below the second main diagonal element to zero. This process is continued until the given matrix has been converted to upper triangular form.

EXAMPLE 31

Reduce the following matrix to upper triangular form and determine its rank.

$$M = \begin{bmatrix} 2 & 1 & 5 & 3 \\ 1 & 0 & 1 & 1 \\ 3 & 2 & 1 & 2 \\ -1 & -1 & 4 & 1 \end{bmatrix}$$

SOLUTION

Following the method described above, the elementary matrix that will reduce the first column is

$$E_1 = \begin{bmatrix} 1 & 0 & 0 & 0 \\ -\dfrac{1}{2} & 1 & 0 & 0 \\ -\dfrac{3}{2} & 0 & 1 & 0 \\ \dfrac{1}{2} & 0 & 0 & 1 \end{bmatrix}.$$

Premultiplying M by this matrix leads to the following result.

$$E_1 M = \begin{bmatrix} 2 & 1 & 5 & 3 \\ 0 & -\dfrac{1}{2} & -\dfrac{3}{2} & -\dfrac{1}{2} \\ 0 & \dfrac{1}{2} & -\dfrac{13}{2} & -\dfrac{5}{2} \\ 0 & -\dfrac{1}{2} & \dfrac{13}{2} & \dfrac{5}{2} \end{bmatrix}$$

The next step is to reduce the last two elements of the second column to zero. This will be accomplished with the following elementary matrix.

$$E_2 = \begin{bmatrix} 1 & 0 & 0 & 0 \\ 0 & 1 & 0 & 0 \\ 0 & 1 & 1 & 0 \\ 0 & -1 & 0 & 1 \end{bmatrix}$$

Premultiplying the previous result by this matrix yields the following.

$$E_2E_1M = \begin{bmatrix} 2 & 1 & 5 & 3 \\ 0 & -\dfrac{1}{2} & -\dfrac{3}{2} & -\dfrac{1}{2} \\ 0 & 0 & -8 & -3 \\ 0 & 0 & 8 & 3 \end{bmatrix}$$

The last required elementary matrix is

$$E_3 = \begin{bmatrix} 1 & 0 & 0 & 0 \\ 0 & 1 & 0 & 0 \\ 0 & 0 & 1 & 0 \\ 0 & 0 & 1 & 1 \end{bmatrix}.$$

Premultiplying the previous result by this elementary matrix gives the desired upper triangular form.

$$E_3E_2E_1M = \begin{bmatrix} 2 & 1 & 5 & 3 \\ 0 & -\dfrac{1}{2} & -\dfrac{3}{2} & -\dfrac{1}{2} \\ 0 & 0 & -8 & -3 \\ 0 & 0 & 0 & 0 \end{bmatrix}$$

From this result, it is seen that the rank of the original matrix is three because of only three nonzero elements on the main diagonal.

EXERCISES

1.*   Consider the following matrix A and elementary matrix E. Evaluate the products EA, E^TA, AE, and AE^T and discuss the results.

$$A = \begin{bmatrix} 2 & -1 & 3 \\ 1 & 2 & 3 \\ -1 & 4 & 5 \end{bmatrix}, \quad E = \begin{bmatrix} 1 & 0 & 0 \\ 0 & 1 & 0 \\ 2 & 0 & 1 \end{bmatrix}$$

2.   Describe and illustrate the difference between premultiplying and postmultiplying an nth-order matrix by the same n^{th}-order elementary matrix.

For the following matrices, reduce them to upper triangular form and determine the rank of the following matrices.

3. $\begin{bmatrix} 1 \\ 0 \end{bmatrix}$ 4. $\begin{bmatrix} -1 \\ 1 \end{bmatrix}$ 5. $\begin{bmatrix} 1 & 2 \\ 3 & 4 \end{bmatrix}$ 6. $\begin{bmatrix} 1 & 2 \\ 2 & 4 \end{bmatrix}$

7. $\begin{bmatrix} 1 & 0 & 1 \\ 0 & 1 & 0 \\ 0 & 0 & 1 \end{bmatrix}$ 8. $\begin{bmatrix} 1 & 2 & -1 \\ -1 & 6 & -3 \\ 1 & -4 & 2 \end{bmatrix}$ 9. $\begin{bmatrix} -2 & 1 & 1 \\ 2 & -3 & 1 \\ 2 & -1 & -1 \end{bmatrix}$

10. $\begin{bmatrix} 1 & -1 & 0 \\ 2 & 0 & 0 \end{bmatrix}$ 11. $\begin{bmatrix} 1 & -1 \\ 2 & 1 \\ 1 & 1 \end{bmatrix}$ 12. $\begin{bmatrix} -2 & 2 & 4 \\ 1 & -1 & 2 \end{bmatrix}$

13. $\begin{bmatrix} 2 & -1 & 1 & 1 \\ 1 & -3 & 2 & 1 \\ 0 & 1 & 2 & 5 \\ -2 & 3 & -1 & 1 \end{bmatrix}$

10. ADDITIONAL SPECIAL TYPES OF MATRICES

Several types of matrices that have special properties or symmetry or other unique characteristics arise in applications of matrices in a number of circumstances. Some of the more common special matrices are listed below for reference purposes.

Conjugate Matrices

In several applications of matrix algebra to physical problems, matrices arise whose elements are complex numbers so that they have the form

$$A = [a_{rs}], \text{ where } a_{rs} = p_{rs} + jq_{rs}, j = \sqrt{-1}.$$

A matrix B that has the same order as A but whose elements are the complex conjugates of the elements of A is called the conjugate matrix of A or, more simply, the conjugate of A. It is usually written in the form

$$B = \overline{A}, \text{ where } b_{rs} = p_{rs} - jq_{rs} = \overline{a}_{rs}.$$

For example, if A is defined as

$$A = \begin{bmatrix} 1+j & -1+3 \\ 3+2j & 2-j \end{bmatrix},$$

then

$$B = \overline{A} = \begin{bmatrix} 1-j & -1-3j \\ 3-2j & 2+j \end{bmatrix}.$$

Involutory Matrix

If $A = A^{-1}$, A is involutory. Examples of involutory matrices are the identity and permutation matrices.

Orthogonal Matrix

Orthogonal matrices hold an important position in matrix algebra and have the special property that if A is orthogonal, then $A^{-1} = A^{T}$. For example, the following matrix can be demonstrated as being orthogonal by evaluating the product AA^{T}.

$$A = \begin{bmatrix} \dfrac{1}{\sqrt{6}} & \dfrac{1}{\sqrt{2}} & \dfrac{1}{\sqrt{3}} \\ \dfrac{2}{\sqrt{6}} & 0 & \dfrac{-1}{\sqrt{3}} \\ \dfrac{1}{\sqrt{6}} & \dfrac{-1}{\sqrt{2}} & \dfrac{1}{\sqrt{3}} \end{bmatrix}$$

Hermitian Matrix

If $\overline{A} = A^{T}$, the matrix A is a Hermitian matrix. The following matrix may be verified as being Hermitian.

$$A = \begin{bmatrix} 1 & 1-j & 2+j \\ 1+j & 1 & 1-2j \\ 2-j & 1+2j & 2 \end{bmatrix}$$

Skew Hermitian Matrix

If $\bar{A} = -A^T$, then A is a skew Hermitian matrix. Again, it may be shown that the following matrix is skew Hermitian.

$$A = \begin{bmatrix} 0 & 1\text{-}j & 2\text{+}j \\ \text{-}1\text{-}j & 0 & 1\text{-}2j \\ \text{-}2\text{+}j & \text{-}1\text{-}2j & 0 \end{bmatrix}$$

EXERCISES

Classify each of the following matrices by all of the following terms that apply: real, complex, symmetric, skew symmetric, upper triangular, lower triangular, orthogonal, Hermitian, skew Hermitian.

1. $\begin{bmatrix} 0 & 1\text{+}j & 2\text{-}3j \\ \text{-}1\text{+}j & 0 & 2\text{-}j \\ \text{-}2\text{-}3j & \text{-}2\text{-}j & 0 \end{bmatrix}$ 2. $\begin{bmatrix} 2 & 0 & 0 \\ \text{-}1 & 1 & 0 \\ 3 & \text{-}2 & 4 \end{bmatrix}$ 3. $\begin{bmatrix} \text{-}3 & 0 & 1 \\ 0 & 2 & 5 \\ 0 & 5 & 3 \end{bmatrix}$

4. $\begin{bmatrix} \text{-}1 & 0 & 3 \\ 0 & 2 & \text{-}2 \\ 3 & \text{-}2 & 1 \end{bmatrix}$ 5. $\begin{bmatrix} \text{-}3 & 1\text{+}j & 1\text{-}j \\ 0 & 2\text{-}j & 1\text{-}2j \\ 0 & 0 & 1 \end{bmatrix}$ 6. $\begin{bmatrix} \frac{1}{\sqrt{2}} & 0 & \frac{\text{-}1}{\sqrt{2}} \\ 0 & 1 & 0 \\ \frac{1}{\sqrt{2}} & 0 & \frac{1}{\sqrt{2}} \end{bmatrix}$

7. $\begin{bmatrix} 0 & 1\text{+}j & 2\text{-}3j \\ 1\text{-}j & 0 & 2\text{-}j \\ 2\text{+}3j & 2\text{+}j & 0 \end{bmatrix}$

8. For the matrix A below, evaluate $A\bar{A}, \bar{A}A^T$ and discuss your results.

$$A = \begin{bmatrix} 1\text{+}j & \text{-}1\text{+}3j \\ 3\text{+}2j & 2\text{-}j \end{bmatrix}$$

Show in general that any real matrix of order n may be expressed as the sum of a symmetric and skew symmetric matrix, both of order n.

ADDITIONAL EXERCISES

1.*   Given the matrix

$$A = \begin{bmatrix} 1 & 1 \\ 1 & 1 \end{bmatrix},$$

show that $A^n = 2^{n-1}A$ where $n = 1, 2, 3, \ldots$

2.   Show that if

$$A = \begin{bmatrix} \cosh\theta & \sinh\theta \\ \sinh\theta & \cosh\theta \end{bmatrix}, \text{ then } A^n = \begin{bmatrix} \cosh(n\theta) & \sinh(n\theta) \\ \sinh(n\theta) & \cosh(n\theta) \end{bmatrix},$$

where $n = \pm 1, \pm 2, \ldots$ and $\cosh\theta$, $\sinh\theta$ are the hyperbolic functions defined as

$$\cosh\theta = \frac{1}{2}\left(e^\theta + e^{-\theta}\right), \qquad \sinh\theta = \frac{1}{2}\left(e^\theta - e^{-\theta}\right).$$

3.   Show that the matrix

$$J = \begin{bmatrix} 0 & -1 \\ 1 & 0 \end{bmatrix}$$

behaves in a manner similar to the unit of imaginaries $j = \sqrt{-1}$ in the theory of complex numbers. That is, $J^2 = -I$, $J^3 = -J$, $J^4 = I$, etc., where I is the second-order unit matrix.

4.   Expand $(A + B)(A - B)$ in general terms. Note that the expansion must contain four and not two terms. Verify that your expansion is correct by evaluating both expressions using a third-order matrix you make up on your own.

5.   Show in general that diagonal matrices of the same order commute.

6.   If $C = AA^T$, show that $C = C^T$.

7.   If X is an nth-order column vector, show that $XX^T = C$, where C is a square matrix with the property that $C = C^T$.

8.*   For the matrix C defined in exercise 7, demonstrate by the use of at least three examples and develop a proof that the rank of C is always one if X is a nonzero vector.

9.   Show that the symmetric matrix

$$A = \begin{bmatrix} a & h \\ h & b \end{bmatrix}$$

may be transformed to a diagonal form by the operation $D = R_\theta A R_\theta^T$, where

$$R_\theta = \begin{bmatrix} \cos\theta & \sin\theta \\ -\sin\theta & \cos\theta \end{bmatrix} \text{ and } \tan(2\theta) = \frac{2h}{(a-b)}$$

This is known as Jacobi's transformation.

10.   Given the same R_θ matrix as in exercise 9 above, what happens when the same operation is applied to the following matrix? Assume that $c \neq h$.

$$A = \begin{bmatrix} a & c \\ h & b \end{bmatrix}$$

11.   The inverse of a general second-order matrix was defined in exercise 1 of section 6. Use algebra to solve the appropriate equations to derive the result stated for the inverse, and state the condition for its existence.

12.*   For the A and E matrices given in exercise 10, section 1 above, determine and discuss what the result is for the operation EAE^T.

13.   It is a fact that any square complex matrix may be expressed as the sum of a Hermitian and a skew Hermitian matrix.

        If $-A = A^T$, then A is a Hermitian matrix. In terms of elements, $-a_{ij} = a_{ji}$. If $-A = -A^T$, A is skew Hermitian. In terms of elements, $-a_{ij} = -a_{ji}$.

For a general n^{th}-order complex matrix, derive expressions by which the Hermitian and skew Hermitian components may be calculated. Determine the Hermitian and skew Hermitian components of the following matrix.

$$A = \begin{bmatrix} 2-j & -1+2j & 3 \\ 1+j & 1+2j & 2-2j \\ -1-j & 4j & 1-2j \end{bmatrix}$$

14.*   For the Hermitian matrix given below, evaluate A^2, A^3, A^4, etc., and discuss your results.

$$A = \begin{bmatrix} 1 & 1+j & 2+j \\ 1+j & 1 & 1-2j \\ 2-j & 1+2j & 2 \end{bmatrix}$$

15.   Show that the inverse of the rotation matrix R_θ is equal to its transpose.

$$R_\theta = \begin{bmatrix} \cos\theta & \sin\theta \\ -\sin\theta & \cos\theta \end{bmatrix}$$

II

LINEAR ALGEBRAIC EQUATIONS AND VECTORS

1. INTRODUCTION

As stated in the introduction to this text, in 1857, the English mathematician Cayley established the basis for the modern theory of matrices. This original work dealt mainly with the utilization of matrices in the solution of linear algebraic and linear differential equations. To the present time, these areas have remained of primary interest in the study of matrices. The reason for this is that either a system of linear algebraic or linear differential equations lies at the heart of solving a large number of problems.

The interested reader will also discover a rich area for further study in the subject of linear spaces that is the basis of the conceptual framework the material of this chapter. Those interested in pursuing this subject will find several good texts available, some of which are listed in the references at the end of this text.

The concept of linear spaces unifies the seemingly diverse topics of vectors and functions into one basic and truly beautiful theory. A detailed development of the theory of linear spaces is not presented as there are several excellent texts available on the subject. The brief introduction presented will be related to vector algebra in the hope that the reader may more easily visualize the basic concepts.

2. LINEAR ALGEBRAIC EQUATIONS

As stated above, linear algebraic equations are encountered in many areas of applied analysis. In chapter I, we employed a system of three linear algebraic equations in three unknowns as a means of introducing the concept of a matrix. They arise with great frequency not only in problems of physics and engineering, but also in economic analysis and a variety of problems in statistics.

As a specific illustration of how a set of such equations arises, consider the resistive electrical circuit shown below.

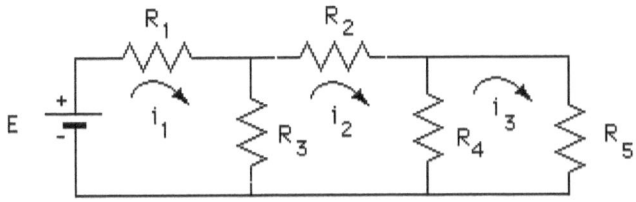

Fig. 2.1

For this circuit, it is desired to determine the three circulating or mesh currents i_1, i_2, and i_3 given values for the dc voltage source and resistances. Assuming that the values for these elements are $E = 12$ V, $R_1 = 5$ Ω, $R_2 = 8$ Ω,$R_3 = 6$ Ω, $R_4 = R_5 =10$ Ω, the three loop equations, found by applying Kirchoff's voltage law[3] are

(2.1)
$$11i_1 - 6i_2 = 12$$
$$-6i_1 + 24i_2 - 10i_3 = 0$$
$$-10i_2 + 20i_3 = 0 .$$

In matrix notation, the coefficient matrix, unknown, and constant vectors would be the following.

$$A = \begin{bmatrix} 11 & -6 & 0 \\ -6 & 24 & -10 \\ 0 & -10 & 20 \end{bmatrix}, \quad X = \begin{bmatrix} i_1 \\ i_2 \\ i_3 \end{bmatrix}, \quad B = \begin{bmatrix} 12 \\ 0 \\ 0 \end{bmatrix}$$

Generally, the coefficients a_{ij} and the quantities b_i are known, and it is required to determine the values of the unknowns x_i. This has resulted in a square system of order 3.

At this point, let us broaden the scope of the problem to a system of m equations and n unknowns without any constraints on either m or n. In particular, this general system has the form

$$a_{11}x_1 + a_{12}x_2 + \ldots + a_{1n}x_n = b_1$$

---

[3]   Readers who are unfamiliar with the details of electrical circuits at this point should not be too concerned. This illustration is being introduced at this point only to show one area where linear algebraic equations arise in practice. In chapter IV, a very direct geometrical method of directly constructing the resulting matrix equations for general circuits will be introduced.

(2.2) $a_{21}x_1 + a_{22}x_2 + \ldots + a_{2n}x_n = b_2$

$$\ldots$$

$$a_{m1}x_1 + a_{m2}x_2 + \ldots + a_{mn}x_n = b_m$$

Equation (2.2) represents m equations for the n unknowns x_j, $j = 1, 2, 3, \ldots, n$. The quantities a_{ij}, $i = 1, 2, \ldots, m$, $j = 1, 2, \ldots, n$, are the known coefficients and the b_i, $i = 1, 2, \ldots, m$, are the known constants. If we utilize matrix notation, equation (2.2) may be written in the convenient form

(2.3) $AX = B$,

where A is the $m \times n$ coefficient matrix,

$$A = \begin{bmatrix} a_{11} & a_{12} & \cdots & a_{1n} \\ a_{21} & a_{22} & \cdots & a_{2n} \\ \vdots & \vdots & \ddots & \vdots \\ a_{m1} & a_{m2} & \cdots & a_{mn} \end{bmatrix}$$

B the mth order constant vector,

$$B = \begin{bmatrix} b_1 \\ b_2 \\ \vdots \\ b_m \end{bmatrix}$$

and X is the nth-order unknown vector,

$$X = \begin{bmatrix} x_1 \\ x_2 \\ \vdots \\ x_n \end{bmatrix}.$$

The problem of solving the general system (2.3) is to determine a set of values x_i, that is, the vector X, which satisfy the given equations. As we shall see in the following sections, if solutions to (2.3) exist, they may or may not be unique.

DEFINITION

Consistent/Inconsistent Systems of Linear Equations

If solutions of any type exist for a general system of equations such as (2.2) above, then the system is said to be consistent. Conversely, if solutions do not exist, the system is called inconsistent.

---

3. NONHOMOGENEOUS SYSTEMS

DEFINITION

Nonhomogeneous/Homogeneous Systems

If the elements of the constant vector B of the general equation (2.3) do not all vanish, i.e., $B \neq 0$, then the system of linear algebraic equations is called nonhomogeneous. By implication, when the vector B is the zero or null vector, then the system is called homogeneous.

DEFINITION

Augmented Matrix

In dealing with systems of linear algebraic equations, we frequently make use of the augmented matrix, which is defined as

$$\text{Aug } A = \begin{bmatrix} A|B \end{bmatrix} = [AB] \ .$$

The augmented matrix is the coefficient matrix A, with the coefficient vector B added as the last column. The order of the augmented matrix for the system (2.3) is $m \times (n + 1)$.

EXAMPLE 1

For the following system of linear algebraic equations, identify the coefficient and augmented matrices.

$$3x_1 + 2x_2 + x_3 = 1$$
$$2x_1 + x_2 - x_3 = 2$$
$$x_1 + 4x_2 = -1$$

SOLUTION

$$A = \begin{bmatrix} 3 & 2 & 1 \\ 2 & 1 & -1 \\ 1 & 4 & 0 \end{bmatrix} = \text{the coefficient matrix}$$

$$\text{Aug } A = \begin{bmatrix} 3 & 2 & 1 & 1 \\ 2 & 1 & -1 & 2 \\ 1 & 4 & 0 & -1 \end{bmatrix}$$

***

We may now state the condition for which system (2.3) is consistent, that is, has a solution.

Condition 1

The system (2.3) of m linear equations and n unknowns is consistent if and only if the coefficient matrix A and the augmented matrix Aug A have the same rank. A related important fact is that when this condition is true, the rank is the number of independent equations.

A set of linear algebraic equations is independent if none of the equations may be represented as a linear combination of any of the other equations.

If the system (2.3) is consistent, the solutions may be categorized into one of the following two cases.

Case 1.1

Rank A = rank Aug $A = r < n$.

   In this case, a complete solution always exists.

   It should be noted that this case includes $r = m < n$ and $r < m, r < n$.

   Also, the rank $r = \min(m, n)$.

   The term complete solution refers to a solution which contains all possible solutions to a given set of equations.

EXAMPLE 2

   Determine if the following system is consistent.

$$\begin{aligned} x_1 - x_2 + x_3 &= 2 \\ 3x_1 - x_2 + 2x_3 &= -6 \\ 3x_1 + x_2 + x_3 &= -18 \end{aligned}$$

In matrix form, this system may be written

(3.1)
$$\begin{bmatrix} 1 & -1 & 1 \\ 3 & -1 & 2 \\ 3 & 1 & 1 \end{bmatrix} X = \begin{bmatrix} 2 \\ -6 \\ -18 \end{bmatrix}.$$

SOLUTION

To check the ranks of the coefficient and augmented matrices, the augmented matrix must be converted to upper triangular form. The augmented matrix is

$$\text{Aug} = \begin{bmatrix} 1 & -1 & 1 & 2 \\ 3 & -1 & 2 & -6 \\ 3 & 1 & 1 & -18 \end{bmatrix}$$

The first elementary matrix needed to obtain the desired form is the following.

$$E_1 = \begin{bmatrix} 1 & 0 & 0 \\ -3 & 1 & 0 \\ -3 & 0 & 1 \end{bmatrix}$$

Premultiplying the augmented matrix by this elementary matrix gives the following modified augmented matrix.

$$Aug_1 = E_1 \, Aug = \begin{bmatrix} 1 & -1 & 1 & 2 \\ 0 & 2 & -1 & -12 \\ 0 & 4 & -2 & -24 \end{bmatrix}$$

The next elementary matrix is

$$E_2 = \begin{bmatrix} 1 & 0 & 0 \\ 0 & 1 & 0 \\ 0 & -2 & 1 \end{bmatrix}.$$

Premultiplying the previous result by this matrix leads to the desired upper triangular form.

$$\text{Augmented matrix for (3.1)} = \begin{bmatrix} 1 & -1 & 1 & 2 \\ 0 & 2 & -1 & -12 \\ 0 & 0 & 0 & 0 \end{bmatrix}$$

From this result, we see that the ranks of the coefficient and augmented matrices are the same and equal to 2 since the coefficient matrix is contained in the augmented matrix and is the leftmost three columns.

Thus, the system (3.1) is consistent; however, it cannot be solved by either Cramer's rule or matrix inversion. The reason for the failure of these methods is that the determinant of the coefficient matrix is zero since its rank is less than its order (has a row of zeros).

***

In cases where a system is consistent but the rank is less than the number of unknowns, the solution may be obtained by operating directly with the equations. As a preview of some of the formal numerical methods by which simultaneous linear equations are solved, which will be presented later, let us apply elementary row operation to the system (3.1). This approach is essentially the same as employed in chapter I, section 10 in evaluating the rank of a matrix. It has been left as an exercise for the reader to verify that performing any of the elementary row operations on the coefficient matrix of a system does not affect the order of unknowns in the unknown vector. In applying elementary operations to the rows of a system of equations such as (3.1), the same operations must be performed on the constant vector. The process is simply the reduction of the system to upper triangular form as done to check the rank in the above example. For the sake of illustration, the steps will be carried out in detail here.

In developing the solution to equation (3.1), we first add minus three times the first row to both the second and third rows in order to reduce the elements in the first column of those rows to zero. The result is

$$\begin{bmatrix} 1 & -1 & 1 \\ 0 & 2 & -1 \\ 0 & 4 & -2 \end{bmatrix} X = \begin{bmatrix} 2 \\ -12 \\ -24 \end{bmatrix} .$$

One should notice that the third row is simply two times the second row, indicating that the last two equations of the set are dependent. The next step is therefore to add minus two times the second row to the third row to obtain

$$(3.2) \qquad \begin{bmatrix} 1 & -1 & 1 \\ 0 & 2 & -1 \\ 0 & 0 & 0 \end{bmatrix} X = \begin{bmatrix} 2 \\ -12 \\ 0 \end{bmatrix} .$$

Clearly, we have arrived at exactly the same form as that given by the augmented matrix, which has been reduced to upper triangular form.

To continue with the solution, let us introduce partitioning of (3.2) in the following way.

$$\left[\begin{array}{cc:c} 1 & -1 & 1 \\ 0 & 2 & -1 \\ \hdashline 0 & 0 & 0 \end{array} \right] \begin{bmatrix} x_1 \\ x_2 \\ x_3 \end{bmatrix} = \begin{bmatrix} 2 \\ -12 \\ \hline 0 \end{bmatrix}$$

Expanding this, we have the two submatrix equations.

$$\begin{bmatrix} 1 & -1 \\ 0 & 2 \end{bmatrix} \begin{bmatrix} x_1 \\ x_2 \end{bmatrix} + \begin{bmatrix} 1 \\ -1 \end{bmatrix} x_3 = \begin{bmatrix} 2 \\ -12 \end{bmatrix}$$

$$[0 \ 0] \begin{bmatrix} x_1 \\ x_2 \end{bmatrix} + 0x_3 = 0$$

Clearly, the second equation is satisfied identically for any value of the solution vector X. The first equation allows us to solve for the subvector $[x_1, x_2]$ by premultiplying by the inverse of its second-order coefficient matrix.

$$\begin{bmatrix} x_1 \\ x_2 \end{bmatrix} = \begin{bmatrix} 1 & -1 \\ 0 & 2 \end{bmatrix}^{-1} \left(\begin{bmatrix} 2 \\ -12 \end{bmatrix} - \begin{bmatrix} 1 \\ -1 \end{bmatrix} x_3 \right)$$

Computing the inverse and performing the multiplications gives the following result.

$$\begin{bmatrix} x_1 \\ x_2 \end{bmatrix} = \frac{1}{2}\begin{bmatrix} 2 & 1 \\ 0 & 1 \end{bmatrix} \left(\begin{bmatrix} 2 \\ -12 \end{bmatrix} - \begin{bmatrix} 1 \\ -1 \end{bmatrix} x_3 \right)$$
$$= -\begin{bmatrix} 4 \\ 6 \end{bmatrix} + \frac{1}{2}\begin{bmatrix} -1 \\ 1 \end{bmatrix} x_3$$

In this solution, we note that the value of x_3 is arbitrary. If we let $x_3 = 2c$, where c is a parameter, then the solution may be expressed as

(3.3)
$$\begin{bmatrix} x_1 \\ x_2 \\ x_3 \end{bmatrix} = \begin{bmatrix} -4-c \\ -6+c \\ 2c \end{bmatrix} = -\begin{bmatrix} 4 \\ 6 \\ 0 \end{bmatrix} + c\begin{bmatrix} -1 \\ 1 \\ 2 \end{bmatrix} .$$

DEFINITION

Complete and Particular Solutions

The solution (3.3) is called a complete or underdetermined solution in that it represents all possible solutions (infinite) as the parameter c ranges over all possible values. Whenever a specific value of c is selected, the solution is called a particular solution. As an illustration if the constant c in the solution (3.3) above is set to one, then a particular solution of equation (3.1) is

$$X = -\begin{bmatrix} 5 \\ 5 \\ -2 \end{bmatrix} .$$

As we will develop later, this complete solution has the form

$$X = X_{nh} + cX_h,$$

where X_{nh} is the nonhomogeneous part and X_h is the homogeneous part. As stated, these terms will be defined later in section 11.

Case 1.2

Rank A = rank Aug $A = r = m = n$.

In this special situation, the complete solution is unique and is given by

(3.4) $X = A^{-1}B.$

Although this last case is a restricted subcase of the general theory, it is perhaps the most frequently encountered case in applied analysis. This particular case is frequently referred to as the nonsingular case.

At this point of our development there have been two ways presented by which the general system of equations may be solved under the situation of case 1.2. One is by the application of Cramer's rule and the other by computation of the inverse of the coefficient matrix for the application of equation (3.4). Whenever the order of the system becomes somewhat large, that is. $n \geq 4$, the amount of computations involved in these methods become cumbersome. In case 1.1 above, the reader was introduced to the beginnings of a method that is the approach taken by one of the most powerful and efficient methods employed to solve large systems. This is the method of Gaussian elimination with back substitution and partial pivoting, which will now be developed in the following sections.

EXERCISES

1.   Show that interchange of any two rows in an augmented matrix does not affect the order of the unknowns in a system of equations such as (2.3).

2.*   Write the augmented matrix for the following system and determine if it is consistent.

$$\begin{bmatrix} 1 & 1 & 1 \\ 2 & -1 & -1 \\ 1 & 2 & -1 \end{bmatrix} X = \begin{bmatrix} 2 \\ 1 \\ 3 \end{bmatrix}$$

3.*   Write the augmented matrix for the following system[4] and determine if it is consistent.

$$\begin{bmatrix} 5 & -1 & 2 \\ 1 & 1 & 1 \\ 9 & -3 & 3 \\ 11 & -1 & 3 \end{bmatrix} X = \begin{bmatrix} -5 \\ 7 \\ 17 \\ -3 \end{bmatrix}$$

***

4.   GAUSSIAN ELIMINATION WITH BACK SUBSTITUTION

The most prevalent method for solving systems of linear algebraic systems is based on the method of Gaussian elimination, the beginnings of which have been employed in some of the examples above. As we shall see, it is a very efficient procedure for solving large systems of equations. With some modifications to be introduced later, it remains the most applicable to a wide range of cases.

To develop the method, let us assume that we have a system of n equations and n unknowns, an nth-order system, which is consistent and of rank n. This is case 1.2 above for which a unique solution exists. Although it is a very restricted example, it does represent a diverse range of applied problems. Let this system be

$$\begin{aligned} a_{11}x_1 + a_{12}x_2 + \ldots + a_{1n}x_n &= b_1 \\ a_{21}x_1 + a_{22}x_2 + \ldots + a_{2n}x_n &= b_2 \\ &\cdots \\ a_{n1}x_1 + a_{n2}x_2 + \ldots + a_{nn}x_n &= b_n \end{aligned}$$

(4.1)

As an additional assumption, let us assume that the equations have been arranged so that none of the diagonal coefficients a_{ii} vanish.

---

[4]   Hohn [1985], p120.

As an overview, the method is to perform elementary operations on the above equations so that they are reduced to upper triangular form. This means that the new system of equations has all zero elements below the main diagonal. Also, rather than working with the equations directly, the operations are applied to the augmented matrix of the system.

(4.2)

$$\text{Aug } A = \begin{bmatrix} a_{11} & a_{12} & \cdots & a_{1n} & b_1 \\ a_{21} & a_{22} & \cdots & a_{2n} & b_2 \\ \cdot & \cdot & \cdots & \cdot & \cdot \\ a_{n1} & a_{n1} & \cdots & a_{nn} & b_n \end{bmatrix}$$

We now seek to reduce all the elements in the first column below a_{11} to zero. For the second row, each element would be replaced by its original value minus a_{21}/a_{11} times the corresponding element in the first row. If we use the superscript [1] to designate this first operation, we could write

(4.3)
$$a^1_{2k} = a_{2k} - \left(\frac{a_{21}}{a_{22}}\right) a_{1k}, k = 1,2,\cdots,n.$$

A brief inspection of this equation will show that the new element in the first row, a^1_{2k}, will be identically zero.

If this procedure is applied to every row, the general operation may be deduced from (4.3) to be

(4.4)
$$a^1_{jk} = a_{jk} - \left(\frac{a_{j1}}{a_{11}}\right) a_{1k}, k = 1,2,\cdots,n, j = 2,3,\cdots,n.$$

After all the rows of the augmented matrix have been transformed, the result is

(4.5)

$$\text{Aug } A_1 = \begin{bmatrix} a_{11} & a_{12} & \cdots & a_{1n} & b_1 \\ O & a^1_{22} & \cdots & a^1_{2n} & b^1_2 \\ \cdot & \cdot & \cdots & \cdot & \cdot \\ O & a^1_{n1} & \cdots & a^1_{nn} & b^1_n \end{bmatrix}$$

This form is achieved through the means discussed in section 10 of chapter I, namely through the use of elementary operations. The reduction of Aug A to Aug A_1 could be achieved by use of the following elementary matrix.

$$
E_1 = \begin{bmatrix}
1 & 0 & 0 & \cdots & 0 \\
-\dfrac{a_{21}}{a_{11}} & 1 & 0 & \cdots & 0 \\
-\dfrac{a_{31}}{a_{11}} & 0 & 1 & \cdots & 0 \\
\vdots & \vdots & \vdots & \ddots & \vdots \\
-\dfrac{a_{n1}}{a_{11}} & 0 & 0 & \cdots & 1
\end{bmatrix}
$$

From the previous discussion in chapter I, we would have

$$\text{Aug } A_1 = E_1 \text{ Aug } A.$$

This matrix multiplication accomplishes all the operations presented formally in equation (4.4) above.

We now repeat the process to reduce all the elements below the main diagonal of the second column to zero. In this case, the general operation is denoted as

(4.6) $\qquad a_{jk}^2 = a_{jk}^1 - \left(\dfrac{a_{j2}^1}{a_{22}^1} \right) a_{2k}^1, k = 2,3,\cdots,n, j = 3,4,\cdots,n.$

In terms of an elementary matrix, these relations are represented by

$$
E_2 = \begin{bmatrix}
1 & 0 & 0 & \cdots & 0 \\
0 & 1 & 0 & \cdots & 0 \\
0 & -\dfrac{a_{32}^1}{a_{22}^1} & 1 & \cdots & 0 \\
\vdots & \vdots & \vdots & \ddots & \vdots \\
0 & -\dfrac{a_{n2}^1}{a_{22}^1} & 0 & \cdots & 1
\end{bmatrix}
$$

The next modified augmented matrix would be given by

$$\text{Aug } A_2 = E_2 \text{ Aug } A_1 = E_2 E_1 \text{ Aug } A.$$

If this process is repeated $n - 1$ times, the final augmented matrix has the following form, which is in row echelon form.

$$
(4.7) \qquad \text{Aug } A_{n-1} = \begin{bmatrix}
a_{11} & a_{12} & a_{13} & \cdots & a_{1n} & b_1 \\
0 & a_{22}^1 & a_{23}^1 & \cdots & a_{2n}^1 & b_2^1 \\
0 & 0 & a_{33}^2 & \cdots & a_{3n}^2 & b_3^2 \\
\cdot & \cdot & \cdot & \cdots & \cdot & \cdot \\
0 & 0 & 0 & \cdots & a_{nn}^{n-1} & b_n^{n-1}
\end{bmatrix}
$$

Since these elementary operations do not alter the solution or change the order of the unknowns, it is convenient to use a simpler notation for the elements of the augmented matrix.

As a means of avoiding cumbersome notation in the remaining development, let us redefine (4.7) as follows.

$$
\text{Aug C} =
\begin{bmatrix}
u_{11} & u_{12} & u_{13} & \cdots & u_{1n} & d_1 \\
O & u_{22} & u_{23} & \cdots & u_{2n} & d_2 \\
O & O & u_{33} & \cdots & u_{3n} & d_3 \\
\cdot & \cdot & \cdot & \cdots & & \\
O & O & O & \cdots & u_{nn} & d_n
\end{bmatrix}
= [U, D]
$$

This means that our original system of equations (4.1) now has the form

(4.8)
$$
\begin{aligned}
u_{11}x_1 + u_{12}x_2 + \ldots + u_{1n}x_n &= d_1 \\
u_{22}x_2 + \ldots + u_{2n}x_n &= d_2 \\
\cdots \\
u_{nn}x_n &= d_n
\end{aligned}
$$

or

$$UX = D,$$

where the coefficient matrix U is upper triangular. It should be noted that each $u_{ii} \neq 0$ since the original coefficient matrix was assumed to be invertable.

From this point, the solution continues by the method of back substitution. From the last equation in (4.8), the variable x_n may be found directly as

$$x_n = d_n/u_{nn}.$$

This result may be back substituted into the preceding equation to give

$$x_{n-1} = (d_{n-1} - u_{n-1,n}x_n)/u_{n-1,n-1}.$$

In general the kth unknown is determined from the following relation by the back substitution process.

$$x_k = \left(d_k - \sum_{r=k+1}^{n} u_{kr}x_r \right)/u_{kk}$$

As each successive unknown is determined by working upward (back substituting) through the equations in order until the last unknown, x_1, is found.

At this point, it is of interest to compare the cost saving of this method of solving a system of simultaneous equations relative to solution by direct inversion. From the discussion in the last chapter of the number of operations required to invert a matrix directly, it can be seen that for a large number of unknowns, the number of operations is approximately $n!(n^2)$. For Gaussian elimination with back substitution, the operation count is $(n^3 + 3n^2 - n)/3$ or approximately $n^3/3$ for large n. If we had a system with 10 unknowns, $n = 10$, direct inversion would require about 3.6×10^8 operations in comparison to 430 for Gaussian elimination with back substitution. This represents a reduction by a factor of nearly a million times fewer operations, a significant saving!

EXAMPLE 3

As a specific example of the Gaussian elimination method, apply it to find the solution to the resistive circuit problem whose equations for the circulating currents are given in equation (2.3).

SOLUTION

For this system, the augmented matrix is

$$\text{Aug } A = \begin{bmatrix} 11 & -6 & 0 & 12 \\ -6 & 24 & -10 & 0 \\ 0 & -10 & 20 & 0 \end{bmatrix}.$$

We now proceed to reduce the elements of the first column below the main diagonal by the row operations defined by equation (4.4). The following elementary matrix represents these operations.

$$E_1 = \begin{bmatrix} 1 & 0 & 0 \\ \dfrac{6}{11} & 1 & 0 \\ 0 & 0 & 1 \end{bmatrix} = \begin{bmatrix} 1 & 0 & 0 \\ 0.546 & 1 & 0 \\ 0 & 0 & 1 \end{bmatrix}$$

A zero appears in the last element of the first column because that element in the original matrix was already zero. At this point, we are also going to switch to floating-point notation and use arithmetic, which is rounded to three decimal places. This is being done to provide an illustration of the effect of round-off error in computations.

The first modified augmented matrix is given[5] by Aug $A_1 = E_1$ Aug A or

$$\text{Aug } A_1 = \begin{bmatrix} 11 & -6 & 0 & 12 \\ 0.006 & 20.724 & -10 & 6.552 \\ 0 & -10 & 20 & 0 \end{bmatrix}.$$

We now move to the second column where we want to reduce the element below the main diagonal to zero. From (4.6), we see that we must subtract the constant multiple of $(-10/20.724)$ times each element of the second row from the third.

(4.9) $$a_{3k}^2 = a_{3k}^1 - \left(\frac{-10}{20.73}\right)a_{3k}^1, \quad k = 2, 3, \cdots, n$$

This is accomplished by premultiplication by the following elementary matrix.

$$E_2 = \begin{bmatrix} 1 & 0 & 0 \\ 0 & 1 & 0 \\ 0 & 0.483 & 1 \end{bmatrix}$$

The final triangularized form of the augmented matrix is now computed from the product $E_2 \text{ Aug } A_1$.

$$\text{Aug } A_2 = \begin{bmatrix} 11 & -6 & 0 & 12 \\ 0.006 & 20.724 & -10 & 6.552 \\ 0.003 & 0.010 & 15.170 & 3.165 \end{bmatrix}$$

This result illustrates the problem of round-off error in the appearance of small nonzero values in the lower triangular portion where zeros should occur. This problem arises due to the computations being made with only three-decimal-digit accuracy. The exact answer is

$$\text{Aug } A_2 = \begin{bmatrix} 11 & -6 & 0 & 12 \\ 0 & \dfrac{228}{11} & -10 & \dfrac{72}{11} \\ 0 & 0 & \dfrac{865}{57} & \dfrac{180}{57} \end{bmatrix}.$$

If we were to carry out the answer to sixteen significant figures and then round the values to the previous three decimal places, we would have the following result.

$$\text{Aug } A_2 = \begin{bmatrix} 11 & -6 & 0 & 12 \\ 0.000 & 20.727 & -10 & 6.546 \\ 0.000 & 0.000 & 15.175 & 3.158 \end{bmatrix}$$

From this augmented matrix, we can see that the original system of equations (2.3) has been transformed to the following matrix equation.

$$(4.10) \qquad \begin{bmatrix} 11 & -6 & 0 \\ 0 & 20.727 & -10 \\ 0 & 0 & 15.175 \end{bmatrix} \begin{bmatrix} i_1 \\ i_2 \\ i_3 \end{bmatrix} = \begin{bmatrix} 12 \\ 6.546 \\ 3.158 \end{bmatrix}$$

We may now proceed with the back substitution process by solving the last equation for the current i_3. The result is $i_3 = 3.158/15.175 = 0.208$ Ampere. The second equation of (4.10) gives

$$20.727 i_2 = 6.546 + 10 i_3,$$

from which we find $i_3 = 0.416$ Ampere. The first equation of (4.10) yields

$$11 i_1 = 12 + 6 i_2.$$

This gives the value of the last unknown as $i_1 = 1.318$ Ampere. The solution vector is

$$X = \begin{bmatrix} i_1 \\ i_2 \\ i_3 \end{bmatrix} = \begin{bmatrix} 1.318 \\ 0.416 \\ 0.208 \end{bmatrix}, \text{ Ampere}$$

***

As stated in the introduction to this text, it is always good practice to check answers once they have been obtained. To check the answer above, we should verify that this solution satisfies the original equation (2.3). To do this, it is convenient to write the general equation (2.3) as follows.

$$AX - B = 0_n$$

If we had obtained the true solution, this equation would indeed vanish upon substitution of the above solution vector. However, this is not the case in almost all real problems. Substituting the current vector obtained above gives the following result.

$$AX - B = \begin{bmatrix} 11 & -6 & 0 \\ -6 & 24 & -1 \\ 0 & -10 & 20 \end{bmatrix} \begin{bmatrix} 1.318 \\ 0.416 \\ 0.208 \end{bmatrix} - \begin{bmatrix} 12 \\ 0 \\ 0 \end{bmatrix} = \begin{bmatrix} 0.001 \\ -0.004 \\ 0.000 \end{bmatrix}$$

This result should not surprise the reader because the solution vector given above was rounded to three decimal places. It should be remembered that this example is only third order, and only a few calculations were involved in finding the solution. In much larger problems, the round-off errors can grow significantly. Since we wanted a three-decimal-place answer, the calculations should have been carried out with several more digits of accuracy and only the final result rounded to the desired three places. Also, the accuracy check above should not have been done with the rounded result but a more accurate nonrounded solution.

In a later chapter on numerical methods, we will discuss how this method may be applied to consistent systems in which the rank is less than the order and when the number of equations does not equal the number of unknowns. For those readers who are interested in programming and would like to write an operation code for this method, one particular note is useful at this point. Although the algorithm is relatively easy to program, it should not be written to apply automatically to all elements lying below the main diagonal. Frequently, many of these will be zero initially, and it is a waste of computational time to recalculate a result that is already known. A more efficient computer code should include a test for a zero value prior to each row operation to save time. Another useful point is to note that the original augmented matrix does not have to be retained. As new element values for each row are computed, they may be stored in the same memory locations or matrix element.

EXAMPLE 4

As a graphic illustration of how the Gaussian elimination method reorients a system of equations being solved, sketch each of the following equations in the two-dimensional solution space x_1, x_2. Repeat the sketches after each step of the process, and discuss what is taking place geometrically. Perform these steps for the following system.

$$\begin{bmatrix} 2 & 1 \\ 1 & 2 \end{bmatrix} X = \begin{bmatrix} 1 \\ 0 \end{bmatrix}$$

SOLUTION

In terms of the elements of the solution vector, the above system may be written as

$$2x_1 + x_2 = 1$$
$$x_1 + 2x_2 = 0.$$

In x_1, x_2 space, these equations may be graphed as shown below in figure 2.2. The solution occurs where the two lines (equations) intersect.

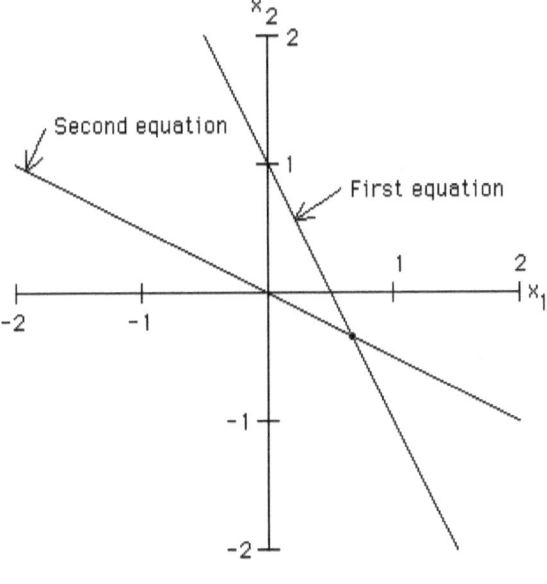

Fig. 2.2

Applying Gaussian elimination requires the transformation of the system of equations to upper triangular form. The resulting equations are

$$2x_1 + x_2 = 1$$
$$\frac{3}{2} x_2 = -\frac{1}{2}.$$

Plotting these in the same manner as above results in the following graph. What this figure indicates is that the process of triangularization has rotated the second equation until it becomes parallel with the x_1 coordinate axis. The point of rotation is the solution point. By applying back substitution, the solution is found to be $x_1 = 2/3$, $x_2 = -1/3$.

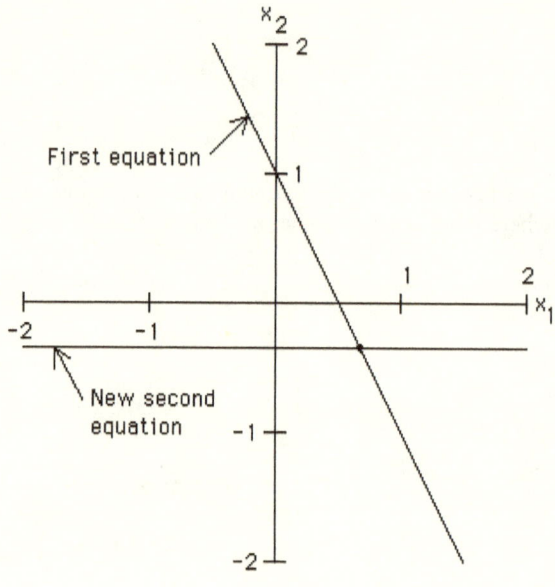

Fig. 2.3

EXERCISES

1.   Consider the following second-order system.

$$\begin{bmatrix} 1 & 1 \\ -2 & 1 \end{bmatrix} X = \begin{bmatrix} 1 \\ -2 \end{bmatrix}$$

   a.   Sketch the lines represented by each equation in x_1, x_2 space.

   b.   Reduce the system to upper triangular form using Gaussian elimination, and sketch the new equations on the same drawing. Discuss what you observe.

2.   Consider the following system of equations.

$$\begin{bmatrix} 1 & 2 & -1 \\ 2 & -1 & 3 \\ 3 & -2 & 3 \end{bmatrix} X = \begin{bmatrix} 3 \\ -1 \\ 2 \end{bmatrix}$$

a.   Sketch the three planes describing each equation in x_1, x_2, x_3 space.

b.   Use Gaussian elimination to reduce the augmented matrix to the following form.

$$\begin{bmatrix} 1 & 2 & -1 \\ 0 & * & * \\ 0 & * & * \end{bmatrix} X = \begin{bmatrix} 3 \\ * \\ * \end{bmatrix}$$

Sketch the planes represented by these three equations, and compare your drawing with that done for part a above.

c.   Again apply Gaussian elimination to reduce the system to the following form.

$$\begin{bmatrix} 1 & 2 & -1 \\ 0 & * & * \\ 0 & 0 & * \end{bmatrix} X = \begin{bmatrix} 3 \\ * \\ * \end{bmatrix}$$

Sketch the planes represented by these three equations, and compare your drawing with those in parts a and b above.

d.   Carefully examine the sketches of the planes in parts a, b, and c above, and describe what the Gaussian elimination procedure has done geometrically in setting up the problem for solution.

3.   Consider the following system.

$$\begin{bmatrix} 1 & 0 & 1 \\ 0 & 1 & -1 \\ 0 & -1 & 1 \end{bmatrix} X = \begin{bmatrix} -1 \\ 2 \\ c \end{bmatrix}$$

What value must the constant c have for the system to be consistent? What is the solution when c takes on this value?

4.*   Solve the following system.

$$\begin{bmatrix} 2 - j & 1 + j \\ 1 + 2j & 2 + j \end{bmatrix} X = \begin{bmatrix} 1 - j \\ 3 - 2j \end{bmatrix}$$

***

5.   IMPROVEMENT BY COLUMN PIVOTING

In most applied problems such as the example covered in the previous section, the diagonal elements are usually those of greatest magnitude for that particular column. However, in some circumstances, this is not the case, and some column element other than the one on the main diagonal has the largest magnitude. These elements are commonly referred to as pivots because they are the basis for further row operations.

To illustrate what happens in such a situation, consider the following second-order system.

$$(5.1) \qquad \begin{bmatrix} 0.01 & 1.00 \\ 10.00 & -0.20 \end{bmatrix} X = \begin{bmatrix} 1.00 \\ -2.50 \end{bmatrix}$$

For simplicity, it will be assumed that a computer or calculator is used that carries only two decimal digits of accuracy in floating-point notation. Examining the first column indicates that the diagonal or first pivot element is much smaller than the a_{21} element. Without further comment, at this point, let us proceed to solve this system by the method of Gaussian elimination presented in the previous section.

The augmented matrix for the system (5.1) is

$$\text{Aug A} = \begin{bmatrix} 1.000 \text{ E-02} & 1.000 \text{ E+00} & 1.000 \text{ E+00} \\ 1.000 \text{ E+01} & -2.000 \text{ E-01} & -2.500 \text{ E+00} \end{bmatrix}.$$

Following the Gaussian elimination procedure reduces this augmented matrix to the following upper triangular form.

$$\text{Aug A}_1 = \begin{bmatrix} 1.000 \text{ E-02} & 1.000 \text{ E+00} & 1.000 \text{ E+00} \\ 0.000 \text{ E+00} & 1.000 \text{ E+03} & 1.002 \text{ E+03} \end{bmatrix}$$

The reader should be sure to note that the assumed three-digit floating-point computations truncate any values after the third digit. Now by applying back substitution, the following solution vector may be found.

$$X = \begin{bmatrix} -2.000 \text{ E-01} \\ 1.002 \text{ E+00} \end{bmatrix}$$

If we substitute this solution into equation (5.1), it is seen that it does not satisfy the original system of equations.

$$AX = \begin{bmatrix} 1.000 \text{ E-02} & 1.000 \text{ E+00} \\ 1.000 \text{ E+01} & -2.000 \text{ E-01} \end{bmatrix} \begin{bmatrix} -2.000 \text{ E-01} \\ 1.002 \text{ E+00} \end{bmatrix} = \begin{bmatrix} 1.000 \text{ E+00} \\ -2.200 \text{ E+00} \end{bmatrix}$$

We see that there is a 12% error in the last element of the computed B vector.

Now let us take the original system but write the equations in the opposite order.

$$\begin{bmatrix} 1.000 \text{ E+01} & -2.000 \text{ E-01} \\ 1.000 \text{ E-02} & 1.000 \text{ E+00} \end{bmatrix} X = \begin{bmatrix} -2.500 \text{ E+00} \\ 1.000 \text{ E+00} \end{bmatrix}$$

Repeating the process leads us to the following solution vector.

$$X = \begin{bmatrix} -2.300 \text{ E-01} \\ 1.000 \text{ E+00} \end{bmatrix}$$

Substitution of this result into (5.1) indicates that it does satisfy the given equations provided we recall that our calculator is only retaining three decimal digits in floating-point notation.

$$AX = \begin{bmatrix} 1.000 \text{ E-02} & 1.000 \text{ E+00} \\ 1.000 \text{ E+01} & -2.000 \text{ E-01} \end{bmatrix} \begin{bmatrix} -2.300 \text{ E-01} \\ 1.002 \text{ E+00} \end{bmatrix} = \begin{bmatrix} 1.000 \text{ E+00} \\ -2.500 \text{ E+00} \end{bmatrix}$$

The problem illustrated by this example is the loss of significance when we are restricted to a fixed number of digits or precision. This is the situation of all calculators and computers and must be attended to when writing numerical algorithms.

The solution by which we obtained a reasonable solution in the above example was to exchange equations so that the lead diagonal element had the largest magnitude of all elements in the first column. This process is called column or partial pivoting. It is implemented in the Gaussian elimination procedure by scanning a given column for the element of greatest magnitude prior to beginning the sequence of calculations to reduce the subdiagonal elements to zero. If the diagonal element is the greatest, then no changes are needed. If an element of greater magnitude is encountered below the diagonal element in question, then the entire two rows in question are interchanged. The reduction process may continue since the new pivot or diagonal element is now the one of largest magnitude. Again it should be recalled that such operations do not disturb the solution in that the order of the quantities in the vector of unknowns is not affected.

EXERCISES

Determine which of the following systems are consistent, and obtain the solution for those that are by using Gaussian elimination. Employ partial pivoting where called for.

1.
$$\begin{bmatrix} 2 & -1 \\ 4 & 3 \end{bmatrix} X = \begin{bmatrix} 5 \\ -1 \end{bmatrix}$$

2.*
$$\begin{bmatrix} 1 & 1 & 1 \\ 2 & -1 & -1 \\ 1 & 2 & -1 \end{bmatrix} X = \begin{bmatrix} 2 \\ 1 \\ 3 \end{bmatrix}$$

3.*[6]
$$\begin{bmatrix} 1 & 1 & 0 \\ 0 & 1 & 6 \\ 2 & 2 & 1 \\ 0 & -1 & 1 \end{bmatrix} X = \begin{bmatrix} 3 \\ 1 \\ 6 \\ -1 \end{bmatrix}$$

4.*[7]
$$\begin{bmatrix} 3 & -1 & 1 \\ 1 & -1 & 0 \\ 2 & 0 & 1 \\ 4 & 2 & 1 \end{bmatrix} X = \begin{bmatrix} 4 \\ 0 \\ 4 \\ 4 \end{bmatrix}$$

5.*[8]
$$\begin{bmatrix} 1 & -1 & 1 & 1 \\ 1 & 1 & -1 & 2 \\ 3 & -1 & 1 & 2 \end{bmatrix} X = \begin{bmatrix} -2 \\ 1 \\ 2 \end{bmatrix}$$

6.*[9]
$$\begin{bmatrix} 11 & 1 & -3 & 1 \\ 2 & 0 & 1 & -1 \\ 1 & 3 & -10 & 4 \end{bmatrix} X = \begin{bmatrix} 2 \\ 0 \\ 6 \end{bmatrix}$$

---

[6]   Campbell [1980], p88.
[7]   Campbell [1980], p97.
[8]   Hohn [1958], p120.
[9]   Campbell [1980], p103.

7.*10
$$\begin{bmatrix} 5 & -1 & 2 \\ 1 & 1 & 1 \\ 9 & -3 & -3 \\ 11 & -1 & 3 \end{bmatrix} X = \begin{bmatrix} -5 \\ 7 \\ 17 \\ -3 \end{bmatrix}$$

8.*
$$\begin{bmatrix} 0 & 0.20 & 0.40 \\ -0.75 & -0.05 & 1.15 \\ -0.25 & -0.15 & 0.45 \end{bmatrix} X = \begin{bmatrix} 0.50 \\ -1.50 \\ 1.00 \end{bmatrix}$$

9.*11
$$\begin{bmatrix} 2 & 0 & -1 \\ 1 & 1 & -2 \\ 3 & -4 & 1 \end{bmatrix} X = \begin{bmatrix} 4 \\ 1 \\ 1 \end{bmatrix}$$

10.*12
$$\begin{bmatrix} 1 & 1 & 1 & 1 \\ 1 & 1 & 1 & -1 \\ 1 & 1 & -1 & -1` \\ 1 & -1 & -1 & -1 \end{bmatrix} X = \begin{bmatrix} 1 \\ 2 \\ 3 \\ 4 \end{bmatrix}$$

***

6.   HOMOGENEOUS SYSTEMS

By the definition given in section 3, if we have $B = 0_n$, then the system in equation (2.3) is said to be homogeneous. In this case, the matrix equation is

(6.1) $AX = 0_n.$

Since B vanishes, the ranks of the coefficient and augmented matrices are identical. Therefore, by condition 1 above, we may state that a homogeneous system is always consistent.

---

[10]   Hohn [1958], p120.
[11]   Campbell [1980], p104.
[12]   Hohn [1958], p120.

DEFINITION

Trivial Solution

After a brief inspection of (6.1), we may note that one solution is $X = 0_n$. This solution is commonly termed the trivial solution, and it clearly always exists.

In some applications, the trivial solution does have a physical interpretation. In problems in mechanics, the unknowns x_i, $i = 1, 2, \ldots, n$ frequently represent displacements of parts of the system from some equilibrium position. The trivial solution therefore represents the system in a state of rest at its equilibrium position.

In most applied problems, it is desired to determine the nonzero or nontrivial solutions or at least the condition for their existence. The condition for the existence of these solutions may be stated in terms of the following:

Condition 2

A necessary and sufficient condition for a nontrivial solution to a system of m homogeneous linear algebraic equations in n unknowns is that the rank of the coefficient matrix be less than the number of unknowns. That is, the rank r of the coefficient matrix is such that $r < n$. If $r = n$, only the trivial solution exists and it is unique.

As in case 1.1 above, a solution may be found for r unknowns in terms of the remaining $n - r$. This means that the r unknowns may be expressed in terms of $n - r$ arbitrary constants in forming the complete solution. Example 2 above is an illustration of this situation.

The above condition implies two important subcases.

Case 2.1

Given a square homogeneous system, i.e., $m = n$, then nontrivial solutions exist if the coefficient matrix is singular. That is, $r < m = n$.

EXAMPLE 5

Prove case 2.1 for the general n^{th}-order system

$$AX = 0_n.$$

SOLUTION

If the rank r of the coefficient matrix A is n, i.e., $r = n$, then the inverse of A exists, and under case 1.2 above, a unique solution exists, which is given by

$$X = A^{-1}0_n = 0_n.$$

Hence, only a trivial solution exists when $r = n$. Thus, for nontrivial solutions to exist, r must be less than n, which results in A being a singular matrix.

***

EXAMPLE 6

Solve the following system.

$$x_1 + 2x_2 - 4x_3 = 0$$
$$-x_1 + x_2 + 3x_3 = 0$$

SOLUTION

In matrix form, the system of equations may be written as

$$\begin{bmatrix} 1 & 2 & -4 \\ -1 & 1 & 3 \end{bmatrix} X = 0_2$$
.

A quick check will show that the rank of the coefficient matrix is two. The system will be solved by Gaussian elimination. Adding the first equation to the second reduces the coefficient matrix to upper triangular form.

$$\begin{bmatrix} 1 & 2 & -4 \\ 0 & 3 & -1 \end{bmatrix} X = 0_2$$

Since the rank of this system is two and there are three unknowns, x_1, x_2, and x_3, two of the unknowns can only be found in terms of the third. One unknown must be left undetermined or as a parameter. For convenience, let x_3 be this unknown. This means that the above system may be rearranged as follows.

$$\begin{bmatrix} 1 & 2 \\ 0 & 3 \end{bmatrix} \begin{bmatrix} x_1 \\ x_2 \end{bmatrix} = \begin{bmatrix} 4 \\ 1 \end{bmatrix} x_3$$

Applying back substitution gives $x_2 = (1/3)x_3$. Substituting this into the first equation and solving for x_1 gives $x_1 = (10/3)x_3$. To clean up the fractions and introduce a more typical constant c, it is convenient to let $x_3 = 3c$. The solution may now be expressed as

$$x_1 = 10c, \; x_2 = c \text{ and } x_3 = 3c,$$

where c is an arbitrary constant. In terms of matrices, a solution vector for the given system is

$$X = \begin{bmatrix} x_1 \\ x_2 \\ x_3 \end{bmatrix} = \begin{bmatrix} 10 \\ 1 \\ 3 \end{bmatrix} c .$$

This solution is easily verified by substitution into the original system.

***

This last example leads us to the following general statement of the second special case.

Case 2.2

If we have a homogeneous system with more equations than unknowns and the rank is less than the number of unknowns, i. e., $m > n > r$, then a nontrivial solution always exists.

It frequently occurs that applied problems involving homogeneous equations usually arise in square form with the coefficients containing an arbitrary parameter. In these situations, we force the system to have nontrivial solutions by requiring the determinant of the coefficient matrix to vanish. This leads to a polynomial equation from which the values of the parameter may be found that guarantee the existence of nontrivial solutions. The following example is an illustration of this type of problem.

EXAMPLE 7

Find the nontrivial solution for the vector X for the following equation, which is typical of applied problems.

$$\begin{bmatrix} 1 & -1 \\ -1 & 2 \end{bmatrix} X - c \begin{bmatrix} 2 & 0 \\ 0 & 1 \end{bmatrix} X = 0_2$$

The constant c is a parameter.

SOLUTION

The given equation is a homogeneous equation that can be seen if it is put into the following form by combining coefficient matrices.

$$\begin{bmatrix} 1-2c & -1 \\ -1 & 2-c \end{bmatrix} X = 0_2$$

As stated above in condition 2, the requirement for a nontrivial solution to exist is that the determinant of the coefficient must vanish. For this example problem, this leads to the following polynomial in the parameter c.

$$\text{Det} \begin{bmatrix} 1-2c & -1 \\ -1 & 2-c \end{bmatrix} = (1-2c)(2-c) - 1$$

$$= 2c^2 - 5c + 1 = 0$$

By applying the quadratic formula, we obtain the two values the constant c must have in order for the determinant to vanish and hence a nontrivial solution to exist. These values are

$$c_1 = \frac{5 + \sqrt{17}}{4}$$

$$c_2 = \frac{5 - \sqrt{17}}{4} \ .$$

$$***$$

This will be introduced later in an important class of problems called eigenvalue problems.

EXERCISES

1.   Consider the following homogeneous system.

$$
\begin{bmatrix} 1 & -1 & 2 \\ -3 & 2 & 0 \\ 2 & -1 & c \end{bmatrix} X = \begin{bmatrix} 0 \\ 0 \\ 0 \end{bmatrix}
$$

What value must the constant c have for a nontrivial solution to exist? What is the solution when c takes on this value?

Determine which of the following systems are consistent, and obtain the solution for those that are by using Gaussian elimination. Employ partial pivoting where called for.

2.*

$$
\begin{bmatrix} 1 & 2 & -1 \\ -1 & 6 & -3 \\ 1 & -4 & 2 \end{bmatrix} X = \begin{bmatrix} 0 \\ 0 \\ 0 \end{bmatrix}
$$

3.*

$$
\begin{bmatrix} 2 & 0 & 3 \\ -1 & 0 & 2 \\ 0 & 1 & 1 \end{bmatrix} X = \begin{bmatrix} 0 \\ 0 \\ 0 \end{bmatrix}
$$

4.*[13]

***

7. SYSTEMS OF VECTORS

Nearly all of the applications considered in this text deal with problems whose solutions are expressed as a system of vectors. The reader should recall that the solution of the various systems of linear algebraic equations discussed in the previous sections are expressed as one or more vectors. In general, we may say that we are dealing with a set of k, n dimensional vectors, $A_1, A_2, \ldots, A_k$ or A_i, $i = 1, 2, \ldots, k$. Each vector is understood as having n elements or components. The entire discussion that follows is equally applicable to row vectors as well as column vectors, and thus, there is no loss of generality by considering only column vectors.

---

[13]   Hohn [1958], p120.

To aid the reader and provide a means of visualizing the material presented, we will draw upon an analogy between the matrix vectors discussed in this text and the more common three-dimensional vectors. Matrix vectors may be used to represent a variety of physical quantities such as distance, velocity, the state of an economic system, etc.

If we consider a three-dimensional vector, the usual notation is

(7.1) $$A = A_x i + A_y j + A_z k,$$

where i, j, and k are unit vectors or vectors of unit length, i.e., the length is one, in the direction of the positive x, y, and z axes respectively. The quantities A_x, A_y, and A_z are the scalar components or projections of A along each of the x, y, and z axes respectively. A graphical representation of the vector A is given in figure 2.4.

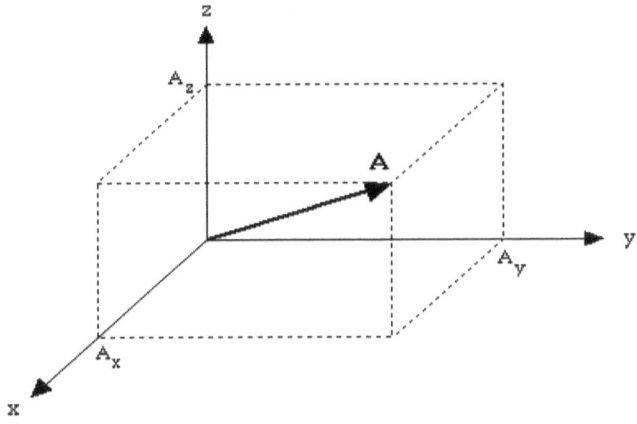

Fig. 2.4

Utilizing matrix notation, we may represent the vector A as follows,

$$A = A = \begin{bmatrix} A_x \\ A_y \\ A_z \end{bmatrix},$$

which illustrates the equivalency of the two notations. From this, one may see the graphical interpretation of a third-order column vector A as well as noting the interpretation of the elements of A as being the projections along the respective axes.

Now let us carry this analogy a step further. Consider a vector B of n

elements, i.e.,

$$B = \begin{bmatrix} b_1 \\ b_2 \\ \vdots \\ b_n \end{bmatrix}.$$

One may now see why this is referred to as a vector of n dimensions and refer to the elements as components. If you will, one may allude to this vector as describing a point in n-dimensional space where the elements represent the components or projections along each of the n coordinate axes respectively.

At this point, we must call the reader's attention to a few fine points of detail concerning the coordinate system itself. In the three-dimensional example illustrated in figure 2.34, use was made of the standard unit vectors i, j, and k. As defined, these are vectors of unit length along three mutually perpendicular directions x, y, and z respectively. Once the coordinates have been agreed upon, these vectors serve to define the coordinate axes. These unit vectors correspond to the orthonormal vectors we will discuss later.

Rather than use the notation of (7.1), we could have used

(7.2) $A = A_x + A_y + A_z,$

which indicates that the vector A may be represented by a linear combination of three vectors, each along a given direction or coordinate. Although it is usually desirable, it is not necessary for the directions or coordinates to be mutually perpendicular.

As we will develop in the following section, either equation (7.1) or (7.2) expresses the vector A as a linear combination of three base vectors. It may be seen in three-dimensional space that any set of three linearly independent vectors will serve as a basis or coordinate system for the space.

8. LINEAR DEPENDENCE

To begin our discussion, let us consider two two-dimensional vectors.

$$(8.1) \qquad\qquad A_1 = \begin{bmatrix} 1 \\ 2 \end{bmatrix}, \quad A_2 = \begin{bmatrix} 2 \\ 4 \end{bmatrix}$$

After a brief inspection, it may be noted that

$$A_2 = 2A_1.$$

This equation expresses the fact that the two vectors are linearly dependent. The equation above may be rearranged as follows:

$$(8.2) \qquad\qquad 2A_1 - A_2 = 0_2$$

As a generalization of this case, it may be stated that two vectors, say B and C, are linearly dependent if one is a scalar multiple of the other, that is,

$$(8.3) \qquad\qquad B = kC$$

or

$$(8.4) \qquad\qquad B - kC = 0_2,$$

where k is a scalar.

Conversely, we may state that if B is not a scalar multiple of C, then B and C are linearly independent.

The concept of linear dependence may now be generalized as stated in the following definition. For convenience, the discussion will be in terms of real vectors. The extension to complex vectors involves only minor notational changes.

DEFINITION

Linear Dependence/Independence

A set of m, n-dimensional nonzero vectors A_i, $i = 1, 2, \ldots, m$, is linearly dependent if and only if a set of constants c_i, $i = 1, 2, \ldots, m$ may be found that satisfy the equation

---

(8.5) $$\sum_{i=1}^{m} c_i A_i = 0$$

with not all of the c_i's equaling zero. If the only solution to equation (8.5) is for all the constants c_i to be zero, then the system of given vectors is termed linearly independent.

As a means of testing a set of vectors for linear dependence, it is convenient to introduce the following matrices.

$$A = \begin{bmatrix} A_1 & A_2 & \cdots & A_m \end{bmatrix}$$

$$C = \begin{bmatrix} c_1 \\ c_2 \\ \vdots \\ c_m \end{bmatrix}$$

Substitution of these into equation (8.5) gives the equivalent matrix form

(8.6) $$AC = 0.$$

In this form, the test for linear dependence may be expressed in terms of the solution to a homogeneous system of equations whose order is n (the dimension of the vectors) by m (the number of vectors). The columns of the matrix A are the vectors under consideration. From the above definition, it may be stated that the set of vectors will be linearly dependent only if a nontrivial solution to equation (8.6) exists.

EXAMPLE 8

Verify that the two vectors given in equation (8.1) are linearly dependent.

SOLUTION

From the above discussion, we may state that this system of two vectors is linearly dependent if a nontrivial solution C to the following homogeneous equation exists.

$$\begin{bmatrix} 1 & 2 \\ 2 & 4 \end{bmatrix} C = \begin{bmatrix} 1 & 2 \\ 2 & 4 \end{bmatrix} \begin{bmatrix} c_1 \\ c_2 \end{bmatrix} = \begin{bmatrix} 0 \\ 0 \end{bmatrix}$$

The rank of this system may be shown to be one. This means that the determinant of the coefficient matrix is zero, and hence, a nontrivial solution for the constant vector C exists. Therefore, the two vectors (the columns of A) are linearly dependent.

In fact, a little exploration of equation (8.1) will lead to the following possibilities for the constants c_1 and c_2. One finds that $c_1 = 2$, $c_2 = -1$ satisfies the definition of linear dependence as does $c_1 = -2$, $c_2 = 1$. More generally, if we set $c_2 = t$, a parameter, then $c_1 = -2t$. The nontrivial constant vector may now be written as

$$C = t \begin{bmatrix} -2 \\ 1 \end{bmatrix}.$$

***

From the theory of homogeneous equations, the following sequence of useful conditions for linear dependence may be deduced.

Condition 3

The m vectors A_i, $i = 1, 2, \ldots, m$ are linearly dependent if and only if the rank of the matrix A of equation (8.6) is less than m. Remember that the given vectors are n^{th} order.

EXAMPLE 9

Determine if the following four three-dimensional vectors are linearly dependent or independent.

$$A_1 = \begin{bmatrix} 1 \\ 0 \\ 0 \end{bmatrix}, \quad A_2 = \begin{bmatrix} 0 \\ 1 \\ 0 \end{bmatrix}, \quad A_3 = \begin{bmatrix} 1 \\ -1 \\ 1 \end{bmatrix}, \quad A_4 = \begin{bmatrix} 1 \\ 1 \\ 1 \end{bmatrix}$$

SOLUTION

The coefficient matrix of equation (8.6) is

$$A = \begin{bmatrix} 1 & 0 & 1 & 1 \\ 0 & 1 & -1 & 1 \\ 0 & 0 & 1 & 1 \end{bmatrix}.$$

The rank of this matrix is clearly three (note it is upper triangular already), and hence, the four given vectors must be linearly dependent.

***

It should be noted that the two vectors given at the beginning of this section also fall under this condition. In fact, the case illustrated here is so useful in applications, it is stated here as a special case of condition 3.

Case 3.1

If $m > n$, the m, n-dimensional vectors A_i, $i = 1, 2, \ldots , m$, are necessarily linearly dependent.

Condition 4

If any k of the n-dimensional vectors A_i, $i = 1, 2, \ldots , m$ such that $k \leq m$ are linearly dependent, then the entire set is linearly dependent.

EXAMPLE 10

Determine whether the following three four-dimensional vectors are linearly independent or dependent.

$$A_1 = \begin{bmatrix} 1 \\ 0 \\ 1 \\ 0 \end{bmatrix}, \quad A_2 = \begin{bmatrix} 0 \\ 1 \\ 0 \\ 1 \end{bmatrix}, \quad A_3 = \begin{bmatrix} -2 \\ 0 \\ -2 \\ 0 \end{bmatrix}$$

SOLUTION

For these vectors, A_1 and A_3 are linearly dependent since $A_3 = -2A_1$. The coefficient matrix of (8.6) is

$$A = \begin{bmatrix} 1 & 0 & -2 \\ 0 & 1 & 0 \\ 1 & 0 & -2 \\ 0 & 1 & 0 \end{bmatrix}.$$

It is a relatively easy matter to show that the rank of this matrix is two, which substantiates that the set of vectors is linearly dependent. An alternative verification would be to note that $2A_1 + 0A_2 + A_3 = 0$.

Case 4.1

Given a set of m linearly independent vectors, all nonempty subsets are also linearly independent. This means that any subcollection of a set of linearly independent vectors must also be linearly independent.

EXAMPLE 11

Consider the following three three-dimensional vectors.

$$A_1 = \begin{bmatrix} 1 \\ 0 \\ 0 \end{bmatrix}, \quad A_2 = \begin{bmatrix} 0 \\ 1 \\ 0 \end{bmatrix}, \quad A_3 = \begin{bmatrix} 1 \\ -1 \\ 1 \end{bmatrix}$$

Determine if they are linearly dependent.

SOLUTION

A quick check of the coefficient matrix will show that the rank is three because it is upper triangular. This means all three vectors are linearly independent. Examination of the above vectors will show that any combination of two of them are also linearly independent, which illustrates this case.

Condition 5

Given a set of m linearly dependent vectors, then at least one of them may be written as a linear combination of one or more of the rest.

EXAMPLE 12

Consider the four three-dimensional vectors given in example 8 above, which have already been shown to be linearly dependent. Verify condition 5 by finding A_4 in terms of A_1, A_2 and A_3.

SOLUTION

From example 8, the given vectors are

$$A_1 = \begin{bmatrix} 1 \\ 0 \\ 0 \end{bmatrix}, \quad A_2 = \begin{bmatrix} 0 \\ 1 \\ 0 \end{bmatrix}, \quad A_3 = \begin{bmatrix} 1 \\ -1 \\ 1 \end{bmatrix}, \quad A_4 = \begin{bmatrix} 1 \\ 1 \\ 1 \end{bmatrix}.$$

For these vectors, we seek the solution to equation (8.6). The augmented matrix for this system is

$$\text{Aug } A = \begin{bmatrix} 1 & 0 & 1 & 1 & 0 \\ 0 & 1 & -1 & 1 & 0 \\ 0 & 0 & 1 & 1 & 0 \end{bmatrix}$$

Since this matrix is already in upper triangular form, we need only to find the c_i's by back substitution. Carrying out this process and setting $c_4 = 1$ yields the particular solution $c_3 = -1$, $c_2 = -2$, and $c_1 = 0$. These results give the following relation between the given vectors.

$$-2A_2 - A_3 + A_4 = 0_3$$

or

$$A_4 = 2A_2 + A_3.$$

This verifies that A_4 is a linear combination of A_2 and A_3; thus, the condition is substantiated.

EXAMPLE 13

Develop a general proof of condition 5.

SOLUTION

Let A_k be a set of m, n^{th} order, linear dependent vectors. Since the set is linearly dependent, we must have

$$c_1A_1 + c_2A_2 + \ldots + c_mA_m = 0_n,$$

where not all the constants c_k are zero. Assume that c_1 is one of the nonzero constants so that we may solve for the vector A_1 as follows.

$$A_1 = -\frac{c_2}{c_1} A_2 - \frac{c_3}{c_1} A_3 - \cdots - \frac{c_m}{c_1} A_m$$

If we now define $d_k = -c_k/c_1$, we may write

$$A_1 = d_2 A_2 + d_3 A_3 + \cdots + d_m A_m = \sum_{k=2}^{m} d_k A_k.$$

This result shows that A_1 has been written as a linear combination of the remaining vectors in the given linearly dependent set.

Condition 6

If any one of the m vectors A_i, $i = 1, 2, \ldots, m$ may be written as a linear combination of any nonempty subset of the remaining vectors, then the entire set of m vectors is linearly dependent. This is just the inverse situation of the previous condition.

Condition 7

If the m vectors A_i, $i = 1, 2, \ldots, m$ are linearly independent but the $m + 1$ vectors A_i, $i = 1, 2, \ldots, m$, B are linearly dependent, then the vector B may be written as a linear combination of the rest.

(8.7)
$$B = \sum_{i=1}^{m} d_i A_i$$
,

where the d_i are constants, not all of which are zero.

This condition has been illustrated by example 12 in which the vectors A_1, A_2, and A_3 are linearly independent but the set A_1, A_2, A_3, and A_4 is dependent. If we let $B = A_4$, then this condition is verified with $B = 2A_2 + A_3$.

Case 8.1

If the m vectors A_i, $i = 1, 2, \ldots, m$ are linearly independent and the vector B cannot be expressed as a linear combination of the A_i as in (8.7), then the $m + 1$ vectors A_i, $(i = 1, 2, \ldots, m)$, B are also linearly independent.

As stated above, these conditions and subcases are quite useful in dealing with solutions to many applied problems that will be handled by the analyst.

EXERCISES

1.   Develop a proof for condition 3.

2.   Develop a proof for case 4.1.

3.   Develop a proof for condition 7.

4.   Show that in three-dimensional space, any two nonzero vectors that are linearly dependent are parallel.

5.*   Determine if the following set of vectors is linearly dependent or independent. If the set is dependent, determine the number of independent vectors and a set of values for the constants c_i that appear in equation (8.5).

$$\begin{bmatrix} 1 \\ 2 \\ 3 \end{bmatrix} \quad \begin{bmatrix} 1 \\ 0 \\ -1 \end{bmatrix} \quad \begin{bmatrix} 1 \\ 2 \\ 1 \end{bmatrix}$$

6.* Determine if the following set of vectors is linearly dependent or independent. If the set is dependent, determine the number of independent vectors and a set of values for the constants c_i that appear in equation (8.5).

$$\begin{bmatrix} 1 \\ 2 \\ 3 \end{bmatrix} \quad \begin{bmatrix} 1 \\ 0 \\ -1 \end{bmatrix} \quad \begin{bmatrix} -1 \\ 2 \\ 5 \end{bmatrix}$$

7.* Determine if the following set of vectors is linearly dependent or independent. If the set is dependent, determine the number of independent vectors and a set of values for the constants c_i that appear in equation (8.5).

$$\begin{bmatrix} 1 \\ 2 \\ -1 \\ 0 \end{bmatrix} \quad \begin{bmatrix} -2 \\ 1 \\ 0 \\ -1 \end{bmatrix} \quad \begin{bmatrix} 2 \\ 0 \\ 1 \\ 3 \end{bmatrix} \quad \begin{bmatrix} 1 \\ 3 \\ 0 \\ 2 \end{bmatrix}$$

8.* Determine if the following set of vectors is linearly dependent or independent. If the set is dependent, determine the number of independent vectors and a set of values for the constants c_i that appear in equation (8.5).

$$\begin{bmatrix} 1 \\ 2 \\ -1 \\ 0 \end{bmatrix} \quad \begin{bmatrix} -2 \\ 1 \\ 0 \\ -1 \end{bmatrix} \quad \begin{bmatrix} 3 \\ 0 \\ -1 \\ 0 \end{bmatrix} \quad \begin{bmatrix} 1 \\ 2 \\ -1 \\ 1 \end{bmatrix}$$

9.* Determine if the following set of vectors is linearly dependent or independent. If the set is dependent, determine the number of independent vectors and a set of values for the constants c_i that appear in equation (8.5).

$$\begin{bmatrix} 1 \\ 2 \\ -1 \\ 0 \end{bmatrix} \quad \begin{bmatrix} -2 \\ 1 \\ 0 \\ -1 \end{bmatrix} \quad \begin{bmatrix} 3 \\ 1 \\ -1 \\ 1 \end{bmatrix} \quad \begin{bmatrix} 0 \\ 5 \\ -2 \\ -1 \end{bmatrix}$$

***

9. BASIS, ORTHOGONAL, AND ORTHONORMAL VECTORS

Solutions to many problems involving n variables are frequently described by a point, line, or surface in n-dimensional space depending upon how many of the variables are independent.

DEFINITION

Basis

A basis or coordinate system for the n-dimensional solution space is defined as any set of n, n^{th}-order vectors that are linearly independent.

If a set of n, n^{th}-order vectors is denoted as B_i and an arbitrary n^{th}-order constant vector by C, then the given set of vectors B_i will constitute a basis only if a nontrivial solution to the homogeneous equation

$$BC = 0_n$$

does not exist by the definition of linear dependence of the previous section. The matrix B is the matrix whose columns are the given vectors B_i. In other words, since only the trivial solution exists, the rank of B is n, and thus, the set of vectors B_i are linearly independent under condition 3 of the forgoing section.

In some problems, the unknowns are the desired basis, and in others, some set of vectors such as the eigenvectors are used. In any situation, problems involving n-independent variables must have a solution space of n dimensions. In this case, solution vectors may be expressed as a linear combination of the coordinates or basis vectors.

The reader should note that a basis is not unique in that there are an infinity of possible linearly independent systems that may serve as a basis. As will be seen later, whatever basis is used, it is frequently convenient to utilize a basis that is orthogonal. Orthogonality of vectors is defined in the following condition. Again we shall restrict our attention to real vectors only as the relations for complex vectors differ only by slight changes in notation.[14]

Condition 8

Two n-dimensional vectors, A_1 and A_2, are orthogonal if and only if

$$A_1^T A_2 = A_2^T A_1 = 0,$$

provided that neither is the null vector.

In condition 8 the matrix product of the two column vectors is termed the inner product, and it is identical with the dot product in vector algebra.

From condition 8, the following condition may be developed.

Condition 9

Any set of m, n-dimensional vectors A_i, $i = 1, 2, \ldots, m$, which are mutually orthogonal, i.e.,

(9.2) $$A_i^T A_j = A_j^T A_i = 0, \; i \neq j; \, i, j = 1, 2, \cdots, m$$

are necessarily linearly independent.

---

[14] If we were considering complex vectors, this relation would be written as $\bar{A}_1^T A_2 = \bar{A}_2^T A_1 = 0_n$ where the bar denotes the complex conjugate.

EXAMPLE 14

Consider the following three three-dimensional vectors. Verify that they are orthogonal and also linearly independent.

$$A_1 = \begin{bmatrix} 1 \\ 0 \\ 1 \end{bmatrix}, \quad A_2 = \begin{bmatrix} 0 \\ 1 \\ 0 \end{bmatrix}, \quad A_3 = \begin{bmatrix} 1 \\ 0 \\ -1 \end{bmatrix}$$

SOLUTION

These three vectors are seen to be orthogonal by evaluating each of the three products indicated in equation (9.2) above. They will also be linearly independent if the rank of the following matrix is three.

$$A = \begin{bmatrix} 1 & 0 & 1 \\ 0 & 1 & 0 \\ 1 & 0 & -1 \end{bmatrix}$$

By subtracting the first row from the last, this matrix is converted to upper triangular form with all ones on the main diagonal. Thus, the rank is three.

***

Before illustrating the usefulness of an orthogonal basis, we will introduce the concept of the length or norm of a vector. As the reader may visualize, a vector may be used to represent the position of a point in space. The position is relative to the origin of a defined coordinate system. From this concept, it may be seen that the following relation represents the length of a vector or the distance between the origin and its end point.

$$\| A \|_2 = \sqrt{A^T A} = \sqrt{a_1^2 + a_2^2 + \cdots + a_n^2}$$

The quantity $\| A \|_2$ is commonly defined as the length or norm of the vector. This expression may also be written as

$$(\| A \|_2)^2 = A^T A,$$

which is another useful relation.

The norm $\| A \|_2$ defined above is known as the Euclidean or $L_2 (A)$ norm. The name Euclidean derives from the fact that the expression above represents the length of a vector in ordinary three-dimensional space. The notation L_2 relates to the definition being based on the squares of the components. In most numerical calculations, the Euclidean norm is not a convenient measure as it requires repeated squaring and the square root operation in its evaluation. There are two other norms that are easier to evaluate computationally and, as a result, are used more frequently in numerical algorithms. These are the L_1 norm and the max-norm L_∞, which are defined below.

$$L_1(A): \quad \| A \|_1 = \sum_{i=1}^{n} |x_i|$$

$$L_\infty(A): \quad \| A \|_\infty = \max_{1 \le i \le n} |x_i|$$

EXAMPLE 15

Compute the L_1, L_2, and L_∞ norms for the following vector.

$$A = \begin{bmatrix} 1 \\ -2 \\ 3 \end{bmatrix}$$

SOLUTION

$$L_1(A): \quad \|A\|_1 = \sum_{i=1}^{n} |x_i| = |1| + |-2| + |3| = 6$$

$$L_2(A): \quad \|A\|_2 = \sqrt{\sum_{i=1}^{n} x_i^2} = \sqrt{1^2 + (-2)^2 + 3^2} = \sqrt{14}$$

$$L_\infty(A): \quad \|A\|_\infty = \max_{1 \le i \le n} |x_i| = \max \{|1|, |-2|, |3|\} = 3$$

EXAMPLE 16

Compute the L_1, L_2, and L_∞ norms for the following vector.

$$A = \begin{bmatrix} 1 + j \\ 2j \\ 1 - j \end{bmatrix}$$

SOLUTION

We recall from the algebra of complex numbers that

$$|a + bj| = \sqrt{a^2 + b^2} = \sqrt{(a + bj)(a - bj)} = \sqrt{z\,\bar{z}},$$

where $z = a + bj$. Now

$$L_1(A) : \quad \|A\|_1 = \sum_{i=1}^{n} |x_i| = |1 + j| + |2j| + |1 - j|$$

$$= \sqrt{2} + 2 + \sqrt{2} = 2 + 2\sqrt{2} = 2(1 + \sqrt{2}).$$

In computing the L_2 norm when the elements are complex, we must take note of the fact that the square of length of a complex value is given by the product of the complex conjugate of the quantity times the quantity itself. In other words, if z is a complex number, then the square of its magnitude (length) is given by $z\bar{z}$.

$$L_2(A) : \quad \|A\|_2 = \sqrt{\sum_{i=1}^{n} |x_i|^2}$$

$$= \sqrt{\overline{(1 + j)}\,(1 + j) + \overline{2j}\,2j + \overline{(1 - j)}\,(1 - j)}$$

$$= \sqrt{(1 - j)\,(1 + j) + (- 2j)\,2j + (1 + j)\,(1 - j)}$$

$$= \sqrt{2 + 4 + 2} = 2\sqrt{2}$$

It is also useful to note that this result may be represented by the following vector representation.

$$L_2(A) : \quad \|A\|_2 = \sqrt{\overline{A}^T A}$$

Now for the last norm, the L_∞ norm.

$$L_\infty(A) : \quad \|A\|_\infty = \max_{1 \le i \le n} |x_i|$$

$$= \max \{|1+j|, |-2j|, |1-j|\}$$

$$= \max \{\sqrt{2}, 2, \sqrt{2}\} = 2$$

***

To illustrate the advantage of an orthogonal basis, let us consider a system of m, n-dimensional linearly independent vectors, which is not orthogonal, say X_i, $i = 1, 2, \ldots, m$. Let us further assume that these vectors will be used as a basis or coordinate system in which we wish to define some given vector A. To do this we would write

(9.3) $A = c_1X_1 + c_2X_2 + \ldots + c_mX_m,$

where c_i, $i = 1, 2, \ldots, m$ are m constants whose meaning will be developed.

The problem of representing the vector A in the given basis is now one of determining the constants c_i. How this is done may be visualized more easily if equation (9.3) is expressed in matrix form.

$$A = XC,$$

where $X = [X_1, X_2, \ldots, X_m]$ and C is the vector of unknown constants. From this expression, we may see that the determination of the constants c_i involves the solution of a system of n nonhomogeneous equations in m unknowns. The process is not easy, but it is simplified if $n = m$, which is the case in many instances. When this occurs, the solution is given by

$$C = X^{-1}A.$$

The inverse exists because the X_i are a basis, which means that they are linearly independent; thus, the rank of X is n. This solution is still an involved process requiring the inversion of an nth-order matrix.

At this point, let us assume that the basis we are working with is orthogonal, i.e.,

(9.4) $X_i^T X_j = 0,\ i \neq j;\ i, j = 1, 2, \ldots, m.$

With this property the c_i's of (9.3) may be found quite easily. To illustrate the procedure, we will solve for c_k in the following manner. Premultiplying (9.3) by X_k^T and applying (9.4) and the L_2 norm gives

$$X_k^T A = c_k \, ||X_k||_2^2$$

From this expression, c_k is easily found and is given by

$$c_k = \frac{X_k^T A}{\| X_k \|_2^2} .$$

This process is valid for all k from 1 to m and hence provides an easy method of determining the values of c_k, $k = 1, 2, \ldots, m$. This illustrates the ease by which the unknown coefficients may be evaluated when the basis is orthogonal.

EXAMPLE 17

Given the three three-dimensional vectors used in example 13,

$$X_1 = \begin{bmatrix} 1 \\ 0 \\ 1 \end{bmatrix}, \; X_2 = \begin{bmatrix} 0 \\ 1 \\ 0 \end{bmatrix}, \; X_3 = \begin{bmatrix} 1 \\ 0 \\ -1 \end{bmatrix},$$

expand the following vector in terms of these three.

$$A = \begin{bmatrix} 2 \\ -1 \\ 3 \end{bmatrix}$$

SOLUTION

The desired expansion has the following form

$$A = \begin{bmatrix} 2 \\ -1 \\ 3 \end{bmatrix} = c_1 X_1 + c_2 X_2 + c_3 X_3$$

$$= c_1 \begin{bmatrix} 1 \\ 0 \\ 1 \end{bmatrix} + c_2 \begin{bmatrix} 0 \\ 1 \\ 0 \end{bmatrix} + c_3 \begin{bmatrix} 1 \\ 0 \\ -1 \end{bmatrix},$$

where it is desired to find the three constants c_1, c_2, and c_3.

As we discussed above, this expansion could be represented by the following matrix equation.

$$XC = A, \quad \begin{bmatrix} 1 & 0 & 1 \\ 0 & 1 & 0 \\ 1 & 0 & -1 \end{bmatrix} C = \begin{bmatrix} 2 \\ -1 \\ 3 \end{bmatrix}$$

The unknown coefficient vector C could be found from inverting the X matrix as previously indicated. However, the unknown c's may be evaluated more easily by use of the fact that the X vectors are orthogonal. From the general expression above, the first coefficient is

$$c_1 = \frac{X_1^T A}{||X_1||_2^2} = \frac{5}{2}$$

This results from premultiplying the expansion above through by X_1^T, which takes advantage of the fact that it is orthogonal with the other two given vectors. Similarly, the remaining two coefficients are found to be

$$c_2 = \frac{X_2^T A}{||X_2||_2^2} = \frac{-1}{1} \, , \quad c_3 = \frac{X_3^T A}{||X_3||_2^2} = \frac{-1}{2} \, .$$

We thus have the required coefficients for the desired expansion.

$$A = \begin{bmatrix} 2 \\ -1 \\ 3 \end{bmatrix} = \frac{5}{2} \begin{bmatrix} 1 \\ 0 \\ 1 \end{bmatrix} - \begin{bmatrix} 0 \\ 1 \\ 0 \end{bmatrix} - \frac{1}{2} \begin{bmatrix} 1 \\ 0 \\ -1 \end{bmatrix}$$

***

The final concept to be introduced in this section is that of an orthonormal set of vectors. An orthonormal set of vectors is one in which the vectors are not only orthogonal, but each also have an L_2 norm or length equal to unity. In other words, a set of m, n-dimensional vectors is orthonormal if

$$X_i^T X_j = \begin{cases} 0, & i \neq j \\ 1, & i = j \end{cases} \, .$$

By this definition, the L_2 or Euclidean norms of the orthonormal X vectors have the value one. Vectors of unit length are also called normalized. From this term, one can see the source of the word *orthonormal* in that it represents a set of orthogonal normalized vectors.

If it is desired to construct a normalized vector from a nonnormalized nonzero vector, the procedure is to divide the elements of the nonnormalized vector by the length of the vector itself. For example, suppose that we are given a nonnormalized vector A out of which we desire to construct a normalized vector $\tilde{A}$. It is easy to show that the vector $\tilde{A}$ is given by

$$\tilde{A} = \frac{A}{\|A\|_2} .$$

EXAMPLE 18

Given the three three-dimensional vectors used in example 13,

$$X_1 = \begin{bmatrix} 1 \\ 0 \\ 1 \end{bmatrix}, X_2 = \begin{bmatrix} 0 \\ 1 \\ 0 \end{bmatrix}, X_3 = \begin{bmatrix} 1 \\ 0 \\ -1 \end{bmatrix},$$

construct an orthonormal set.

SOLUTION

The L_2 norms of these three vectors are $\sqrt{2}$, 1, and $\sqrt{2}$. The orthonormal vectors are obtained by multiplying each given vector by the reciprocal of its norm. The result is

$$\tilde{X}_1 = \begin{bmatrix} \frac{1}{\sqrt{2}} \\ 0 \\ \frac{1}{\sqrt{2}} \end{bmatrix}, \tilde{X}_2 = \begin{bmatrix} 0 \\ 1 \\ 0 \end{bmatrix}, \tilde{X}_3 = \begin{bmatrix} \frac{1}{\sqrt{2}} \\ 0 \\ \frac{-1}{\sqrt{2}} \end{bmatrix}.$$

***

Although an orthonormal set is convenient due to the unit length of all the vectors, it is not frequently used in actual applied problems because of the additional work required to normalize an orthogonal set to produce an orthonormal one. However, with the increasing availability of matrix computational software packages that easily generate sets of orthonormal vectors, this situation is changing. In problems involving large matrices where computational time is important, the other two norms are usually the preferred choice.

Based on the concept of orthonormal vectors is the orthogonal matrix, which holds a very important position in matrix applications. When all the columns of a square matrix comprise an orthornormal set of vectors, the matrix is called orthogonal. As a result, we have the following definition.

DEFINITION

Orthogonal Matrix

If $A^{-1} = A^T$, then A is an orthogonal matrix.

EXAMPLE 19

Show that the following matrix is orthogonal.

$$A = \begin{bmatrix} \dfrac{1}{\sqrt{6}} & \dfrac{1}{\sqrt{2}} & \dfrac{1}{\sqrt{3}} \\ \dfrac{2}{\sqrt{6}} & 0 & \dfrac{-1}{\sqrt{3}} \\ \dfrac{1}{\sqrt{6}} & \dfrac{-1}{\sqrt{2}} & \dfrac{1}{\sqrt{3}} \end{bmatrix}$$

SOLUTION

By the above definition, A will be orthogonal if $AA^T = I_3$.

$$AA^T = \begin{bmatrix} \dfrac{1}{\sqrt{6}} & \dfrac{1}{\sqrt{2}} & \dfrac{1}{\sqrt{3}} \\ \dfrac{2}{\sqrt{6}} & 0 & \dfrac{-1}{\sqrt{3}} \\ \dfrac{1}{\sqrt{6}} & \dfrac{-1}{\sqrt{2}} & \dfrac{1}{\sqrt{3}} \end{bmatrix} \begin{bmatrix} \dfrac{1}{\sqrt{6}} & \dfrac{2}{\sqrt{6}} & \dfrac{1}{\sqrt{6}} \\ \dfrac{1}{\sqrt{2}} & 0 & \dfrac{-1}{\sqrt{2}} \\ \dfrac{1}{\sqrt{6}} & \dfrac{-1}{\sqrt{3}} & \dfrac{1}{\sqrt{3}} \end{bmatrix} = \begin{bmatrix} 1 & 0 & 0 \\ 0 & 1 & 0 \\ 0 & 0 & 1 \end{bmatrix}$$

EXERCISES

1.   Develop a proof for condition 9.

2.   Show that the three vectors given below are linearly independent and hence form a basis for three-dimensional space.

$$V_1 = \begin{bmatrix} 0 \\ 0 \\ 1 \end{bmatrix}, \ V_2 = \begin{bmatrix} 0 \\ 1 \\ 1 \end{bmatrix}, \ V_3 = \begin{bmatrix} 1 \\ 1 \\ 1 \end{bmatrix}$$

3.*   Given the arbitrary third-order (dimensional) vector V below, write it as a linear combination (expansion) of the three basis vectors in exercise 2 above.

$$V = \begin{bmatrix} 2 \\ 1 \\ 3 \end{bmatrix} = c_1 V_1 + c_2 V_2 + c_3 V_3$$

Evaluate the L_1, L_2, and $L_{\hat{Y}}$ norms for each of the following vectors.

4. $\begin{bmatrix} 1 \\ -1 \\ 1 \end{bmatrix}$ 5. $\begin{bmatrix} 1 + j \\ 2 - j \\ 1 - 2j \end{bmatrix}$ 6. $\begin{bmatrix} 4 \\ 0 \\ -5 \\ 1 \end{bmatrix}$ 7. $\begin{bmatrix} 2 \\ -4 \\ 1 \\ 3 \end{bmatrix}$

8.*   Using the vectors given in exercise 2 above, construct an orthonormal set and show that the matrix, for which the resulting vectors are the columns, is orthogonal.

***

10. GRAM-SCHMIDT ORTHOGONALIZATION

     In some cases in applied problems, the analyst is given or has obtained a nonorthogonal set of vectors. From these it may be desired to construct a set of orthogonal vectors with the same span. By *span* we mean that every vector of the new set is expressible as a linear combination of the vectors of the original set. The procedure by which this is accomplished is called the Gram-Schmidt orthogonalization process. To illustrate this method, consider the following situation.

Assume that a linearly independent set of n real[15] vectors, X_i, $i = 1, 2, \ldots, n$, are given and that they are not orthogonal, i. e.,

$$X_i^T X_j \neq 0 \, , \, i \neq j$$

From this given set, it is desired to construct a set of n vectors, V_i, $i = 1, 2, \ldots, n$, which is orthogonal, i. e.,

$$V_i^T V_j = 0 \, , \, i \neq j$$

To accomplish this, begin as follows. Set $V_1 = X_1$. With V_1 thus defined, we now set

(10.1) $$V_2 = X_2 - c_2^1 V_1 \qquad .$$

This equation indicates that V_2 will be constructed from the vector X_2 by subtracting a portion of the known vector V_1. The amount to be subtracted is controlled by the unknown coefficient c_2^1. The notation for this coefficient is that the subscript denotes the vector from which the new one is to be formed, and the superscript denotes the new orthogonal vector it relates to.

To determine this constant, we require V_2 to be orthogonal to V_1, that is,

(10.2) $$V_1^T V_2 = V_1^T X_2 - c_2^1 V_1^T V_1 = 0 \qquad .$$

From this equation, the constant c_2^1 may be found to be

(10.3) $$c_2^1 = \frac{V_1^T X_2}{V_1^T V_1} = \frac{V_1^T X_2}{\| V_1 \|_2^2} \, ,$$

where the L_2 norm is used.

We may now proceed to determine V_3. In a similar fashion, we develop this vector from X_3 by seeking the appropriate fractions of V_1 and V_2 to subtract from X_3 so that V_3 is orthogonal to each of them. Specifically, we set

$$V_3 = X_3 - c_3^1 V_1 - c_3^2 V_2$$

---

[15]   As before, we shall restrict the discussion to real vectors. Complex vectors may be handled by a slight change in notation as reflected in the orthogonality relation $X_i^T X_j = 0$, $i \neq j$.

The two necessary equations from which the constants $c_3{}^1$ and $c_3{}^2$ are found are obtained from the requirement that V_3 be orthogonal with V_1 and V_2, that is,

(10.4)
$$V_1^T V_3 = V_1^T X_3 - c_3^1 V_1^T V_1 - c_3^2 V_1^T V_2 = 0$$
$$V_2^T V_3 = V_2^T X_3 - c_3^1 V_2^T V_1 - c_3^2 V_2^T V_2 = 0 \quad.$$

Since V_1 and V_2 are known and are orthogonal to one another, we have

$$V_1^T V_2 = V_2^T V_1 = 0$$
$$V_1^T V_1 = ||V_1||_2^2, \quad V_2^T V_2 = ||V_2||_2^2 \quad.$$

Applying these equations to (10.4) and solving for the coefficients gives

$$c_3^1 = \frac{V_1^T X_3}{||V_1||_2^2}, \quad c_3^2 = \frac{V_2^T X_3}{||V_2||_2^2} \quad.$$

This procedure may be generalized for the determination of the kth orthogonal vector V_k. For this vector we write

(10.5)
$$V_k = X_k - \sum_{i=1}^{k-1} c_k^i V_i \quad.$$

From orthogonality, we have the requirement

$$V_j^T V_k = 0 \quad \text{for } i \le j < k.$$

Since also,

$$V_i^T V_j = \begin{cases} 0, & i \ne j \\ ||V_i||_2^2, & i = j \end{cases},$$

we may determine the coefficients c_k^i in (10.5) as

$$c_k^i = \frac{V_i^T X_k}{||V_i||_2^2} \quad.$$

This result may be substituted into (10.5) to yield the general expression by which the kth orthogonal vector is generated.

$$(10.6) \qquad V_k = X_k - \sum_{i=1}^{k-1} \frac{V_i^T X_k}{|| V_i ||_2^2} V_i$$

In the later chapter on numerical methods, a more efficient method of generating an orthogonal set of vectors will be presented.

EXAMPLE 20

Apply the Gram-Schmidt method to develop an orthogonal set from the following three three-dimensional vectors that are linearly independent.

$$X_1 = \begin{bmatrix} 1 \\ 0 \\ 0 \end{bmatrix}, X_2 = \begin{bmatrix} 1 \\ 1 \\ 0 \end{bmatrix}, X_3 = \begin{bmatrix} 1 \\ 1 \\ 1 \end{bmatrix}$$

SOLUTION

Following the Gram-Schmidt procedure we take $V_1 = X_1$. From equation (10.1), the next vector is given by

$$V_2 = X_2 - c_2^1 V_1 = \begin{bmatrix} 1 \\ 1 \\ 0 \end{bmatrix} - c_2^1 \begin{bmatrix} 1 \\ 0 \\ 0 \end{bmatrix}.$$

Requiring this vector to be orthogonal to V_1 gives

$$V_1^T V_2 = V_1^T X_2 - c_2^1 V_1^T V_1 = \begin{bmatrix} 1 \\ 0 \\ 0 \end{bmatrix}^T \begin{bmatrix} 1 \\ 1 \\ 0 \end{bmatrix} - c_2^1 \begin{bmatrix} 1 \\ 0 \\ 0 \end{bmatrix}^T \begin{bmatrix} 1 \\ 0 \\ 0 \end{bmatrix} = 0$$

or $1 - c_2^1 = 0$ from which $c_2^1 = 1$. Knowing this constant allows us to construct the second orthogonal vector

$$V_2 = \begin{bmatrix} 1 \\ 1 \\ 0 \end{bmatrix} - c_2^1 \begin{bmatrix} 1 \\ 0 \\ 0 \end{bmatrix} = \begin{bmatrix} 0 \\ 1 \\ 0 \end{bmatrix}.$$

We have gone through all the steps to emphasize the principles involved. The second vector could have been determined more directly from the application of equation (10.6).

$$V_2 = X_2 - \frac{V_1^T X_2}{||V_1||_2^2} V_1 = \begin{bmatrix} 1 \\ 1 \\ 1 \\ 0 \end{bmatrix} - \frac{\begin{bmatrix} 1 \\ 0 \\ 0 \end{bmatrix}^T \begin{bmatrix} 1 \\ 1 \\ 0 \end{bmatrix}}{\begin{bmatrix} 1 \\ 0 \\ 0 \end{bmatrix}^T \begin{bmatrix} 1 \\ 0 \\ 0 \end{bmatrix}} \begin{bmatrix} 1 \\ 0 \\ 0 \end{bmatrix} = \begin{bmatrix} 0 \\ 1 \\ 0 \end{bmatrix}$$

Rather than repeating the steps at this point, the third orthogonal vector will be computed from (10.6) directly.

$$V_3 = X_3 - \frac{V_1^T X_3}{||V_1||_2^2} V_1 - \frac{V_2^T X_3}{||V_2||_2^2} V_2 =$$

(10.6)

$$\begin{bmatrix} 1 \\ 1 \\ 1 \end{bmatrix} - \frac{\begin{bmatrix} 1 \\ 0 \\ 0 \end{bmatrix}^T \begin{bmatrix} 1 \\ 1 \\ 1 \end{bmatrix}}{\begin{bmatrix} 1 \\ 0 \\ 0 \end{bmatrix}^T \begin{bmatrix} 1 \\ 0 \\ 0 \end{bmatrix}} \begin{bmatrix} 1 \\ 0 \\ 0 \end{bmatrix} - \frac{\begin{bmatrix} 0 \\ 1 \\ 0 \end{bmatrix}^T \begin{bmatrix} 1 \\ 1 \\ 1 \end{bmatrix}}{\begin{bmatrix} 0 \\ 1 \\ 0 \end{bmatrix}^T \begin{bmatrix} 0 \\ 1 \\ 0 \end{bmatrix}} \begin{bmatrix} 0 \\ 1 \\ 0 \end{bmatrix} = \begin{bmatrix} 0 \\ 0 \\ 1 \end{bmatrix}$$

EXERCISES

1.*   Repeat the problem in example 20 above except take X_2 as the starting vector, i.e., $V_1 = X_2$.

Apply the Gram-Schmidt process to develop orthogonal sets from the following sets of given vectors.

2.* $\begin{bmatrix} 1 \\ 1 \\ 1 \end{bmatrix}, \begin{bmatrix} 1 \\ 0 \\ 1 \end{bmatrix}, \begin{bmatrix} 1 \\ 1 \\ 0 \end{bmatrix}$ 3.* $\begin{bmatrix} 1 \\ 1 \\ 0 \end{bmatrix}, \begin{bmatrix} 1 \\ 0 \\ 1 \end{bmatrix}, \begin{bmatrix} 0 \\ 1 \\ 1 \end{bmatrix}$

4.* $\begin{bmatrix} 1 \\ 0 \\ 1 \end{bmatrix}, \begin{bmatrix} 0 \\ 0 \\ 1 \end{bmatrix}, \begin{bmatrix} 1 \\ 1 \\ 1 \end{bmatrix}$ 5.* $\begin{bmatrix} 1+j \\ 2 \\ j \end{bmatrix}, \begin{bmatrix} 1 \\ 1-j \\ -1 \end{bmatrix}, \begin{bmatrix} 2-j \\ 2j \\ 2+j \end{bmatrix}$

6.* $\begin{bmatrix} 1 \\ 0 \\ 0 \\ 0 \end{bmatrix}, \begin{bmatrix} 1 \\ 0 \\ 1 \\ 0 \end{bmatrix}, \begin{bmatrix} 0 \\ 1 \\ 0 \\ 1 \end{bmatrix}, \begin{bmatrix} 0 \\ 0 \\ 1 \\ 1 \end{bmatrix}$

7.*   For the vectors of example 4 above, construct an orthonormal set. Use the resulting vectors as the columns of a third-order matrix, and show that it is orthogonal.

8.*   Construct an orthonormal set of vectors from those given in exercise 5 above. Use the resulting vectors as the columns of a third-order matrix, and show that it is orthogonal.

9.*   Use the vectors given in example 6 above to form an orthonormal set. Show that the matrix whose columns are the resulting vectors is orthogonal.

10.   Establish the equivalent form of equation (10.6) for use when the given vectors are complex.

***

11. LINEAR ALGEBRAIC EQUATIONS REVISITED

As a conclusion to this chapter, it will be useful to pull several of the concepts presented together in a brief discussion of the solution of a system of linear algebraic equations. As was stated earlier, the most common method of solving linear algebraic systems is by the method of Gaussian elimination with partial pivoting and back substitution. Our discussion began by considering a system of m equations and n unknowns represented by equation (2.3), which is repeated here for reference.

$$(11.1) \qquad\qquad AX = B$$

At this point, the case in which the number of equations is greater than the number of unknowns will be left for special consideration in the next chapter. We will concentrate our attention on the situation $m \leq n$. As we observed in section 4, the method of Gaussian elimination was used to convert this original system to an upper triangular one of the form given in equation (4.10) or

$$(11.2) \qquad\qquad UX = D.$$

In partitioned form, the augmented matrix of (11.2) will appear in the somewhat symbolic form where r is the rank of the system.

$$(11.3) \qquad \text{Aug } U = \left[\begin{array}{c|c|c} U_{r,r} & C_{r,n-r} & D_{r,1} \\ \hline [0]_{m-r,r} & [0]_{m-r,n-r} & [0]_{m-r,1} \end{array} \right]$$

If the system were not consistent, that is, the rank of Aug A > rank A, then the lower right-hand column vector would not be a zero vector but would contain some nonzero elements. What this notation attempts to show is that the Gaussian elimination procedure reduces the system of equations to one in which there appears a square upper triangular matrix whose order is the rank of the system. All the remaining rows will be filled with zero elements.

In the special case that $r = m = n$, we have the nonsingular case (case 1.2 of section 3) where a unique solution exists. In this particular instance, Gaussian elimination would have produced

$$(11.4) \qquad \text{Aug } U = \left[\begin{array}{c|c} U_{n,n} & D_{n,1} \end{array} \right].$$

The solution could be obtained directly by the back substitution process.

Should we encounter a situation in which $r < n$, we need to examine the solution to see how we should proceed with the Gaussian elimination process. In case 1.1 of section 3 of this chapter, the following example problem (example 2) was considered.

$$\begin{bmatrix} 1 & -1 & 1 \\ 3 & -1 & 2 \\ 3 & 1 & 1 \end{bmatrix} X = \begin{bmatrix} 2 \\ -6 \\ -18 \end{bmatrix}$$

The solution was found to be equation (3.3).

$$X = \begin{bmatrix} -4-c \\ -6+c \\ 2c \end{bmatrix} = - \begin{bmatrix} 4 \\ 6 \\ 0 \end{bmatrix} + c \begin{bmatrix} -1 \\ 1 \\ 2 \end{bmatrix}$$

The first vector in this solution is called the nonhomogeneous solution in that it is obtained by solving $AX = B$. This is the part of the complete or general solution that produces the constant vector B when it is premultiplied by the coefficient matrix A. In other words,

$$\begin{bmatrix} 1 & -1 & 1 \\ 3 & -1 & 2 \\ 3 & 1 & 1 \end{bmatrix} \begin{bmatrix} -4 \\ -6 \\ 0 \end{bmatrix} = \begin{bmatrix} 2 \\ -6 \\ -18 \end{bmatrix}.$$

The nonhomogeneous solution is unique and, as such, is a point in the three-dimensional space of the problem variables $x_1 x_2 x_3$.

The second vector in the complete solution above is the homogeneous solution, that is, the solution to $AX = 0$. As there are three unknowns and the rank of the system is two, there is only one ($nr = 3 - 2 = 1$) linearly independent nontrivial solution to $AX = 0$. It should also be noted that the homogeneous and nonhomogeneous solutions are linearly independent. The existence of one nontrivial solution to the homogeneous equation means that the complete solution is a line (one-dimensional) in the variable space $x_1 x_2 x_3$. This result is depicted in figure 2.5 below. The coordinate axes are the three problem variables x_1, x_2, and x_3.

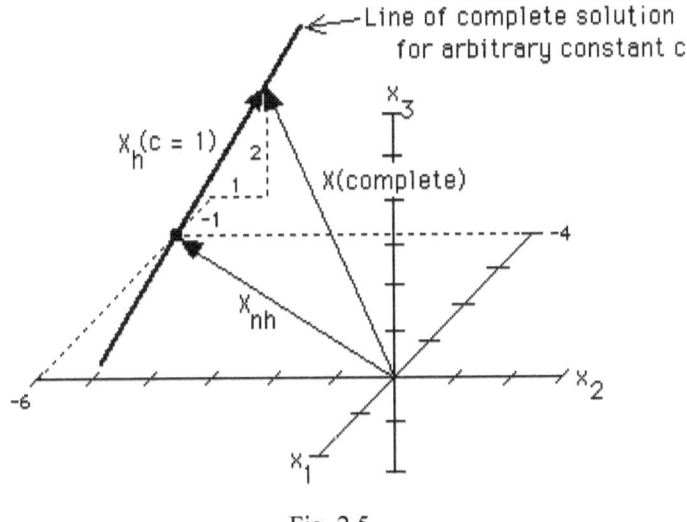

Fig. 2.5

The nonhomogeneous solution is the point $(-4, -6, 0)$, which is the vector X_{nh}. The homogeneous solution X_h passes through this point, which occurs when $c = 0$. The homogeneous solution is the vector shown from the point of the nonhomogeneous solution along the line whose relative coordinates are $-1, 1, 2$.

These relative coordinates are for a particular homogeneous solution that is given when the value of the constant c is taken to be one.

The complete solution is the sum of the nonhomogeneous and homogeneous solutions as indicated by the particular solution in figure 2.4 for the value $c = 1$. All solutions comprise the one-dimensional line through the point given by the nonhomogeneous solution along the direction given by the single nontrivial solution vector of the homogeneous equation $AX = 0$.

Let us now try to summarize the implications of the previous third-order example. The nonhomogeneous solution is a unique point in the space of the three problem variables, $x_1 x_2 x_3$. Since the rank of the system is two, only two of the original equations are independent. The difference between the order of the system and the rank of the coefficient matrix tells us the number of linearly independent nontrivial solution vectors to the homogeneous equation $AX = 0$ that exist. In this example this difference was one ($3 - 2 = 1$). This means the complete solution is one-dimensional or a line in the space of the problem's variables.

If the rank had been three, then only a unique nonhomogeneous solution would have existed. The complete solution would be just the unique point in the $x_1 x_2 x_3$ space.

If the rank had been one, then there would have been two linearly independent nontrivial solution to the homogeneous equation $AX = 0$. The existence of these two solutions would have meant that the homogeneous solution would be described by a plane (two-dimensional solution) containing the point given by the nonhomogeneous solution.

In general, it may now be stated that when the rank r of a system of equations $AX = B$ is less than the number of unknowns n, the general solution has the following form.

(11.5)
$$X = X_{nh} + \sum_{k=1}^{n-r} c_k X_k$$

Here, X_{nh} is the nonhomogeneous solution X_k, $k = 1, 2, \ldots, n - r$, the homogeneous solutions and c_k, $k = 1, 2, \ldots, n - r$, arbitrary coefficients. The complete set of vectors X_{nh}, X_k with $k = 1, 2, \ldots e$, $n - r$ is linearly independent.

It is also important to note the situation that occurs when one is dealing with a homogeneous system. In such cases, the nonhomogeneous solution vector X_{nh} is zero. Equation (11.3) then states that there will be only $n - r$ linearly independent nontrivial solution vectors. These linearly independent vectors constitute a basis for the solution to the given linear algebraic equations. The space defined by this basis is called the solution space in that it contains all the solutions to a given equation. In other words, every point in the solution space is a particular solution to the given equation, and the space itself is the complete solution.

EXAMPLE 21

Solve following system of equations.

$$\begin{bmatrix} 3 & 4 & -1 & 2 \\ -1 & 2 & 2 & 1 \\ 1 & 8 & 3 & 4 \\ 2 & 6 & 1 & 3 \end{bmatrix} X = \begin{bmatrix} 5 \\ -3 \\ -1 \\ 2 \end{bmatrix}$$

SOLUTION

The augmented matrix is

$$\begin{bmatrix} 3 & 4 & -1 & 2 & 5 \\ -1 & 2 & 2 & 1 & -3 \\ 1 & 8 & 3 & 4 & -1 \\ 2 & 6 & 1 & 3 & 2 \end{bmatrix}.$$

Following the Gaussian elimination procedure, this may be reduced to the following upper triangular matrix.

(11.6)
$$\begin{bmatrix} 3 & 4 & -1 & 2 & 5 \\ 0 & 10 & 5 & 5 & -4 \\ 0 & 0 & 0 & 0 & 0 \\ 0 & 0 & 0 & 0 & 0 \end{bmatrix}$$

Clearly, the rank of the system is two, and it is consistent. Based on equation (11.5) above, the general solution must include the nonhomogeneous solution plus two linearly independent homogeneous solution vectors since $n - r = 2$.

To determine these solution vectors, it is convenient to rewrite (11.6) in the form of a partitioned matrix equation as in (11.3) above.

$$
\begin{bmatrix}
3 & 4 & -1 & 2 \\
0 & 10 & 5 & 5 \\
\hline
0 & 0 & 0 & 0 \\
0 & 0 & 0 & 0
\end{bmatrix}
\begin{bmatrix} x_1 \\ x_2 \\ x_3 \\ x_4 \end{bmatrix}
=
\begin{bmatrix} 5 \\ -4 \\ \hline 0 \\ 0 \end{bmatrix}
$$

Expanding the nontrivial portion of this equation gives

$$
\begin{bmatrix} 3 & 4 \\ 0 & 10 \end{bmatrix}
\begin{bmatrix} x_1 \\ x_2 \end{bmatrix}
+
\begin{bmatrix} -1 & 2 \\ 5 & 5 \end{bmatrix}
\begin{bmatrix} x_3 \\ x_4 \end{bmatrix}
=
\begin{bmatrix} 5 \\ -4 \end{bmatrix}
$$

or

(11.7)
$$
\begin{bmatrix} 3 & 4 \\ 0 & 10 \end{bmatrix}
\begin{bmatrix} x_1 \\ x_2 \end{bmatrix}
=
\begin{bmatrix} 5 \\ -4 \end{bmatrix}
-
\begin{bmatrix} -1 & 2 \\ 5 & 5 \end{bmatrix}
\begin{bmatrix} x_3 \\ x_4 \end{bmatrix}
$$

To determine the nonhomogeneous solution, the lower portion of the unknown vector $[x_3, x_4]$ is set to zero and the remaining equation solved by back substitution for $[x_1, x_2]$. The result for the nonhomogeneous solution vector is

$$
X_{nh} =
\begin{bmatrix} 11/5 \\ -2/5 \\ 0 \\ 0 \end{bmatrix}
$$

For the homogeneous solution, we proceed as follows. Since we want $AX = 0$, we replace $[5, -4]$ in equation (11.7) by $[0, 0]$ to obtain

$$
\begin{bmatrix} 3 & 4 \\ 0 & 10 \end{bmatrix}
\begin{bmatrix} x_1 \\ x_2 \end{bmatrix}
=
-\begin{bmatrix} -1 & 2 \\ 5 & 5 \end{bmatrix}
\begin{bmatrix} x_3 \\ x_4 \end{bmatrix}
$$

At this point, it is important to recall that solutions to homogeneous systems may be determined only up to an arbitrary constant, and since the rank of the example problem is two, there will be two ($n - r = 2$) linearly independent solution vectors. To obtain the first vector, we set $[x_3, x_4] = [1, 0]$ and solve for

$[x_1, x_2]$ by back substitution. This gives the first homogeneous solution vector to be

$$X_1 = \begin{bmatrix} 1 \\ -1/2 \\ 1 \\ 0 \end{bmatrix}.$$

Now setting $[x_3, x_4] = [0, 1]$ and again applying back substitution to find $[x_1, x_2]$ gives

$$X_2 = \begin{bmatrix} 0 \\ -1/2 \\ 0 \\ 1 \end{bmatrix}.$$

The general solution to the original system of equations is

$$X = \begin{bmatrix} 11/5 \\ -2/5 \\ 0 \\ 0 \end{bmatrix} + c_1 \begin{bmatrix} 1 \\ -1/2 \\ 1 \\ 0 \end{bmatrix} + c_2 \begin{bmatrix} 0 \\ -1/2 \\ 0 \\ 1 \end{bmatrix}.$$

***

This example illustrates how Gaussian elimination with back substitution may be used to determine solutions for any type of system of simultaneous linear algebraic equations.

In summary, let us review how the Gaussian elimination procedure with back substitution is applied to cases in which the rank r of a consistent system is less than the number of unknowns n. As noted above, the Gaussian elimination method reduces a given system such as (11.1) to an upper triangular form shown schematically in (11.3). The upper r elements of the nonhomogeneous solution are found by applying back substitution to the subsystem

$$U_{r,r} (X^*_{r,1})_{nh} = D_{r,1},$$

where we have used subscripts to explicitly indicate the order of each matrix. The nonhomogeneous solution vector is given in partitioned form by

$$X_{nh} = \begin{bmatrix} \left(X^{*}_{r,\,1}\right)_{nh} \\ \cdots \\ O_{n-r} \end{bmatrix}_{nh}.$$

The remaining $n - r$ homogeneous solutions are developed by applying back substitution to the subsystem

$$(11.8) \qquad U_{r,r}(X_{r,1})_h = -B_{r,n-r} \begin{bmatrix} X_{r+1,1} \\ \vdots \\ X_{n,1} \end{bmatrix},$$

where $\begin{bmatrix} X_{r+1,1} \\ \vdots \\ X_{n,1} \end{bmatrix}$ is set respectively equal to $\begin{bmatrix} X_{r+1,1} \\ \vdots \\ X_{n,1} \end{bmatrix} \begin{bmatrix} 1 \\ 0 \\ \vdots \\ 0 \end{bmatrix}, \begin{bmatrix} 0 \\ 1 \\ \vdots \\ 0 \end{bmatrix}, \cdots, \begin{bmatrix} 0 \\ 0 \\ \vdots \\ 1 \end{bmatrix}.$

Actually, the postmultiplication by this series of constructed lower portions of the homogeneous vectors is not required. Inspection of equation (11.8) will indicate that each desired homogeneous vector is obtained by back substitution of (11.8) when the right-hand side becomes each respective column of the submatrix $-B_{r,\,n-r}$.

12. A GEOMETRIC INTERPRETATION

In concluding this chapter, it may be helpful for the reader in understanding systems of linear equations to make recourse to a geometric interpretation. For ease of visualization, consider the case of a nonhomogeneous system of third order.

$$a_{11}x_1 + a_{12}x_2 + a_{13}x_3 = b_1$$

$$a_{21}x_1 + a_{22}x_2 + a_{23}x_3 = b_3$$

$$a_{31}x_1 + a_{32}x_2 + a_{33}x_3 = b_3$$

If we consider the variables x_1, x_2, and x_3 as coordinate axes in three-dimensional space, then each equation above describes a plane in that space. The first equation would be a plane whose intercepts with the three respective axes are b_1/a_{11}, b_1/a_{12}, and b_1/a_{13}. This plane is shown in figure 2.6 below.

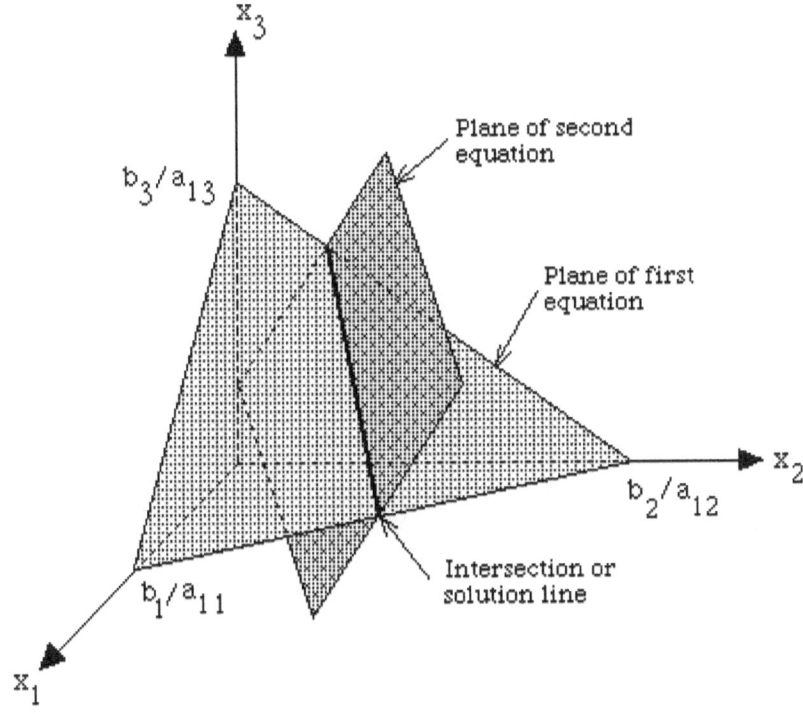

Fig. 2.6

The plane of the second equation is also depicted in figure 2.6 along with its line of intersection with the first plane (equation). If the system included only the first two equations, then this intersection line would represent all possible solutions. Clearly, a unique solution (point) does not exist as any point on the intersection line lies on (satisfies) both planes (equations). The general solution may be seen as a linear combination of two vectors. One is a vector from the origin of the coordinate system to a specific point on the intersection line (the nonhomogeneous solution). The other is a vector lying along the intersection line from the above point to an arbitrary point (the homogeneous solution).

If we now include the third equation, it is a third plane that may be pierced by the intersection line (solution) of the first two. In this situation, only the single point exists, which lies on all three planes. In other words, there is only one unique solution to the system of equations, which is a point in the solution space as shown in figure 2.7.

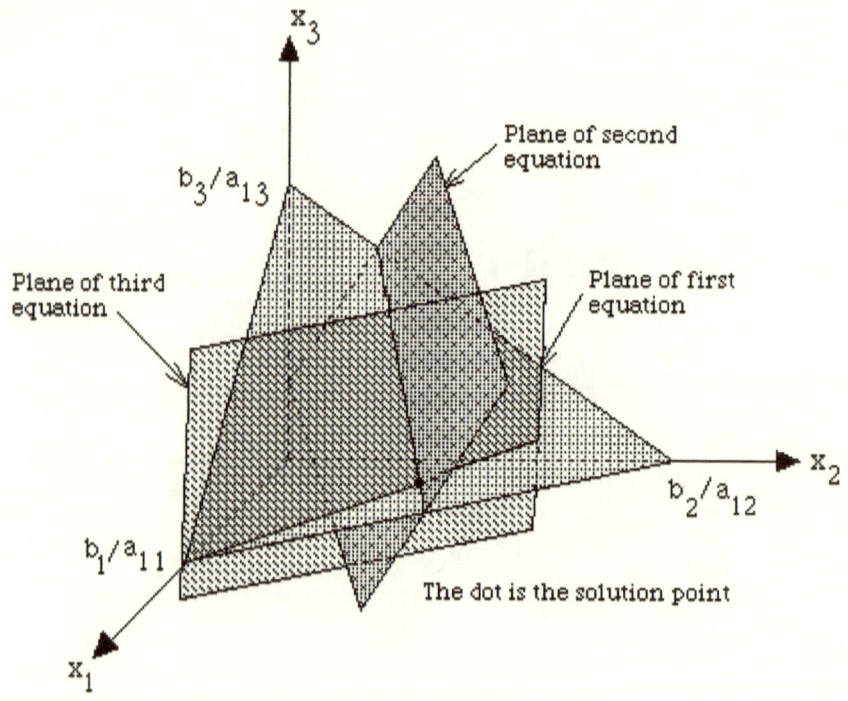

Fig. 2.7

This situation is also indicated by the rank of the augmented matrix of the system having the value three.

Should any two planes be parallel, we would have a case of the two equations being linearly dependent, and a unique solution would not exist. This situation also occurs when the third equation is parallel to the line of intersection of the first two but not parallel to either plane.

As the reader can hopefully see, each particular case or type of solution to a system of linear algebraic equations may be visualized by means of this graphic interpretation.

Ill-Conditioned Systems

In certain situations of solving systems of linear algebraic equations, the rank of the coefficient matrix may equal the rank of the augmented matrix (a consistent system), but the determinant of the coefficient matrix can be very small. This gives rise to solutions with a very high degree of uncertainty and can differ greatly from the "true" solution. When this situation occurs, the system of equations is termed ill conditioned.

As a simple illustration, consider the case of two equations and two unknowns. The equations are depicted as the two lines in figure 2.7 below. It is also assumed that there is an amount of uncertainty present due to unavoidable numerical accuracy problems as well as the possibility of experimental accuracy if the values are derived from scientific measurements. The total uncertainty is represented by the shaded zones around the equation lines in the following figures. The situation of a well-conditioned problem is illustrated in figure 2.8.

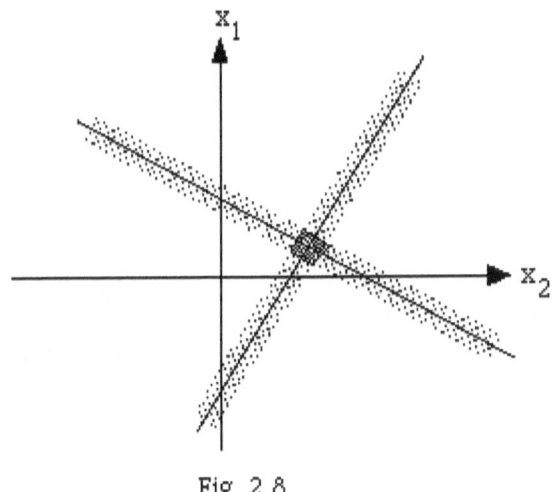

Fig. 2.8

In figure 2.8, the zone in which a calculated solution would occur is indicated by the region of heavier shading. It has about the same uncertainty tolerance as each equation has.

When the equations are ill conditioned, the equation lines become very close to parallel but not quite. If the lines were exactly parallel, then the equations would be linearly dependent and the rank of the system would be one. An ill-conditioned system is depicted in figure 2.9.

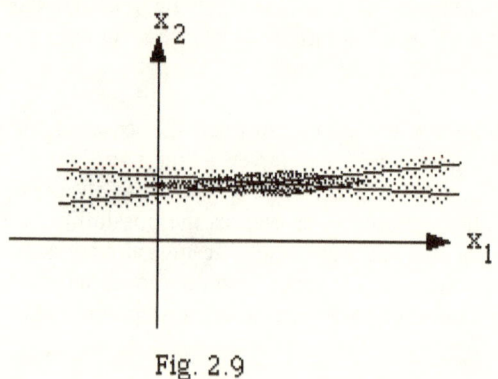

Fig. 2.9

As may be seen in figure 2.9, the uncertainty zone or region in which the computed solution will lie is much greater than for the well-conditioned case. This means that in the case of ill-conditioned equations, the uncertainty or accuracy of the computed solution is quite large.

A logical question at this point is whether there is a way to tell if a system of equations is well conditioned or not. The answer is that there is a way of telling, and that is by the use of the condition number. The condition number is defined as the norm[16] of the coefficient matrix times the norm of the inverse of the coefficient matrix.

$$\text{cond}(A) = \| A \| \, \| A^{-1} \|$$

The calculation of the norm of the inverse matrix is an involved process, and hence, the methods by which the condition number is calculated will not be presented. Most commonly available matrix calculator software programs include the evaluation of the condition number as a built-in function. The condition number is close to unity for well-conditioned problems. A system is usually considered ill conditioned when the condition number is on the order of 100 or larger. However, in any situation, it is always good practice to check computed solutions by back substitution into the original equations for verification.

---

[16] The definition of the norm of a matrix is given in section 7 of chapter IX.

ADDITIONAL EXERCISES

1.*   Consider the following system[17] and determine whether it is consistent
       or not. If it is consistent, state the form of the solution and find it.

$$\begin{bmatrix} 2 & 1 & 0 & 1 & 3 \\ 1 & 1 & 1 & 1 & 1 \\ 2 & 1 & 1 & 2 & 1 \\ 2 & 2 & 1 & 1 & 4 \end{bmatrix} \begin{bmatrix} x_1 \\ x_2 \\ x_3 \\ x_4 \\ x_5 \end{bmatrix} = \begin{bmatrix} 3 \\ 3 \\ 3 \\ 6 \end{bmatrix}$$

2.*   Consider the column vector $V = [1, 2, -1]$.

   a. Given the set of three three-dimensional linearly independent
       vectors below, expand the vector V in terms of them.

$$X_1 = \begin{bmatrix} 1 \\ 0 \\ 1 \end{bmatrix}, \; X_2 = \begin{bmatrix} 1 \\ 1 \\ 0 \end{bmatrix}, \; X_3 = \begin{bmatrix} 0 \\ 1 \\ 1 \end{bmatrix}$$

       In other words, find the constants c_i such that

$$V = \sum_{i=1}^{3} c_i X_i \; .$$

   b. Given the set of orthogonal vectors

$$V_1 = \begin{bmatrix} 1 \\ 0 \\ -1 \end{bmatrix}, \; V_2 = \begin{bmatrix} 1 \\ 1 \\ 1 \end{bmatrix}, \; V_3 = \begin{bmatrix} 1 \\ -2 \\ 1 \end{bmatrix},$$

       expand the given vector V in terms of them. Use orthogonality to
       find the constants d_i such that

$$V = \sum_{i=1}^{3} d_i V_i \; .$$

   c. Discuss the advantage of using an orthogonal system of vectors as a
       basis or coordinate system for representing vectors.

---

[17]   Marcus and Miinc [1965], p107.

d. Discuss the further advantage of using an orthonormal system of vectors as a basis or coordinate system for representing vectors.

3.*   Consider the following system.

$$\begin{bmatrix} 2 & 0 & -1 \\ 1 & 1 & -2 \\ 3 & -4 & 1 \end{bmatrix} X = B_i$$

Find the solutions X_i for each of the following three B_i vectors.

$$B_1 = \begin{bmatrix} 1 \\ 0 \\ 0 \end{bmatrix}, \ B_2 = \begin{bmatrix} 0 \\ 1 \\ 0 \end{bmatrix}, \ B_3 = \begin{bmatrix} 0 \\ 0 \\ 1 \end{bmatrix}$$

If these solutions are placed in order as the columns of a matrix A, what does this matrix represent? Verify your answer.

4.*   Attempt to solve the following system of equations by Gaussian elimination, and discuss what happens.

$$\begin{bmatrix} 1 & 1 & -1 \\ 2 & -1 & 3 \\ 3 & -3 & 7 \end{bmatrix} X = \begin{bmatrix} 1 \\ 4 \\ 7 \end{bmatrix}$$

5.*   The following system[18] is called tridiagonal. It is also termed sparse because of the many zero elements. Show that it remains sparse when reduced to upper triangular form by Gaussian elimination.

$$\begin{bmatrix} 2 & -1 & 0 & 0 & 0 & 0 \\ -1 & 2 & -1 & 0 & 0 & 0 \\ 0 & -1 & 2 & -1 & 0 & 0 \\ 0 & 0 & -1 & 2 & -1 & 0 \\ 0 & 0 & 0 & -1 & 2 & -1 \\ 0 & 0 & 0 & 0 & -1 & 2 \end{bmatrix} X = \begin{bmatrix} 1 \\ 1 \\ 1 \\ 1 \\ 1 \\ 1 \end{bmatrix}$$

[18]   Dorn and McCracken, 1972, p. 206.

6.   Suppose you wished to solve two sets of simultaneous linear algebraic systems of equations[19] that had identical coefficient matrices but different constant vectors such as the following.

$$AX = B_1 , AX = B_2$$

Develop an extension to the Gaussian elimination method that would solve both sets with only one elimination process.

7.*   Consider the following fourth-order system of linear algebraic equations of rank four.

$$AX = \begin{bmatrix} 2 & 6 & 1 & 6 \\ 6 & 3 & -4 & 2 \\ 1 & -4 & 0 & 3 \\ 6 & 2 & 3 & -8 \end{bmatrix} X = B$$

Solve this system when the constant column B is set equal to each of the columns of A. What do you observe? Can you prove your finding? Does it matter that A is symmetric?

8.   As an extension of exercise 7 above, consider the general case of an nth-order system of rank n and a general constant vector B.

$$AX = B$$

Let B be expanded as a linear combination of the columns of A.

$$B = \sum_{k=1}^{n} c_k A_k , \quad A_k = k^{th} \text{ column of A}$$

What, if any, is the relation between the constants c_k in the above expansion and the solution vector X?

9.   As discussed in section 4, verify that in solving a system on n equations and n unknowns requires $(n_3 + 3n_2 - n)/3$ operations not counting additions or subtractions.

---

[19]   Dorn and McCracken, 1972, p. 207.

III

APPLICATION OF LINEAR ALGEBRAIC EQUATIONS

1. INTRODUCTION

In this chapter, the reader will be introduced to a wide variety of applications of the theory of linear algebraic equations and related topics. Specifically, we will cover the method of least squares, linear electrical circuits, Leontief economic models, solution of potential problems, difference equations, and the equilibrium of systems of forces.

No attempt will be made to provide all the underlying theory of these methods as there is not sufficient space or time to cover each subject in detail. The object is to provide a sufficient amount of information so that the reader may set up the appropriate governing equations for a simple problem in each given area and exercise. The theory of linear algebraic equations developed in the last chapter will then be employed to solve these equations.

2. THE METHOD OF LEAST SQUARES

One of the most common and widely used applications of linear algebraic equations is that of least squares curve fitting. In the area of statistics, it is known as linear or multiple regression. In other areas, it is characterized by problems in which it is desired to find an equation to fit or represent a large number of data points. Most frequently, a linear or straight-line fit is desired.

In this section, underlying theory will be developed, and then the problem will be formulated in terms of matrix equations representing a system of linear algebraic equations.

To begin with, let us assume that we have a problem in which a number of data points or observations are given. Suppose that there are m pairs of data points (x_1, y_1), (x_2, y_2), . . . , (x_m, y_m). We further assume that x_i are the independent variables or values and y_i the dependent variables or values. It is desired to fit or represent these data by a polynomial of degree n, which will be a function of x.

$$(2.1) \qquad p_n(x) = a_0 + a_1 x + a_2 x^2 + \cdots + a_n x^n = \sum_{i=0}^{n} a_i x^i$$

156

In an application, this polynomial would be used to produce a curve that would represent the behavior of the data. In other situations, $p_n(x)$ would be used to predict the value of the dependent variable y at any x or the specific points x_j. The method of least squares is the most common method of evaluating the unknown coefficients of the polynomial (2.1) and provides the "best fit" in the least squares sense.

Figure 3.1 below provides a graphic illustration of this problem.

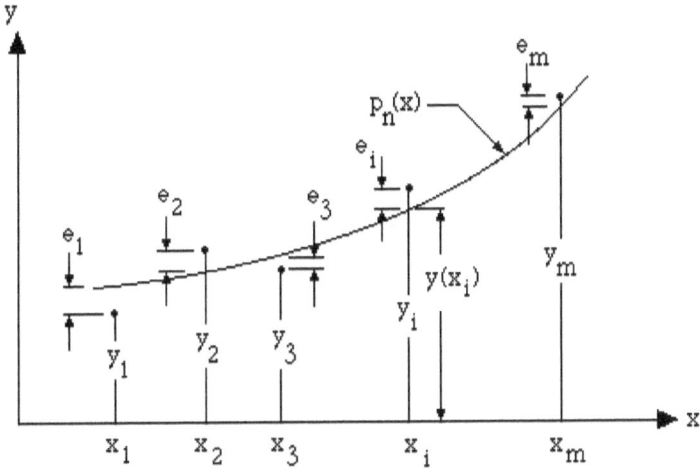

Fig. 3.1

In this figure, the data points (x_i, y_i), $i = 1, 2, \ldots, m$ are shown by dots, and the error or amount of discrepancy between the polynomial curve and the data points by e_i. At each data value x_i, the polynomial may be used to predict a y value from $y_i = y(x_i) = p_n(x_i)$. The error of fit at each point is given by

$$e_i = y_i - y(x_i) = y_i - p_n(x_i).$$

By examining the polynomial, it may be seen that it is controlled by $n + 1$ arbitrary constants a_i. The problem is to select these constants so that all the errors are minimized. The least squares error technique does this by requiring that the sum of the squares of all the individual errors must be a minimum. This produces a set of $n + 1$ equations for the $n + 1$ unknowns a_i, $i = 0, 1, \ldots, n$.

The total square error may be written as

$$E(a_k) = \sum_{i=1}^{m} e_i^2 = \sum_{i=1}^{m} [y_i - y(x_i)]^2 = \sum_{i=1}^{m} [y_i - p_n(x_i)]^2$$

$$= \sum_{i=1}^{m} \left[y_i - \sum_{k=0}^{n} a_k x_i^k \right]^2 .$$

In this result, the notation $E(a_k)$ has been employed to clearly indicate that the total error is a function of the unknown constants a_k. To minimize the error, we set

$$\frac{\partial E}{\partial a_j} = 2 \sum_{i=1}^{m} e_i \frac{\partial e_i}{\partial a_j} = 0 , j = 0, 1, 2, \dots , n ,$$

which represents a set of $n + 1$ equations for the $n + 1$ unknown constants.

$$\frac{\partial E}{\partial a_j} = 2 \sum_{i=1}^{m} \left[y_i - \sum_{k=0}^{n} a_k x_i^k \right] \left(-x_i^j \right) = 0$$

Dropping the constant factor of two and rearranging this equation gives the following result

(2.2) $$\sum_{k=0}^{n} \left(\sum_{i=1}^{m} x_i^{j+k} \right) a_k = \sum_{i=1}^{m} x_i^j y_i , j = 0, 1, 2, \dots , n ,$$

which represents the $n + 1$ linear algebraic equations for the $n + 1$ unknown constants. The coefficients in this problem are calculated from the data values.

As an aid to seeing these equations, consider a case in which it is desired to fit a straight line to the given m data points. For this case, $n = 1$ and the two equations will be

$$m a_0 + \left(\sum_{i=1}^{m} x_i \right) a_1 = \sum_{i=1}^{m} y_i$$

$$\left(\sum_{i=1}^{m} x_i \right) a_0 + \left(\sum_{i=1}^{m} x_i^2 \right) a_1 = \sum_{i=1}^{m} x_i y_i .$$

In matrix form, these equations would appear as

$$\begin{bmatrix} m & \left(\sum_{i=1}^{m} x_i\right) \\ \left(\sum_{i=1}^{m} x_i\right) & \left(\sum_{i=1}^{m} x_i^2\right) \end{bmatrix} \begin{bmatrix} a_0 \\ a_1 \end{bmatrix} = \begin{bmatrix} \sum_{i=1}^{m} y_i \\ \sum_{i=1}^{m} x_i y_i \end{bmatrix}.$$

or $CA = B$, where

$$C = \begin{bmatrix} m & \left(\sum_{i=1}^{m} x_i\right) \\ \left(\sum_{i=1}^{m} x_i\right) & \left(\sum_{i=1}^{m} x_i^2\right) \end{bmatrix}, \quad A = \begin{bmatrix} a_0 \\ a_1 \end{bmatrix}, \quad B = \begin{bmatrix} \sum_{i=1}^{m} y_i \\ \sum_{i=1}^{m} x_i y_i \end{bmatrix}.$$

Due to measurement uncertainties as well as computer round off, it would be a highly unusual situation for the coefficient matrix C to be singular. Thus, the unknown constants for the best fit equation would be found from $C = A^{-1}B$.

When it is desired to fit higher-order polynomials to a set of data points, equation (2.2) could be used to calculate all the elements of the matrices C and B. However, there is a better approach utilizing matrices from the beginning that provides the same result. Let us return to the general polynomial we wish to fit to the data as given by equation (2.1). If all the data values are respectively substituted into this equation, one would have the following system .

(2.3)
$$p_n(x_1) = a_0 + a_1 x_1 + a_2 x_1^2 + \cdots + a_n x_1^n = y_1$$
$$p_n(x_2) = a_0 + a_1 x_2 + a_2 x_2^2 + \cdots + a_n x_2^n = y_2$$
$$\cdots$$
$$p_n(x_m) = a_0 + a_1 x_m + a_2 x_m^2 + \cdots + a_n x_m^n = y_m$$

Since the data values are known, this system represents a set of m equations for the $n + 1$ unknowns a_i. Actually, these equations cannot all be satisfied exactly unless the polynomial passed through every data point. What we are seeking is a set of coefficients a_0, a_1, etc., which will cause the polynomial to satisfy all the above equations as closely as possible by minimizing the errors in the least squares sense. Using matrix notation, equation (2.3) may be written as $XA = Y$ or

$$(2.4) \quad \begin{bmatrix} 1 & x_1 & x_1^2 & \cdots & x_1^n \\ 1 & x_2 & x_2^2 & \cdots & x_2^n \\ \cdot & \cdot & \cdot & \cdots & \cdot \\ 1 & x_m & x_m^2 & \cdots & x_m^n \end{bmatrix} \begin{bmatrix} a_0 \\ a_1 \\ a_2 \\ \vdots \\ a_n \end{bmatrix} = \begin{bmatrix} y_1 \\ y_2 \\ \vdots \\ y_m \end{bmatrix} .$$

In nearly all situations, the degree of the desired polynomial is much smaller than the number of data points such that $m > n + 1$. This system falls under case 1.1 of the last chapter, and although a solution always exists, it is not unique. We will now show how (2.4) may be transformed by matrix operations into a square form of order $n + 1$ and a unique solution found, which is in fact the least squares solution.

We begin by premultiplying (2.4) by X^T to obtain

$$X^T X A = X^T Y.$$

Examining the product $X^T X$ indicates that it is a square matrix of order $n + 1$. The desired constant vector A may be found if we now multiply by $(X^T X)^{-1}$. The final result is

$$A = (X^T X)^{-1} X^T Y.$$

In comparing this result to the linear fit developed above, it should be noted that $C = X^T X$ and $B = X^T Y$. It is left as an exercise for the reader to show that in the general case, the matrices C and B as developed from X and Y give exactly the least squares equations (2.2).

The reader might be tempted to expand the inverse of the product $X^T X$ to obtain the product of inverses in the reverse order. This would appear to result in $A = X^{-1} Y$, which is meaningless in this situation because X is not a square matrix. The point to be made is that $(AB)^{-1} = B^{-1} A^{-1}$ only when B^{-1} and A^{-1} exist.

Rather than solve for the constant vector A by direct inversion, it should be clear that the use of the Gaussian elimination method previously developed is the more efficient method.

EXAMPLE 1

Consider the following set of data[20], which has been obtained from measuring the heights and weights of nine men of fairly consistent body structure (average build). In place of the X and Y matrices, we will use H and W to relate to the height and weight values. The object of this problem will be to obtain a linear equation from which one could predict the weight of a man of average build based on knowing his height.

Height (in.), h_i	72	68	66	74	62	70	64	76	64
Weight (lb), w_i	174	152	154	180	135	161	140	174	157

SOLUTION

The H, W, and A matrices for this case are

$$X = \begin{bmatrix} 1 & 72 \\ 1 & 68 \\ 1 & 66 \\ 1 & 74 \\ 1 & 62 \\ 1 & 70 \\ 1 & 64 \\ 1 & 76 \\ 1 & 64 \end{bmatrix}, \ Y = \begin{bmatrix} 174 \\ 152 \\ 154 \\ 180 \\ 135 \\ 161 \\ 140 \\ 174 \\ 157 \end{bmatrix}, \ A = \begin{bmatrix} a_0 \\ a_1 \end{bmatrix}$$

Calculating the matrices C and B from $X^T X$ and $X^T Y$ gives

$$C = \begin{bmatrix} 9 & 616 \\ 616 & 42{,}352 \end{bmatrix}, \ B = \begin{bmatrix} 1{,}427 \\ 98{,}220 \end{bmatrix}.$$

The solution for the desired constant vector is

$$A = \begin{bmatrix} a_0 \\ a_1 \end{bmatrix} = \begin{bmatrix} 9 & 616 \\ 616 & 42{,}352 \end{bmatrix}^{-1} \begin{bmatrix} 1{,}427 \\ 98{,}220 \end{bmatrix} = \begin{bmatrix} -39.2617 \\ 2.8902 \end{bmatrix}.$$

---

[20]   The data for this example was taken from a problem in the text by Dorn and McCracken [1972], p. 320.

This result tells us that the desired linear equation from which the dependent variable y can be predicted, based on least squares error, is

$$w = -39.2617 + 2.8902h$$

***

The reader should note that the analyst must determine the degree of the polynomial chosen to fit the data in the previous example. In many situations, there is some idea of what the degree should be based on an understanding of the theory underlying the system from which the measurements were taken. In any situation, one would not want to attempt to fit a polynomial whose degree is close to the number of data points. In fact, if the degree was chosen such that $n = m - 1$, a polynomial would be obtained, which passes exactly through every data point producing zero error at each point. Although this might seem to be desirable, if the polynomial were plotted, it would be seen to wander wildly through the points and not indicate any trend.

In most situations, it is desired to not have the polynomial reflect the small dispersions of the data but provide a smoothing effect. This is frequently achieved by using a small value of n. Many least squares fitting programs do provide some statistical measure of the degree of fit as determined from an analysis of the individual errors at each point. When available, the analyst must be familiar with the measures provided to avoid any misunderstanding of what the particular values mean.

EXERCISES

1.*   Find the best straight line in the least squares sense that fits the following data.

x	0.0	0.1	0.2	0.3	0.4	0.5
y	-1.1	-0.2	0.4	1.0	2.3	2.9

2.*   Determine the best

    a.   quadratic
    b.   cubic

least squares that fit the following data.

x	-1	0	1	2	3
y	-22	-1	4	5	14

3.*   You have just started a new business, and your earnings for each of the first six months of operation are $1,485, $2,010, $2,505, $2,980, $3,450 and $3,890.

      a.   Find the best linear least squares equation to fit this information, and use it to predict your earnings for the seventh month of operation.

      b.   Find the best quadratic least squares equation to fit this information, and use it to predict your earnings for the seventh month of operation.

      c.   Use the linear and quadratic fits developed in parts a and b above, and predict your earnings for the ninth month using each approximation.

4.*   A test run has been made of a turbine-powered vehicle on the Bonneville Salt Flats in Utah. The results of six time trials are given below.

Time, t (s)	10.0	20.0	30.0	40.0	50.0	60.0
Speed, v (m/s)	15.1	32.2	63.4	84.5	118.0	139.0

Determine the best least squares linear equation to predict the speed s on the basis of the time t.

***

3. LINEAR ELECTRICAL CIRCUITS (STEADY STATE)

As we stated in the introduction, this text will not attempt to present the extensive theory behind the various areas of application that will be cited. This is particularly true in the area of electrical circuit theory. Only a brief superficial introduction to one aspect will be provided as a means of illustrating how complex circuits reduce to the solution of a system of linear algebraic equations. Sufficient information will be given so that the reader may set up and solve a system of equations for simple steady state circuits.

To begin with, let us consider a general circuit that is illustrated in figure 3.2 below. This symbolic sketch is fairly typical of how a circuit is drawn. The symbol S^{21} denotes voltage or potential sources, the Zs are branch impedances,

---

[21.]   The reader should note that in this section, capital letters are being used to designate scalar circuit elements as well as matrix quantities. This notation is being followed as it is the standard notation of nearly all texts on circuit analysis.

and the i's are the circulating mesh currents. It is assumed that mesh currents are assumed to all flow in the same direction, i.e., all clockwise or all counterclockwise. The complete circuit is visualized as a network of individual mesh not unlike a fish net. Each branch or side of a mesh can contain a voltage source and/or an impedance. Each mesh is assumed to have a circulating current in some arbitrary direction. For convenience, all the mesh currents in this general example have been assumed to flow clockwise.

If a branch is common to two mesh, the current flowing through the branch is the difference of the two adjacent mesh currents. For example, the impedance Z_4 below has a current (branch current) $i_4 - i_3$ flowing from left to right. On the other hand, the impedance Z_3 has only i_3 flowing from left to right.

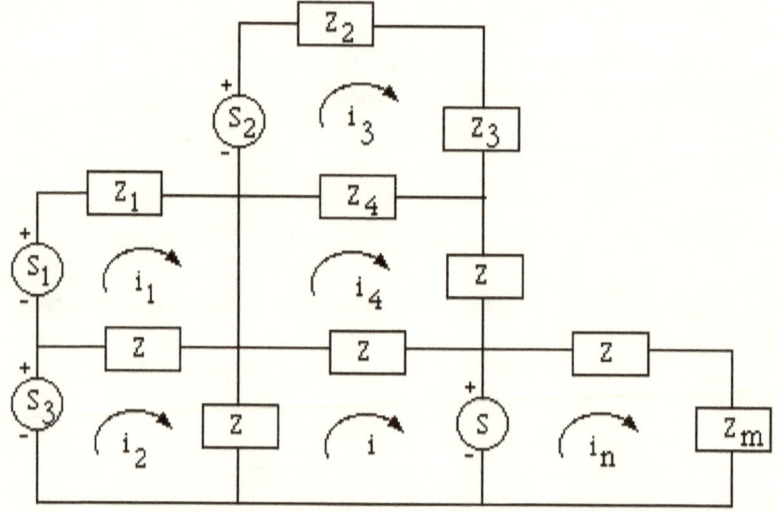

Fig. 3.2

The branch impedances may be composed of any combination of three electrical components: resistance, capacitance, and inductance. The impedance of both capacitances and inductances are frequency dependent, which must be reflected in writing expressions for the impedance. Without delving into too much circuit theory, figure 3.3 below illustrates the algebraic form an impedance would have if it were a series connection of a resistance R, capacitance C, and inductance L.

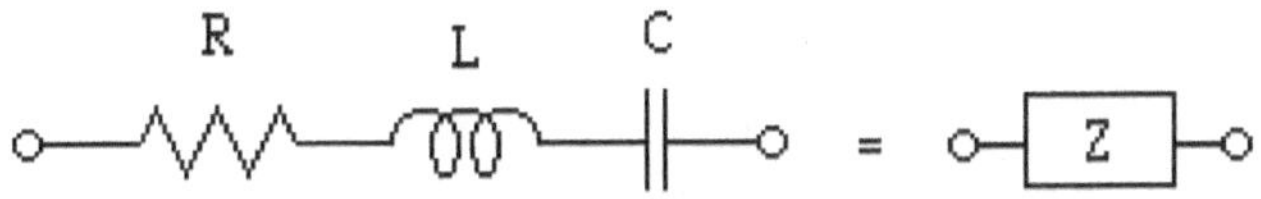

$$Z = R + j\omega L - j\frac{1}{\omega C} \quad , \quad j = \sqrt{-1}$$

Fig. 3.3

In a circuit problem, a consistent set of units is volts (V) for voltage, amperes (A) for current, ohms (Ω) for impedance and resistance, farads (F) for capacitance, and henrys (H) for inductance.[22] The frequency ω is angular frequency with units of inverse seconds or s^{-1}. For standard household voltage, the frequency is 60 hertz (Hz) for which ω is 377 s^{-1}.

It must be clear that this discussion is restricted to steady state problems only. This means that we are dealing with only two types of problems. One is problems of direct current (DC) for which $\omega = 0$ and the impedances are resistive only. The other type of problems are those for which the frequency of the driving voltage sources is a constant alternating current (DC), and we are seeking the steady state solution for the mesh currents.

In terms of an impedance matrix Z, unknown current vector I, and voltage source vector E, we may write the following system of linear equations for a general circuit such as the one depicted in figure 3.2 above.

(3.1) $ZI = E,$

where

(3.2) $Z = \begin{bmatrix} Z_{11} & -Z_{12} & \cdots & -Z_{1n} \\ -Z_{21} & Z_{22} & \cdots & -Z_{2n} \\ \cdot & \cdot & \cdots & \cdot \\ -Z_{n1} & -Z_{n2} & \cdots & Z_{nn} \end{bmatrix}, \quad I = \begin{bmatrix} i_1 \\ i_2 \\ \vdots \\ i_n \end{bmatrix}, \quad E = \begin{bmatrix} e_1 \\ e_2 \\ \vdots \\ e_n \end{bmatrix} .$

In practice, one would typically set up the above equations by the application of Kirchoff's voltage law to each individual mesh. It is convenient that the result is such that the above matrices may be written down directly by examining a

---

[22]   In terms of volts (V), amperes (A), and seconds (s), an ohm (Ω) is one volt per ampere (V/A), a Farad (F) is one ampere-second per volt (A-s/V), and a henry (H) is one volt-second per ampere (V-s/A).

specific given circuit. First it should be apparent that the unknown current vector contains the n unknown mesh currents for the general circuit. In the voltage vector, the element e_i represents the sum of all the voltage rises going around the mesh in the same direction as the assigned mesh current.

The elements of the impedance matrix are determined in the following manner. The diagonal elements, Z_{ii}, represent the sum of all the impedances common to the ith mesh. The off diagonal elements are always zero or negative as shown by the negative signs in (3.2) above. The value of Z_{ij} is the sum of the impedances of the ith mesh, which are common to a neighboring j^{th} mesh. Also, Z_{ij} is always nonnegative, and the impedance matrix is always symmetric.

As an illustration, consider the third mesh of the general circuit shown above. For this mesh i_3 is the mesh or circulating current. The source voltage term e_3 would be $+S_2$ as it is a voltage rise when traversing the mesh in the same direction as the assigned current. The self impedance Z_{33} is $Z_2 + Z_3 + Z_4$. The only nonzero coupling impedance is Z_4, which is common to mesh four, thus $Z_{34} = Z_4$.

EXAMPLE 2

As a specific example, recall the simple resistive electrical circuit that was introduced in chapter II. Following the rules described above, construct the voltage, current, and impedance matrices that describe the circuit.

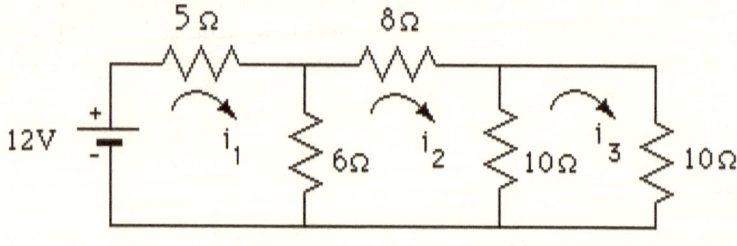

Fig. 3.4

SOLUTION

By following the above discussion, the voltage and unknown mesh current vectors are

$$E = \begin{bmatrix} 12 \\ 0 \\ 0 \end{bmatrix}, \quad I = \begin{bmatrix} i_1 \\ i_2 \\ i_3 \end{bmatrix}.$$

Following the rules for constructing the impedance matrix given above, we may write the following.

$$Z_{11} = 5 + 6 = 11$$
$$Z_{12} = Z_{21} = 6$$
$$Z_{13} = Z_{31} = 0$$
$$Z_{22} = 6 + 8 + 10 = 24$$
$$Z_{23} = Z_{32} = 10$$
$$Z_{33} = 10 + 10 = 20$$

Assembling these values in the circuit impedance matrix gives

$$Z = \begin{bmatrix} 11 & -6 & 0 \\ -6 & 24 & -10 \\ 0 & -10 & 20 \end{bmatrix}.$$

These results are the same as presented in section 2, chapter II.

***

As a further example, let us examine an AC circuit that has a similar structure to the resistive DC one just analyzed.

EXAMPLE 3

For the circuit shown in figure 3.5 below, determine the voltage, current, and impedance matrices. Also solve for the circulating currents that have been defined in figure 3.5.

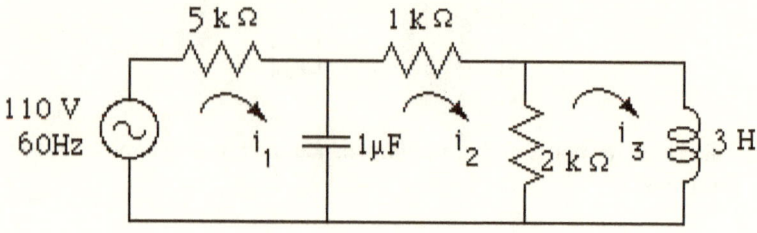

Fig. 3.5

SOLUTION

For this circuit, the steady state impedance of the capacitor at 60 Hz is $Z_C = -j(1/\omega C) = -2.65jk\Omega$, and the impedance of the inductor is $Z_L = j\omega L = 1.13jk\Omega$. Here, j is $\sqrt{-1}$ as these impedances are pure imaginary values. The driving voltage in the first mesh is $e_1 = 110$ volts, which is assumed to be at a zero phase angle. For those readers who are familiar with phasor notation,[23] this source voltage would be written as $e_1 = 110\underline{/\,0°}$. The phase angles of the circulating mesh currents will all be relative to the phase of the source voltage.

The known voltage and unknown current vectors are

$$E = \begin{bmatrix} 100\angle 0 \\ 0 \\ 0 \end{bmatrix}, V \quad \text{and} \quad I = \begin{bmatrix} i_1 \\ i_2 \\ i_3 \end{bmatrix}$$

---

[23]   Phasor notation is essentially the same as polar notation for complex numbers. If z is a complex number with real part a and complex part b, we write $z = a + bj$. The standard polar notation for this number is $z = |z|\exp(j\theta)$ where $|z| = R(a^2 + b^2)$ and $\theta = \arctan(b/a)$. In phasor notation, this would be written as $z = |z| \underline{/\ \theta}$.

The reader may verify that the impedance matrix for this circuit is the following.

$$Z = \begin{bmatrix} 5\text{-}2.65j & 2.65j & 0 \\ 2.65j & 3\text{-}2.65j & -2 \\ 0 & -2 & 2+1.13j \end{bmatrix}, k\Omega$$

The solution for the circulating mesh currents is

$$I = Z^{-1}E.$$

The solution[24] is

$$I = \begin{bmatrix} 15.85+0.75j \\ 14.45\text{-}10.85j \\ 6.30\text{-}14.41j \end{bmatrix}, mA = \begin{bmatrix} 15.87\angle 2.7° \\ 18.07\angle\text{-}36.9° \\ 15.73\angle\text{-}66.4° \end{bmatrix}, mA$$

The answers for the current are in milliamperes since the impedances were in kilo-ohms and the values are presented in both complex Cartesian and polar or phasor forms.

EXERCISES

1.*   Write the matrix equation for the three circulating mesh direct (DC) currents in the circuit below, and solve for them.

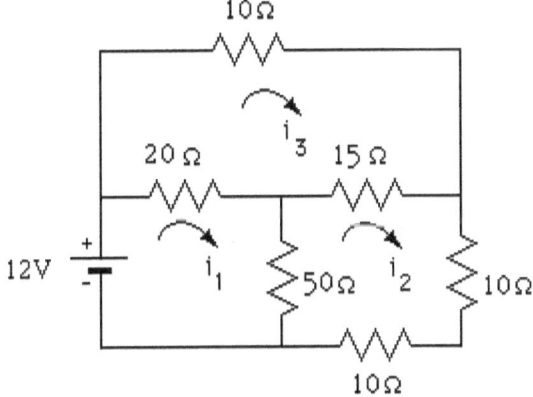

2.*   Write the matrix equation for the four circulating mesh direct currents in the circuit below, and solve for them. All resistance values are in ohms.

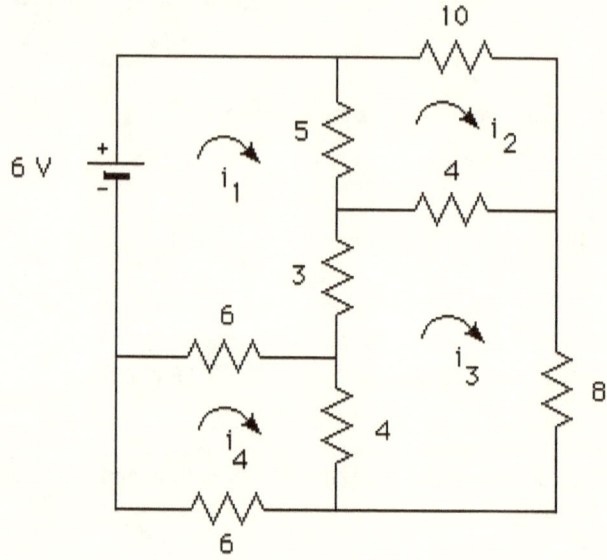

3.*   Write the matrix equation for the four circulating mesh direct currents in the circuit below, and solve for them.

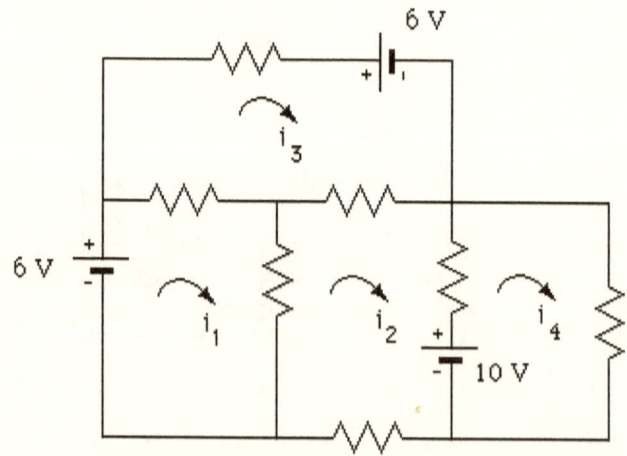

Assume all resistances are 10 ohms.

***

4. INPUT-OUTPUT ECONOMIC ANALYSIS

In the area of economic analysis, a widely known application of matrix algebra is that of Leontief input-output model. Essentially, it is a supply-and-demand model that may be used to study the effect of changes in demand on supply or production among others. The model may be applied to specific industries or to broader scale segments of an economy.[25]

In establishing this model, we define x_i as the productivity or output of the ith industry or segment of an economy. For standardization purposes, x_i is measured in units of value or money such as dollars. The output of the ith segment may be utilized by many of the other segments in order to produce their respective outputs. In other words, a segment j may produce a demand for the output of segment i.

Algebraically, this demand may be expressed by the term $c_{ij}x_j$. Here, x_j represents the production or output of the jth segment, and the coefficient c_{ij} is the output of the ith segment used to produce one unit of the jth segment.

Demand can arise not only from other producing segments of the particular economy being analyzed, it may also arise from outside or external sources. The external demand for the ith product is represented by d_i. If we now assume that demand equals supply and that there are n segments or industries in the economy under study, the following demand equals supply equation may be written for the ith segment.

$$c_{i1}x_1 + c_{i2}x_2 + \cdots + c_{in}x_n + d_i = x_i$$

Since there are n segments to the economy, there will be a similar equation for each segment. All these equations may be represented by the following matrix equation.

(4.1) $CX + D = X,$

where C is the coefficient matrix of internal demands, D the external demand vector, and X the total production vector. If we define a matrix A such that $A = I_n$

---

25.   For example, the interested reader is referred to the article "The Structure of the U.S. Economy" by Wassily W. Leontief, *Scientific American*, vol. 212, no. 4, pp. 25–35, April 1965.

$- C$, where I_n is the nth-order identity matrix, equation (4.1) may be written in the following form.

$$AX = D$$

This shows how the Leontief economic model is represented by a system of n linear algebraic equations and n unknowns. If the demand is known, then the production vector may be found, which meets the stated demand.

As an illustration, consider a simple economy in which there are four identifiable segments—agriculture (x_1), mining (x_2), manufacturing (x_3), and transportation (x_4). The table in figure 3.6 gives values for the production and consumption of the output of these segments.

Produced For ↑ / Consumed From →	Agriculture	Mining	Manufacturing	Transportation	External Demand	Total Production
Agriculture	22	0	41	50	10	123
Mining	0	2	16	2	3	23
Manufacturing	7	2	200	5	181	395
Transportation	3	2	28	8	31	72

Input-Output Table for Example Economy
[Billions (10^9) Dollars]

Fig. 3.6

From this table we may see the external demand vector as being

$$D = \begin{bmatrix} 10 \\ 3 \\ 181 \\ 31 \end{bmatrix}$$

The entries for the segments in each row represent the demand for the output of the segment at the left. As indicated, the last column is the total demand, and hence, the total production under the assumption of supply equals demand. To normalize the coefficients of the C matrix for a per-unit basis, each value for a given segment must be divided by the total production. This means the C matrix will be the following.

$$C = \begin{bmatrix} 22/123 & 0/23 & 41/395 & 50/72 \\ 0/123 & 2/23 & 16/395 & 2/72 \\ 7/123 & 2/23 & 200/395 & 5/72 \\ 3/123 & 2/23 & 28/395 & 8/72 \end{bmatrix}$$

If we employ three significant figure arithmetic, the resulting A matrix will be

$$A = I_4 - C = \begin{bmatrix} 0.821 & 0.000 & -0.104 & -0.694 \\ 0.000 & 0.913 & -0.0405 & -0.0278 \\ -0.0569 & -0.0870 & 0.494 & -0.0694 \\ -0.0244 & -0.0870 & -0.0709 & 0.889 \end{bmatrix}$$

For the given demand vector, the resulting production vector is

$$X = A^{-1} D = \begin{bmatrix} 123.314 \\ 23.010 \\ 395.046 \\ 72.272 \end{bmatrix}.$$

It should be noted that the inverse of the coefficient matrix, A, is used above to only symbolically indicate the solution. In actuality, the solution is determined by Gaussian elimination.

The above answer is not completely correct due to the round-off errors resulting from computing the A matrix only to three significant figures of accuracy. The exact answer should be

$$X = \begin{bmatrix} 123 \\ 23 \\ 395 \\ 72 \end{bmatrix},$$

which is the total production as given in the data table of figure 3.6.

EXAMPLE 4

Consider the example economy just presented. Assuming that the system constants, the C matrix, remain the same, what is the production vector if the external demand for manufacturing decreases by 10% to a level of 163 units? Also compute the total output of the economy, which could be called its GNP.

SOLUTION

The new external demand vector is

$$D = \begin{bmatrix} 10 \\ 3 \\ 163 \\ 31 \end{bmatrix}.$$

The resulting total production is

$$X = A^{-1}D = \begin{bmatrix} 116 \\ 21 \\ 357 \\ 69 \end{bmatrix}.$$

In this result, the answers have been rounded to the nearest integer or whole billion-dollar value.

The total output of the economy is the sum of all the individual output values. Since the production vector has four elements, their sum may be obtained by the following matrix product.

$$\text{Total output} = \text{GNP} = [\,1\ 1\ 1\ 1\,]\,X$$

$$\text{GNP} = [\,1\ 1\ 1\ 1\,] \begin{bmatrix} 116 \\ 21 \\ 357 \\ 69 \end{bmatrix} = 563$$

EXAMPLE 5

In example 4 above, rather than study the effect of a 10% drop in external demand for manufacturing, assume a boycott occurs, which reduces the external demand for manufactured goods to zero. Again, assume all the system constants do not change, and determine the effect on the total production of the economy as well as its GNP.

SOLUTION

The new demand vector representing a total boycott on manufactured goods is

$$D = \begin{bmatrix} 10 \\ 3 \\ 0 \\ 31 \end{bmatrix}.$$

The resulting total production vector and GNP for this demand are

$$X = \begin{bmatrix} 45 \\ 5 \\ 11 \\ 38 \end{bmatrix}$$

$$\text{GNP} = [\,1\ 1\ 1\ 1\,] \begin{bmatrix} 45 \\ 5 \\ 11 \\ 38 \end{bmatrix} = 99$$

EXERCISES

1.*   In the example economic system examined in this section, determine the effect on production if the external demand for manufacturing goods increased by 20%. Assume that the system coefficients are affected by this change in the demand.

2.*   Repeat exercise 1 above if the external demand for manufacturing goods fell by 50%.

3.*   Consider a medium-size energy company that owns wells from which it pumps its own oil, produces gasoline, and generates electricity. Suppose that to produce one unit of oil, the company consumes

        0 units of crude oil,
        1 unit of gasoline,
        1.5 units of electricity.

To produce one unit of gasoline, the company consumes

        2 units of crude oil,
        0.3 unit of gasoline,
        0.5 unit of electricity.

To produce one unit of electricity, the company consumes

        0.3 unit of crude oil,
        0.5 unit of gasoline,
        0.2 unit of electricity.

If the external demand is for 70 units each of crude oil, gasoline, and electricity, what should the total production of the company be in order to meet the internal and external demand?

***

5. SOLUTION OF POTENTIAL PROBLEMS

In the field of engineering and other applied sciences, a large class of problems are classified as continuum or distributed variable problems. The electric circuit problems discussed in section 4 above are classified as lumped parameter problems. The reason is that the individual circuit elements and their behavior are considered as acting at a single point. The currents and voltages that are the variables are functions of time only in the general case. On the other hand, a problem of heat flowing through a solid or the transmission of electromagnetic waves are distributed problems in that the variables are function of both time and space. In the most general situations, a dependent variable would be a function of four independent variables—three spatial dimensions and time. In these cases, the mathematical models that are used to define a problem are partial differential equations. For example, the equation that describes the motion of a vibrating string is the one-dimensional wave equation

$$\frac{\partial^2 u(x,t)}{\partial x^2} = c^2 \frac{\partial^2 u(x,t)}{\partial t^2}.$$

In this equation, $u(x, t)$ represents the deflection of the string from its equilibrium position.

In many applications of distributed value problems, it is desired to find the steady state solution. Such problems are termed potential problems and are characterized by Laplace's equation.

(5.1) $$\nabla^2 \phi(x,y,z) = \frac{\partial^2 \phi}{\partial x^2} + \frac{\partial^2 \phi}{\partial y^2} + \frac{\partial^2 \phi}{\partial z^2} = 0$$

Here, $\phi(x, y, z)$ is the potential function that may represent steady state problems of heat conduction, fluid flow, electrostatics, or magnetostatics, among others.

To solve a potential problem, the object is to find the function $\phi(x, y, z)$, which satisfies equation (5.1) throughout the region of the problem and matches specified values on the boundary. A common numerical method of solving this type of problem is that of finite differences. In this method, the region is considered to be subdivided into uniform squares (two dimensional) or cubes (three dimensional) with points at each corner. Figure 3.7 below illustrates such a two-dimensional problem in which the region in which a solution is desired is the area enclosed by the boundary curve.

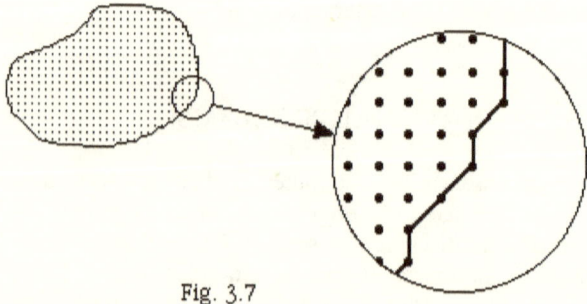

Fig. 3.7

The method requires that the boundary pass through the grid of points as depicted in the enlarged view of figure 3.7. In most real problems, the boundaries are smooth continuous curves, which means that they must be approximated by straight-line segments as indicated above. Clearly, the finer the grid, the better a curved boundary will be represented. However, this also means a very large increase in the size of the numerical problem as can be seen in the following development.

Solutions for the unknown function are found for each grid point within the boundary based on known values at the boundary points. Based on first-order approximations, it may be shown that the function $\phi(x, y)$ that satisfies the two-dimensional Laplace's equation is determined at any grid point by the average of the values at the four nearest grid points. This is depicted in figure 3.8 below.

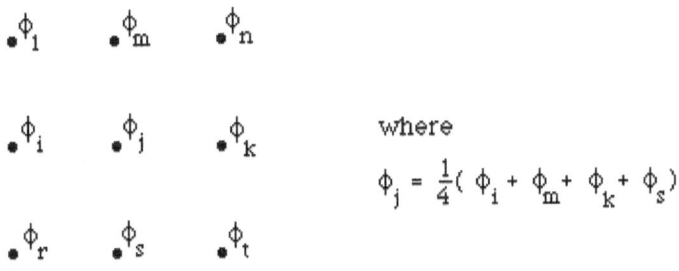

where

$$\phi_j = \frac{1}{4}(\phi_i + \phi_m + \phi_k + \phi_s)$$

Fig. 3.8

If the problem is such that there are n grid points internal to the region, then there will be n unknown values of the potential function ϕ_i.

Application of the equation shown in figure 3.8 and the values of the function at the boundary points will yield a system of n linear algebraic equations for the n unknowns ϕ_i.

To illustrate the procedure, let us consider the problem shown in figure 3.9 of determining the temperature distribution in an L-shaped region whose boundary temperatures are known.

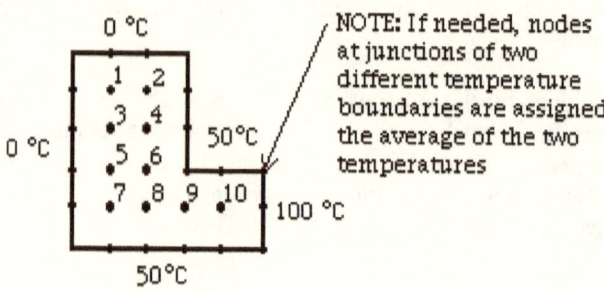

Fig. 3.9

In this problem, there are ten internal nodes at which we desire to find the temperatures. The unknown potential function is the temperature, and it will be designated by t rather than ϕ as in the general discussion above. From the basic equation for calculating the solution at a given grid point, we may write for the first grid point of the problem.

(5.2)
$$t_1 = \frac{1}{4}(0 + 0 + t_2 + t_3)$$

For the temperature at node four, we have

(5.3)
$$t_4 = \frac{1}{4}(50 + t_2 + t_3 + t_6)$$

Once all the node temperature equations have been written out, they are written in matrix form as follows,

(5.4)
$$T = (1/4)CT + B,$$

where T is the vector of unknown node temperatures, B a vector of averaged known boundary temperatures at each node, and C a coefficient matrix that relates how the temperature at a given node depends on the temperatures of the surrounding four closest nodes. The constant value of 1/4 has been factored out of the coefficient matrix as a matter of convenience. For our particular example, the T, C, and B matrices are

$$
T = \begin{bmatrix} t_1 \\ t_2 \\ t_3 \\ t_4 \\ t_5 \\ t_6 \\ t_7 \\ t_8 \\ t_9 \\ t_{10} \end{bmatrix}, \quad B = \left(\frac{1}{4}\right) \begin{bmatrix} 0 \\ 50 \\ 0 \\ 50 \\ 0 \\ 50 \\ 50 \\ 50 \\ 100 \\ 200 \end{bmatrix}
$$

$$
C = \begin{bmatrix}
0 & 1 & 1 & 0 & 0 & 0 & 0 & 0 & 0 & 0 \\
1 & 0 & 0 & 1 & 0 & 0 & 0 & 0 & 0 & 0 \\
1 & 0 & 0 & 1 & 1 & 0 & 0 & 0 & 0 & 0 \\
0 & 1 & 1 & 0 & 0 & 1 & 0 & 0 & 0 & 0 \\
0 & 0 & 1 & 0 & 0 & 1 & 1 & 0 & 0 & 0 \\
0 & 0 & 0 & 1 & 1 & 0 & 0 & 1 & 0 & 0 \\
0 & 0 & 0 & 0 & 1 & 0 & 0 & 1 & 0 & 0 \\
0 & 0 & 0 & 0 & 0 & 1 & 1 & 0 & 1 & 0 \\
0 & 0 & 0 & 0 & 0 & 0 & 0 & 1 & 0 & 1 \\
0 & 0 & 0 & 0 & 0 & 0 & 0 & 1 & 0 & 0
\end{bmatrix}.
$$

It is helpful to note that the B and C matrices may be constructed by inspection of the given problem. An element b_i of B represents one-fourth the sum of the known boundary temperatures of the four nodes nearest to the ith node. If none of the four nearest nodes are boundary nodes, then the value of b_i is automatically set to zero. An element c_{ij} of C has the value one only if node j is not a boundary node and is one of the four nodes nearest to node i. Following these rules, the reader may verify the values in the B and C matrices above. For example, node 1 has two neighboring nodes, number 2 and number 3. This means that elements c_{12} and c_{13} in the C matrix are ones. Because the C matrix must be symmetrical, c_{21} and c_{31} must be equal to one. It is an instructive exercise to verify that the first and fourth rows of equation (5.4) yield equations (5.2) and (5.3) respectively.

If we define a new matrix A as follows,

$$A = I_{10} - (1/4)C,$$

where I_{10} is a tenth-order identity matrix, equation (5.4) may be transformed into the standard form for linear algebraic equations, which is

$$AT = B.$$

Constructing the matrix A and solving[27] for the temperature vector gives

$$T = A^{-1}B = \begin{bmatrix} 9.3 \\ 22.5 \\ 14.9 \\ 30.7 \\ 19.4 \\ 35.3 \\ 27.6 \\ 40.9 \\ 50.9 \\ 62.7 \end{bmatrix}.$$

As an aid to visualizing this solution, we have replotted figure 3.9 as figure 3.10 below with the nodal temperatures shown at the node locations.

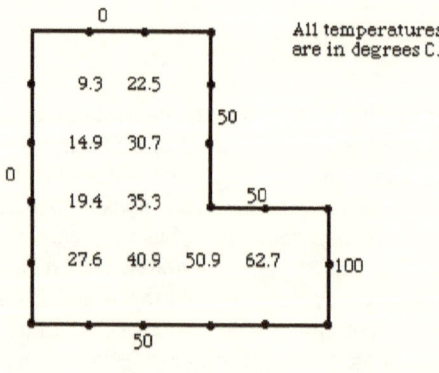

Fig. 3. 10

---

[27]   This solution was obtained by using the built-in matrix functions in an EXCEL spreadsheet program.

EXERCISES

1.* Solve for the temperature distribution (node temperatures) for the following geometry.

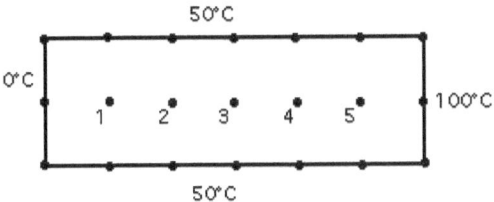

2.* Solve for the temperature distribution (node temperatures) for the following geometry.

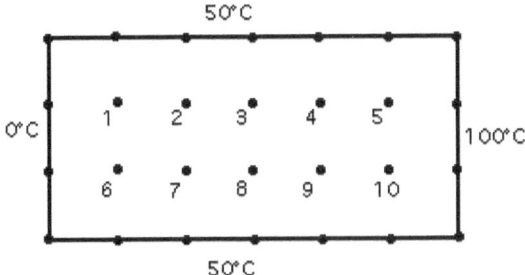

3.* Solve the example problem of this section (figure 3.9) if the left-hand edge is at 20°C rather than 0°C.

***

6. JACOBI ITERATION

From the examples in the two previous sections, it may be seen that the order of the matrices involved in some systems of linear algebraic equations can become quite large. This is particularly true in the solution of potential problems when the boundaries become complex and the grid size is made small to provide greater detail. In some cases, it is not uncommon to encounter matrices of order 5,000 to 10,000 or more. In these situations, even the Gaussian elimination method of solution can require substantial computation time. There is another method that is available for use in such cases, which may be faster provided the coefficient matrices are sparse or nearly all the elements are less than one in magnitude. To be sparse means that a matrix contains only a small percentage of elements whose values are other than zero. This is clearly the situation in the case of finite difference solutions of potential problems as can be seen in the previous example. The Leontief economic model of section 4 is one in which all the elements are less than one although a few diagonal elements might be unity.

In these situations, the method that may produce a solution more rapidly is that of Jacobi iteration. This procedure requires that the problem be stated in the following general form.

(6.1) $X = AX + B,$

where X is an nth-order unknown vector, A an nth-order coefficient matrix of known values, and B a known n^{th}-order constant vector. The applied problems of the two previous sections indeed match this situation, as can be seen from equations (4.1) and (5.4).

The Jacobi iteration method proceeds as follows. A trial solution vector, X_0, is guessed. The trial or first guess vector is substituted into the right-hand side of (6.1) above. This generates a second trial vector on the left side. In other words,

$$X_1 = AX_0 + B.$$

The process is then repeated to give

$$X_2 = AX_1 + B.$$

If this operation is repeated n times, the general result is

$$X_n = AX_{n-1} + B.$$

The process is stopped when it has converged or $X_n = X_{n-1}$ to whatever accuracy is desired. If the coefficient matrix A is sparse or all its elements are less than one, this method will converge to a constant answer quite rapidly. Unfortunately, there is no reliable way to measure how rapidly the convergence will occur. As a result, one cannot tell ahead of time for a given large problem whether to use this method or Gaussian elimination. The decision must be based on experience or trial and error.

The best measure of testing convergence is to test the difference between the norms of two successive trial vectors or $\|X_n\| - \|X_{n-1}\|$. When this difference is zero to a preselected number of decimal places, the process may be halted. Since the standard norm involves summing the squares of all the elements in a vector, it can add a significant amount of computation time. It is usually recommended to use either the L_1 or L_∞ norms for the above test.

It should be noted that if the magnitudes of the solution vector are small, then the norms will be small, and the above criteria for testing convergence would require a very small tolerance. As an alternative, a fractional convergence measure could be employed, such as

$$(\|X_n\| - \|X_{n-1}\|)/\|X_{n-1}\|.$$

When the Jacobi iteration method was applied[28] to the examples in section 4 and section 5 above, the number of iterations required to converge to the number of decimal places given was 15 and 20 respectively. For illustration, the example problem from section 4 was the Leontief economic model expressed by equation (4.1), which is repeated here with the specific values included. It has also been written in proper form for Jacobi iteration.

$$X_{n+1} = \begin{bmatrix} 22/123 & 0/23 & 41/395 & 50/72 \\ 0/123 & 2/23 & 16/395 & 2/72 \\ 7/123 & 2/23 & 200/395 & 5/72 \\ 3/123 & 2/23 & 28/395 & 8/72 \end{bmatrix} X_n + \begin{bmatrix} 10 \\ 3 \\ 181 \\ 31 \end{bmatrix}$$

The initial trial vector was $X_0 = [1, 1, 1, 1]$. The tolerance on the convergence test, $\|X_n\|_1 - \|X_{n-1}\|_1$, was set to 0.49 to insure the result would be accurate to the nearest integer value. As stated, the procedure required 15 iterations to converge to the previous answer.

$$X_{15} = \begin{bmatrix} 123 \\ 23 \\ 395 \\ 72 \end{bmatrix}$$

To conclude this section, it should be mentioned how a standard system of linear algebraic equations, namely

(6.2) $AX = B$

is put in a form appropriate for the Jacobi iteration procedure. This may be easily accomplished by defining a new matrix C as follows.

(6.3) $C = I - A$ or $A = I - C$,

where I is an identity matrix of appropriate order. Substitution of (6.3) into (6.2) leads to the standard form of the equations for the application of Jacobi iteration.

[28.]   The application of Jacobi iteration to the two example problems were carried out using an EXCEL spreadsheet program.

At this point, a word of caution should be interjected. When Jacobi iteration is applied to solve systems of linear algebraic equations, the process might not converge. The reason for divergence is that some or all the eigenvalues of the C matrix have magnitudes greater than one. This should become clear after the concept of eigenvalues has been introduced in chapter 5. Given a problem that diverges, it sometimes can be made to converge by appropriate scaling of the system equations.

To illustrate this situation, consider the electrical circuit problem given in section 2 of chapter II. In the form of equation (6.2), this problem is to solve the equation

$$\begin{bmatrix} 11 & -6 & 0 \\ -6 & 24 & -10 \\ 0 & -10 & 20 \end{bmatrix} X = \begin{bmatrix} 12 \\ 0 \\ 0 \end{bmatrix},$$

where X is the unknown current vector.

If this problem is set up directly for Jacobi iteration as discussed above, one quickly discovers that the process diverges in that each succeeding X vector becomes larger and larger. However, if the equations above are scaled by a sufficiently large factor, the process should converge. The question is how large should the factor be. This cannot be answered at present and must be left until a later chapter. For the moment, we will use a scale factor of 30.

If we divide the system equations by this factor, the above matrix equation becomes

$$\begin{bmatrix} 0.3667 & -0.2000 & 0 \\ -0.2000 & 0.8000 & -0.3333 \\ 0 & -0.3333 & 0.6667 \end{bmatrix} X = \begin{bmatrix} 0.4000 \\ 0 \\ 0 \end{bmatrix}.$$

The values have been shown in rounded form to four decimal places. Using equation (6.3) to convert this to the appropriate form for iteration, we obtain the following result.

$$X_{n+1} = \begin{bmatrix} 0.6333 & 0.2000 & 0 \\ 0.2000 & 0.2000 & 0.3333 \\ 0 & 0.3333 & 0.3333 \end{bmatrix} X_n + \begin{bmatrix} 0.4000 \\ 0 \\ 0 \end{bmatrix}$$

If we now take the initial trial vector to be $X_0 = [1,1,1]$, the process converges to the answer that is accurate to four decimal places after 34 iterations.

$$X_{34} = \begin{bmatrix} 1.3179 \\ 0.4162 \\ 0.2081 \end{bmatrix}$$

EXERCISES

1.*   Set up the following heat-conduction problem for the steady state temperatures at the indicated nodes, and solve by Jacobi iteration.

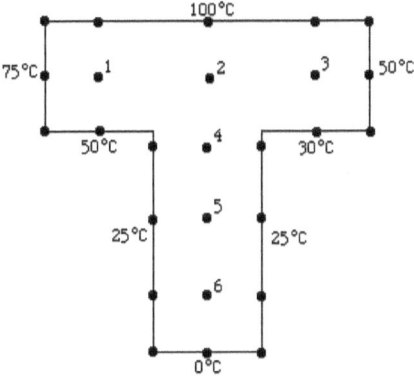

2.*   Set up the following heat-conduction problem for the steady state temperatures at the indicated nodes, and solve by Jacobi iteration. Compare the solution to that of exercise 1 above.

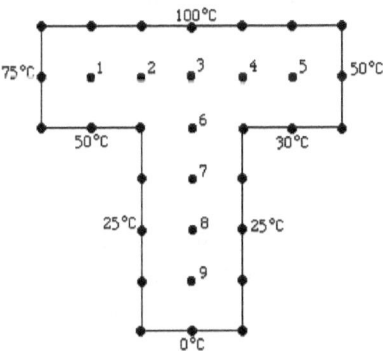

3*.   Set up the example problem of figure 3.9 for solution by Jacobi iteration, and find the solution.

***

7.   DIFFERENCE EQUATIONS

Matrix difference equations can arise in a wide variety of applied areas. Only two will be presented in this section. The first area will be the analysis of random walk problems. As an analytical tool, this approach is used to study a number of problems that involve a random or stochastic process. As an illustration, consider the classic problem of an inebriated person attempting to cross a bridge. We shall assume that the guard rails are sufficient to prevent the possibility of falling off the sides. The bridge is short in that it is only five paces across. For the purpose of analysis, a state vector of six elements will be introduced. Also, to aid in setting up this problem, it is useful to visualize the bridge as shown in figure 3.11 below.

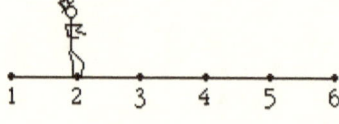

Fig. 3. 11

The bridge has been divided into the five paces with a location point at each end of each interval. Point one is the left end, and thus in crossing the bridge, there are six possible places at which the person can be at. Due to the physical condition of the person, we shall assume that in attempting to cross from left to right, there is one chance of stepping back, two chances of moving forward, and two chances of staying in the same place. Each position on the bridge will be called a state and the vector of all these states a state vector X_i, that is,

$$X_i = \begin{bmatrix} x_1 \\ x_2 \\ \vdots \\ x_6 \end{bmatrix}_i .$$

The subscript $_i$ designates the state the person is in after $_i$ attempts to take a step. In other words, the value of x_2 in the vector above would be the probability the person would be at the second position after having attempted i steps. If we know the state vector at some point i, then the state the person will be in when another step is attempted is given by

(7.1) $$X_{i+1} = PX_i.$$

Equation (7.1) is known as a difference equation, and the P matrix is frequently referred to as a transfer matrix. For our particular example, the matrix P is a probability coefficient matrix whose elements are the probabilities of being in state i knowing the person is in state j. For example, p_{23} is the probability the person will wind up in position 2 given that they are in position 3. This would be a step backward from position 3 to position 2. As stated above, at each position, there are five possibilities—one of stepping back, two of staying put, and two of stepping ahead. This means that the value of p_{23} must be 1/5 or 0.2. On the other hand p_{33} and p_{43} must be 2/5 or 0.4.

The situations at the ends of the bridge need special consideration. At the left end, there are only two possibilities, either stepping ahead or staying put. A step back cannot be allowed. From the statement of the problem, these are equally probable, so we must have $p_{11} = p_{21} = 0.5$. The reader should verify that the values for p_{56} and p_{66} for the right end are 1/3 and 2/3 respectively.

The assembled P matrix will now be

$$P = \begin{bmatrix} 1/2 & 1/5 & 0 & 0 & 0 & 0 \\ 1/2 & 2/5 & 1/5 & 0 & 0 & 0 \\ 0 & 2/5 & 2/5 & 1/5 & 0 & 0 \\ 0 & 0 & 2/5 & 2/5 & 1/5 & 0 \\ 0 & 0 & 0 & 2/5 & 2/5 & 1/3 \\ 0 & 0 & 0 & 0 & 2/5 & 2/3 \end{bmatrix},$$

which is known as a Markov matrix. One characteristic of a Markov matrix is that every column sums to a value of one because the elements are probability values. To help in understanding the example problem, let us examine the third equation given by equation (7.1).

$$(x_3)_{i+1} - \frac{2}{5}(x_2)_i + \frac{2}{5}(x_3)_i + \frac{1}{5}(x_4)_i$$

This equation states that to be in the third position after $i + 1$ steps, the probability is two-fifths that the person got there from the second position, two-fifths that they were already there, and one-fifth that they stepped back from the fourth position.

The solution proceeds as follows. As the problem statement indicated, the person is attempting to cross the bridge from left to right, so the initial state vector is

$$X_0 = \begin{bmatrix} 1 \\ 0 \\ \vdots \\ 0 \end{bmatrix}.$$

After the first attempted step, we would have

$$X_1 = \begin{bmatrix} 0.5 \\ 0.5 \\ 0 \\ 0 \\ 0 \\ 0 \end{bmatrix}.$$

At this point, we must clarify that this state vector is a probability vector, or in other words, it only tells us the probabilities of being at each possible position after each step. If the process is continued indefinitely, we find that the state vector tends to a steady state or constant value. This is a characteristic of a Markov process. For our example problem, the state vector converges to four significant digits after seventy-three steps. The tenth, thirtieth, and seventy-third state vectors[29] are

$$X_{10} = \begin{bmatrix} 0.0843 \\ 0.1641 \\ 0.2008 \\ 0.2040 \\ 0.1939 \\ 0.1528 \end{bmatrix}, X_{30} = \begin{bmatrix} 0.0204 \\ 0.0482 \\ 0.0884 \\ 0.1636 \\ 0.3116 \\ 0.3678 \end{bmatrix}, X_{73} = \begin{bmatrix} 0.0160 \\ 0.0400 \\ 0.0800 \\ 0.1600 \\ 0.3200 \\ 0.3840 \end{bmatrix}.$$

By converge, we mean that $X_{74} = X_{73}$ to within four significant digits.

---

[29] This iteration process was performed using an EXCEL spreadsheet program as are the others for the examples in this section.

It is also of interest to note that from the repeated application of equation (7.1) the following result may be demonstrated.

$$X_n = P^n X_0,$$

which indicates that repeated iterations simply raises the coefficient or transfer matrix to higher powers. The convergence of the process implies that as the power n increases, the matrix P^n converges to a constant matrix. Another important observation is that it does not matter what the initial vector is; in a sufficient number of steps, the process will converge to the same result.

As the second example of difference equations, we will consider a problem of the growth of a bacteria. Let us begin by considering a specific case of a bacteria that has an average life span of four hours. The life span will be considered to be subdivided into hour segments so that we can specify a population vector of four elements, namely b_1 for the number of bacteria of age zero to one-hour old, b_2 for one to two hours, etc. In terms of reproduction, we assume that the b_1's do not produce any offspring, the b_2's produce two per individual on the average in the one-hour span, and the b_3's and b_4's one each on the average. In terms of survivability, 70% survive the first hour, 85% the second, and 35% the third.

This growth problem may now be stated in matrix form as follows.

(7.2) $B_{i+i} = GB_i,$

where B_i is the bacteria population vector at the ith hour and G the fourth-order growth matrix. Specifically, the G and B_i matrices will be

$$B_i = \begin{bmatrix} b_1 \\ b_2 \\ b_3 \\ b_4 \end{bmatrix}_i , \quad G = \begin{bmatrix} 0 & 2 & 1 & 1 \\ 0.70 & 0 & 0 & 0 \\ 0 & 0.85 & 0 & 0 \\ 0 & 0 & 0.35 & 0 \end{bmatrix}.$$

In population studies, the G matrix is called a Leslie matrix. If the population is known at any time, equation (7.2) is employed to determine how it changes with time. For example, assume that we have an initial population of 1,000 new (b_1) bacteria, that is,

$$B_0 = \begin{bmatrix} 1.00E{+}03 \\ 0 \\ 0 \\ 0 \end{bmatrix}.$$

Because this model is based on the assumption of unlimited food, the solution will grow without bound. The population vectors after 50 and 100 generations (hours) are

$$B_{50} = \begin{bmatrix} 1.98E+08 \\ 1.07E+08 \\ 3.70E+07 \\ 5.71E+06 \end{bmatrix}, \quad B_{100} = \begin{bmatrix} 8.54E+13 \\ 4.61E+13 \\ 1.60E+13 \\ 2.47E+12 \end{bmatrix}.$$

We will return to this problem in the chapter on eigenvalues and eigenvectors, which will allow us to extract some long-term behavior of the system.

Let us now define the general form the Leslie matrix has for a general population of n groups. The notation c_i represents the number of offspring each member of the ith group has and s_i the probability individuals in the ith group have of surviving to the next group. In these terms, the general Leslie matrix has the form

$$G = \begin{bmatrix} c_1 & c_2 & c_3 & \cdots & c_{n-1} & c_n \\ s_1 & 0 & 0 & \cdots & 0 & 0 \\ 0 & s_2 & 0 & \cdots & 0 & 0 \\ \cdot & \cdot & \cdot & \cdots & \cdot & \cdot \\ 0 & 0 & 0 & \cdots & s_{n-1} & 0 \end{bmatrix}.$$

EXAMPLE 6

Consider a species of bacteria that has a life span of three hours, which is divided into three equal stages. In the second stage, an individual produces a single offspring. In the third stage, each individual divides into two new bacteria. Conditions are harsh, so the probability of an individual surviving the first stage is only 55%. The survival rate for those in the second stage is 70%. If a colony begins with only 100 bacteria in the first stage of life, how many hours will it take for the total population to exceed 10,000, and how will it be distributed over the three stages?

SOLUTION

For this problem, the Leslie growth matrix is

$$G = \begin{bmatrix} 0 & 1 & 2 \\ 0.55 & 0 & 0 \\ 0 & 0.70 & 0 \end{bmatrix}$$

The initial population matrix is

$$X = \begin{bmatrix} 100 \\ 0 \\ 0 \end{bmatrix}$$

The total number of bacteria in the population at any hour is given by the following matrix product, which produces the summing operation.

$$\text{total population} = [1\ 1\ 1]\,X$$

The problem is solved by repeating the iteration operation of equation (7.2) above until the total population reaches a value of 10,000 or larger. After forty-six hours (iterations), the total population reaches 10,152 with the following distribution.

$$X = \begin{bmatrix} 5,628 \\ 2,778 \\ 1,746 \end{bmatrix}$$

***

To conclude this section, we shall consider another example of a Markov problem. Suppose that you are in Las Vegas playing a game for which the bet is set at a value of $25. At each play, the probability of winning is 0.25, of staying 0.30 and losing 0.45, and we desire to express this game in matrix form. Clearly, the game will end when all your money is gone, and let us assume you are going to play conservatively and quit when your winnings total $100.

We begin by defining a state or probability vector that describes each possible state of your winnings. The first state is that of no money, which indicates that the game will end. The last state (fifth) is when you have won $100, and again the game will end. The intermediate states indicate the holding of increments of $25. The state vector is defined as

$$M_i = \begin{bmatrix} \text{Probability of no money or losing} \\ \text{Probability of holding } \$25 \\ \text{Probability of holding } \$50 \\ \text{Probability of holding } \$75 \\ \text{Probability of holding } \$100 \text{ or winning} \end{bmatrix}_i.$$

The subscript i designates the number of times the game has been played.

Each play of the game is described by the standard finite difference or transfer equation

$$M_{i+1} = GM_i.$$

An element g_{ij} of the transfer matrix G represents the probability of moving to state i having just been in state j. For example, g_{32} would be the probability of holding $50 having just bet your last $25. As this is a case of winning, the value of g_{ij} is 0.25. The value of g_{11} must be one since the game ends when you lose all you money. It is left to the reader to verify the G matrix is the following.

$$G = \begin{bmatrix} 1 & 0.45 & 0 & 0 & 0 \\ 0 & 0.30 & 0.45 & 0 & 0 \\ 0 & 0.25 & 0.30 & 0.45 & 0 \\ 0 & 0 & 0.25 & 0.30 & 0 \\ 0 & 0 & 0 & 0.25 & 1 \end{bmatrix}$$

In general, the following rule may be stated for constructing a G matrix. Let p_l be the probability of losing, p_s the probability of staying or holding, and p_w the probability of winning on a given play of the game. All elements of the G matrix are zero except $g_{11} = g_{nn} = 1$ and

$$g_{ij} = \begin{cases} p_l \text{ when } i = j - 1, \ 2 \le j \le n - 1 \\ p_s \text{ when } i = j, \ 2 \le j \le n - 1 \\ p_w \text{ when } i = j + 1, \ 2 \le j \le n - 1. \end{cases}$$

At this point, we should observe that since you are a diligent player, the game will end when you have either lost all your money or won some predetermined amount. Since G is a Markov matrix, the process will converge to a steady state.

If we let p be the probability of eventually losing, the steady state solution vector will have the following form.

$$M_\infty = \begin{bmatrix} p \\ 0 \\ 0 \\ 0 \\ 1-p \end{bmatrix}$$

EXAMPLE 7

Consider the specific game described above. Assuming that you start the game with $50, determine your probability of winning. Repeat the problem if you begin with $75.

SOLUTION

Your initial state vector will be

$$M(\$50)_0 = \begin{bmatrix} 0 \\ 0 \\ 1 \\ 0 \\ 0 \end{bmatrix}.$$

If we seek a solution that is accurate to two decimal digits, the nearest cent, 25 iterations of the basic equation

$$M_{i+1} = GM_i$$

is required, and the result is

$$M(\$50)_{25} = \begin{bmatrix} 0.76 \\ 0.00 \\ 0.00 \\ 0.00 \\ 0.24 \end{bmatrix}.$$

This indicates that you only have about one chance in four of winning with the given odds. Should you have \$75 to begin with, your starting and final state vectors are

$$M(\$75)_0 = \begin{bmatrix} 0 \\ 0 \\ 0 \\ 1 \\ 0 \end{bmatrix}, \quad M(\$75)_{21} = \begin{bmatrix} 0.49 \\ 0.00 \\ 0.00 \\ 0.00 \\ 0.51 \end{bmatrix}.$$

In this situation, your odds of winning or losing are about even.

EXERCISES

1.*   Consider the fourth-order population matrix used as an illustration in this section. With the same initial population vector, i.e., 1,000 bacteria in the first age group, determine how many hours it will take for the total population to just exceed 1,000,000 bacteria. Also state what the population distribution (vector) is at this time.

2.*   For the gambling problem in this section, what are you odds of winning if you have only \$25 to begin the game with?

3.*   For the gambling problem in this section, change the odds to 0.30 for winning, 0.35 for staying, and 0.35 for losing.

   a.   Determine the transfer or Markov matrix for this game.
   b.   What are you odds of winning if you begin play with \$25?
   c.   With \$50?
   d.   With \$75?
   e.   How could you have solved parts b, c, and d with a single iteration sequence?

4.*   You are a "high roller" and play a game in which the odds of winning are 0.33, staying 0.30, and losing 0.37. The bet is set at \$250, and you will only quit the game if you lose all you money or win \$2,000.

   a.   Construct the Markov matrix for this problem.
   b.   If you begin the game with \$1,000, what are the odds that you will win?

***

8.   EQUILIBRIUM OF SYSTEMS OF FORCES (TRUSSES)

As a final example of the occurrence of linear algebraic equations, we shall examine the equilibrium of force systems. The approach is similar for two- and three-dimensional force systems, so we will focus on a two-dimensional example for convenience.

From basic mechanics, if one studies a body that has a number of forces, say n, acting on it, then static equilibrium requires that the sum of all the forces must be zero. By their nature, forces are vector quantities since they possess both magnitude and direction. Let us designate the ith force by F_i. Static equilibrium now requires that

$$\sum_{i=1}^{n} \overline{F_i} = \overline{0} .$$

In two-dimensional problems, this vector equilibrium equation implies that there are two scalar equations, that is, the sum of all force components in the x and y directions must vanish or

(8.1) $$\sum_{i=1}^{n} (F_i)_x = 0 , \quad \sum_{i=1}^{n} (F_i)_y = 0 .$$

To illustrate the application of linear algebraic equations to the solution of a system of forces, we will examine the solution of the two-dimensional truss problem shown in figure 3.12 below.

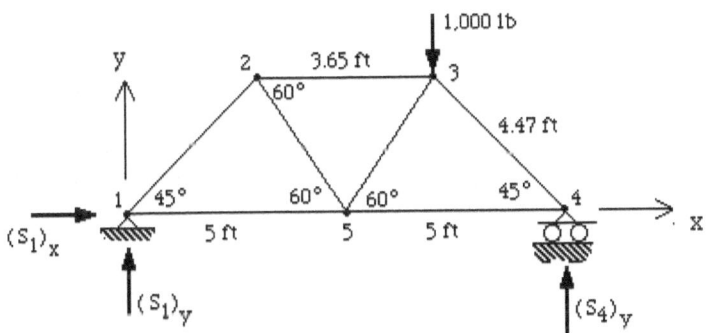

Fig. 3.12

The unknown support forces are designated as $(S_1)_x$, $(S_1)_y$. and $(S_4)_y$. Two support forces are required at joint number 1 as it is a pin connection, and only one is needed at joint number 4 because it is a roller that is free to move in the x direction. The object of the problem is to determine the force in each of the members (there are seven) and the three support forces. The total of unknowns is ten, and hence, ten independent equations are required. From figure 3.12, it is seen that there are five joints, and the equilibrium of each will yield two scalar equations of the form (8.2). This means that equilibrium will provide the needed ten equations.

In order to write the equilibrium equations for each joint in the truss above, let us consider a general situation depicted in figure 3.13 below.

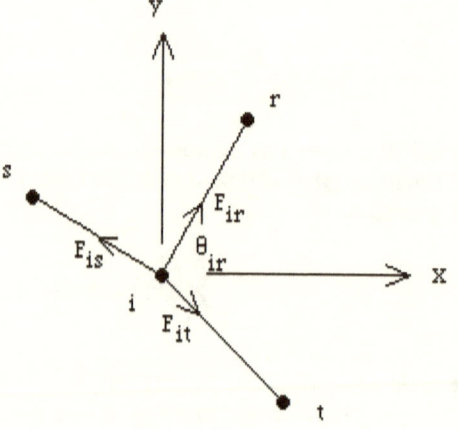

Fig. 3.13

In this figure, we have shown a joint i and three neighboring joints r, s, and t to which the ith joint is connected by members. As a standard approach, it is assumed that the force in each member is in tension, which means that it pulls away from each joint. This action on the ith joint is shown by the arrows on the members in figure 3.13. Once a solution is obtained, any negative values for the member forces indicates that the member is in compression rather than tension. The member forces are designated by two subscripts, which are the joint numbers to which the member is connected.

At the ith joint, the direction of the member force F_{ij} is defined by the symbol θ_{ij}, which is the angle the member force vector makes with the positive x axis. This angle is measured in a counterclockwise sense from the x axis to the line of the member force.

With these definitions, we may write the two equilibrium equations for the general joint of figure 3.13 above. Applying equations (8.1), the following result is obtained.

$$\sum_{i=1}^{n} (F_i)_x = F_{ir}\cos\theta_{ir} + F_{is}\cos\theta_{is} + F_{it}\cos\theta_{it} = 0$$

$$\sum_{i=1}^{n} (F_i)_y = F_{ir}\sin\theta_{ir} + F_{is}\sin\theta_{is} + F_{it}\sin\theta_{it} = 0$$

Should any applied loads or support forces occur at this joint, they would also have to be included in the above equations.

The equilibrium equations for the example truss of figure 3.12 may now be written. For illustration, only two will be written in detail so the above information may be verified by the reader. Summing forces in the x direction for joint number 1, we have

$$\sum (F_1)_x = F_{12}\cos 45° + F_{15}\cos 0° + (S_1)_x = 0$$

The sum of forces in the y direction at joint number 3 would yield

$$F_{32}\sin 180° + F_{34}\sin 315° + F_{35}\sin 240° - 1000 = 0$$

Once all ten equations have been written, they may be expressed in the following matrix form.

(8.2) $AX + L = 0$ or $AX = -L,$

where X is the vector of unknown member and support forces and L the applied load vector. For the example problem under consideration, these vectors would be

$$X = \begin{bmatrix} F_{12} \\ F_{15} \\ F_{23} \\ F_{25} \\ F_{34} \\ F_{35} \\ F_{45} \\ (S_1)_x \\ (S_1)_y \\ (S_4)_y \end{bmatrix}, \quad L = \begin{bmatrix} 0 \\ 0 \\ 0 \\ 0 \\ 0 \\ -1000 \\ 0 \\ 0 \\ 0 \\ 0 \end{bmatrix}.$$

$$
A =
\begin{bmatrix}
 & F_{12} & F_{15} & F_{23} & F_{25} & F_{34} & F_{35} & F_{45} & (S_1)_x & (S_1)_y & (S_4)_y \\
 & \cos45° & 1 & 0 & 0 & 0 & 0 & 0 & 1 & 0 & 0 \\
 & \sin45° & 0 & 0 & 0 & 0 & 0 & 0 & 0 & 1 & 0 \\
 & \cos225° & 0 & 1 & \cos300° & 0 & 0 & 0 & 0 & 0 & 0 \\
 & \sin225° & 0 & 0 & \sin300° & 0 & 0 & 0 & 0 & 0 & 0 \\
 & 0 & 0 & -1 & 0 & \cos315° & \cos240° & 0 & 0 & 0 & 0 \\
 & 0 & 0 & 0 & 0 & \sin315° & \sin240° & 0 & 0 & 0 & 0 \\
 & 0 & 0 & 0 & 0 & \cos315° & 0 & -1 & 0 & 0 & 0 \\
 & 0 & 0 & 0 & 0 & \sin135° & 0 & 0 & 0 & 0 & 1 \\
 & 0 & -1 & 0 & \cos120° & 0 & \cos60° & 1 & 0 & 0 & 0 \\
 & 0 & 0 & 0 & \sin120° & 0 & \sin60° & 0 & 0 & 0 & 0
\end{bmatrix}
$$

In the coefficient equation above, the member and support force labels have been included at the heading of each column so it will be easier to determine the coefficients for each individual equilibrium equation.

At this point, the solution for the unknown member and support forces could be obtained directly by solving the given system of linear algebraic equations. However, it is of interest to see how partitioning can be used to reduce the system to lower order and separate the solution for the member forces from the support forces.

To apply partitioning in order to separate out the support forces, the eighth equation of (8.2) must be moved up to become the third equation. By doing this, the support forces appear only in the first three equations of the full system. This may be accomplished by using an appropriate permutation matrix. All the other equations may be left in their same relative order. Once this is done, the revised (8.2) may be partitioned in the following schematic way.

$$
(8.3) \qquad
\begin{bmatrix} B_{3,7} & I_3 \\ \hline C_7 & O_{7,3} \end{bmatrix}
\begin{bmatrix} F_{7,1} \\ \hline S_{3,1} \end{bmatrix}
= -
\begin{bmatrix} O_{3,1} \\ \hline L^{*}_{7,1} \end{bmatrix} ,
$$

where the 0s designate null matrices, I an identity matrix, and

$$
B = \begin{bmatrix}
\cos 45^\circ & 1 & 0 & 0 & 0 & 0 & 0 \\
\sin 45^\circ & 0 & 0 & 0 & 0 & 0 & 0 \\
0 & 0 & 0 & 0 & \sin 135^\circ & 0 & 0
\end{bmatrix}
$$

$$
C = \begin{bmatrix}
\cos 225^\circ & 1 & 1 & \cos 300^\circ & 0 & 0 & 0 \\
\sin 225^\circ & 0 & 0 & \sin 300^\circ & 0 & 0 & 0 \\
0 & 0 & -1 & 0 & \cos 315^\circ & \cos 240^\circ & 0 \\
0 & 0 & 0 & 0 & \sin 315^\circ & \sin 240^\circ & 0 \\
0 & 0 & 0 & 0 & \cos 315^\circ & 0 & -1 \\
0 & -1 & 0 & \cos 120^\circ & 0 & \cos 60^\circ & 1 \\
0 & 0 & 0 & \sin 120^\circ & 0 & \sin 60^\circ & 0
\end{bmatrix}
$$

$$
F = \begin{bmatrix}
F_{12} \\ F_{15} \\ F_{23} \\ F_{25} \\ F_{34} \\ F_{35} \\ F_{45}
\end{bmatrix},
\quad
S = \begin{bmatrix}
(S_1)_x \\ (S_1)_y \\ (S_4)_y
\end{bmatrix}
$$

By expanding equation (8.3), it may be shown that the following equations are obtained for the unknown force and support vectors.

$$F = -C^{-1}L,* \qquad\qquad S = -BF$$

Solving these equations gives the following result.

$$F = \begin{bmatrix} -448 \\ 317 \\ -500 \\ 366 \\ -966 \\ -366 \\ 683 \end{bmatrix}, \qquad S = \begin{bmatrix} 0 \\ 317 \\ 683 \end{bmatrix}$$

This solution may be verified by checking the support forces through the application of equilibrium to the entire truss as a free or single body.

EXAMPLE 8

Determine the internal forces in each of the three members of the truss shown below in figure 3.14.

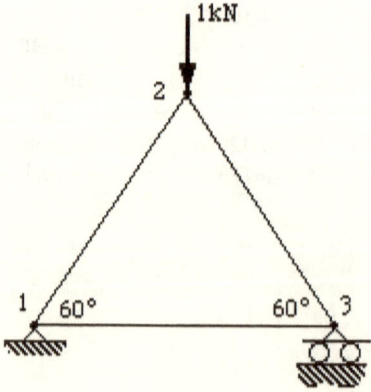

Fig. 3.14

SOLUTION

As a means to indicate the internal member forces and the support forces more clearly, figure 3.14 has been redrawn as figure 3.15 below with the member forces shown assuming they are in tension. The support symbols have been removed, and only the forces produced by the supports acting on the truss (support forces) shown.

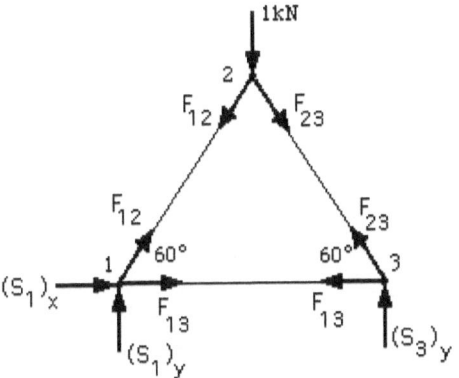

Fig. 3.15

For this truss, the elements of the unknown force vector will be defined as follows.

$$
F = \begin{bmatrix} F_{12} \\ F_{13} \\ F_{23} \\ (S_1)_x \\ (S_1)_y \\ (S_3)_y \end{bmatrix}
$$

Based on this order, the coefficient A matrix may be constructed directly from the truss diagram above. Each node is taken in order with the x and y components respectively to form each row. The column entries correspond to each of the elements in the unknown force vector respectively. Figure 3.16 below gives the A

matrix, and each column has been headed with the unknown force it corresponds to, and each row is indicated by its appropriate node and coordinate direction.

	F_{12}	F_{13}	F_{23}	$(S_1)_x$	$(S_1)_y$	$(S_3)_y$
Node one x direction →	$\cos60°$	1	0	1	0	0
Node one y direction →	$\sin60°$	0	0	0	1	0
Node two x direction →	$\cos240°$	0	$\cos300°$	0	0	0
Node two y direction →	$\sin240°$	0	$\sin300°$	0	0	0
Node three x direction →	0	-1	$\cos120°$	0	0	0
Node three y direction →	0	0	$\sin120°$	0	0	1

Fig. 3.16

The first element of A, a_{11}, is $\cos60°$ because member 12 is connected to node 1, and the direction it makes with the positive x axis is $+60°$. Likewise, member 13 in which the unknown force is F_{13}, is attached to node 1, and it is at an angle of $0°$ with the positive x axis, which gives a value for a_{12} of 1 since $\cos0° = 1$. Because node 1 is a pin support, the two components of the unknown support force must also be included. These are $(S_1)_x$ in the positive x direction, which gives $a_{14} = \cos0° = 1$, and $(S_1)_y$ in the positive y direction, which gives $a_{15} = \cos90° = 0$. Actually, one does not have to be concerned with whether or not a given member is in tension or compression. The entries in the A matrix are just the cosine and sine of the angle each member attached to a particular node makes with the positive x axis.

The required load vector in Newtons is

$$
L = \begin{bmatrix} 0 \\ 0 \\ 0 \\ -1,000 \\ 0 \\ 0 \end{bmatrix}
\begin{array}{l}
\leftarrow \text{ Node one} \\ \text{ x direction} \\
\leftarrow \text{ Node one} \\ \text{ y direction} \\
\leftarrow \text{ Node two} \\ \text{ x direction} \\
\leftarrow \text{ Node two} \\ \text{ y direction} \\
\leftarrow \text{ Node three} \\ \text{ x direction} \\
\leftarrow \text{ Node three} \\ \text{ y direction}
\end{array}
$$

The solution for the unknown forces from equation (8.2) is

$$
F = \begin{bmatrix} -577.4 \\ 288.7 \\ -577.4 \\ 0.0 \\ 500.0 \\ 500.0 \end{bmatrix}, N
$$

EXERCISES

1.*   Find the member forces for the truss shown below.

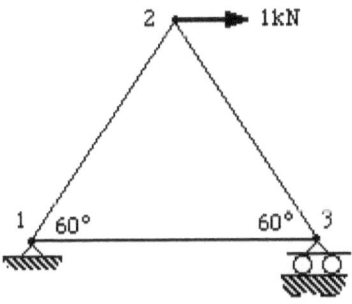

2.*   Consider the sign support truss shown below. Note that the pin support at joint number 1 sustains only a horizontal support force. Determine the forces in all members.

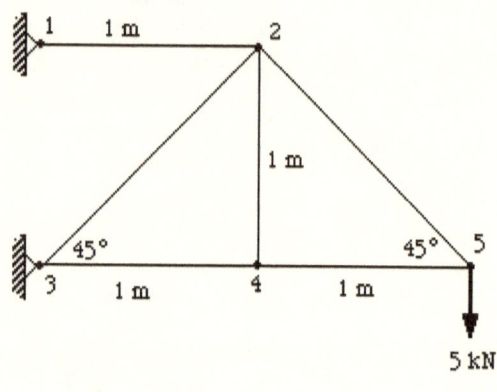

***

9. PARTITIONED MATRICES

From several of the above applications of matrices, the reader should see that in many problems, the order of the matrices encountered may become quite large. Even with today's large computer memory, some problems may be still too big to store. For this reason and others, it is sometimes convenient to partition or subdivide a matrix into smaller matrices. This is particularly useful when dealing with problems involving very large matrices and especially those in which only a small fraction of the elements of a matrix have nonzero values. Such a matrix is said to be sparse.

As an illustration, consider the matrix A of order 3×3:

(9.1)
$$A = \begin{bmatrix} 2 & -1 & 0 \\ 4 & 2 & -3 \\ 1 & 6 & 5 \end{bmatrix} = \begin{bmatrix} P & Q \\ R & S \end{bmatrix}$$

This matrix has been partitioned into the four submatrices P, Q, R, and S, where

(9.2)
$$P = \begin{bmatrix} 2 & -1 \\ 4 & 2 \end{bmatrix}, \quad Q = \begin{bmatrix} 0 \\ -3 \end{bmatrix}$$
$$R = \begin{bmatrix} 1 & 6 \end{bmatrix}, \quad S = \begin{bmatrix} 5 \end{bmatrix}$$

In this case, the matrix A has been partitioned in such a manner that the submatrices P and S along the diagonal are square, and the partitioning is diagonally symmetrical. Now let B be another square matrix of the third order that is similarly partitioned.

(9.3)
$$B = \left[\begin{array}{cc:c} -1 & 0 & 7 \\ 3 & 8 & -2 \\ \hdashline 0 & 1 & 4 \end{array} \right] = \left[\begin{array}{c:c} P_1 & Q_1 \\ \hdashline R_1 & S_1 \end{array} \right]$$

Now it can easily be seen that the sum $C = A + B$ may be expressed in terms of the above submatrices in the form

(9.4)
$$C = A + B = \left[\begin{array}{c:c} P+P_1 & Q+Q_1 \\ \hdashline R+R_1 & S+S_1 \end{array} \right] = \left[\begin{array}{cc:c} 1 & -1 & 7 \\ 7 & 10 & -5 \\ \hdashline 1 & 7 & 9 \end{array} \right]$$

This indicates that when two matrices of the same order are partitioned similarly, their sum may be obtained by adding their various submatrices as if they were individual matrices.

Multiplication in Terms of Submatrices

As a consequence of the fundamental rule for the multiplication of matrices, a rectangular matrix B may be premultiplied by another rectangular matrix A provided the two matrices are conformable—that is, the number of rows of B is equal to the number of columns of A. Now if A and B are both partitioned into submatrices such that the grouping of columns in A agrees with the grouping of rows in B, it can be shown that the product AB may be obtained by treating the submatrices as ordinary matrices and proceeding according to the multiplication rule.

In the case of the matrices A and B discussed above, the partitioning is such that the product $D = AB$ may be carried out by treating the submatrices of (9.1) and (9.3) as if they were ordinary elements and thus obtaining the following:

(9.5)

$$D = AB = \begin{bmatrix} P & Q \\ \hline R & S \end{bmatrix} \begin{bmatrix} P_1 & Q_1 \\ \hline R_1 & S_1 \end{bmatrix}$$

$$= \begin{bmatrix} (PP_1+QR_1) & (PQ_1+QS_1) \\ \hline (RP_1+SR_1) & (RQ_1+SS_1) \end{bmatrix}$$

$$= \begin{bmatrix} \begin{bmatrix} 2 & -1 \\ 4 & 2 \end{bmatrix}\begin{bmatrix} -1 & 0 \\ 3 & 8 \end{bmatrix} + \begin{bmatrix} 0 \\ -3 \end{bmatrix}[0\ 1] & \begin{bmatrix} 2 & -1 \\ 4 & 2 \end{bmatrix}\begin{bmatrix} 7 \\ -2 \end{bmatrix} + \begin{bmatrix} 0 \\ -3 \end{bmatrix}[4] \\ \hline [1\ 6]\begin{bmatrix} -1 & 0 \\ 3 & 8 \end{bmatrix} + [5][0\ 1] & [1\ 6]\begin{bmatrix} 7 \\ -2 \end{bmatrix} + [5][4] \end{bmatrix}$$

$$= \begin{bmatrix} -5 & -8 & 16 \\ 2 & 13 & 12 \\ \hline 17 & 53 & 15 \end{bmatrix}$$

EXAMPLE 9

For the matrices A and B given above, compute the product AB if A is partitioned as follows.

$$A = \begin{bmatrix} 2 & -1 & 0 \\ \hline 4 & 2 & -3 \\ 1 & 6 & 5 \end{bmatrix}$$

SOLUTION

The first decision that must be made is how to partition the matrix B so that all the products of the submatrices are conformable. If we used the same notation for the submatrices of A as above, namely P, Q, R, and S, then we have

$$P = [2\ \text{-}1],\quad Q = [0],\quad R = \begin{bmatrix} 4 & 2 \\ 1 & 6 \end{bmatrix},\quad S = \begin{bmatrix} -3 \\ 5 \end{bmatrix}$$

In terms of orders, P is 1×2, Q is 1×1, R is 2×2, and S is 2×1. As indicated in equation (9.5) above, the product must have the following form in terms of the individual submatrices.

$$AB = \begin{bmatrix} P & Q \\ \hline R & S \end{bmatrix} \begin{bmatrix} P_1 & Q_1 \\ \hline R_1 & S_1 \end{bmatrix} = \begin{bmatrix} (PP_1 + QR_1) & (PQ_1 + QS_1) \\ \hline (RP_1 + SR_1) & (RQ_1 + SS_1) \end{bmatrix}$$

For the product PP_1 to be conformable, P_1 must have two rows. This means that the horizontal partition of the matrix B must fall between the second and third rows.

We are now left with determining the position of the vertical partition line for the matrix B. After some checking, it should become clear that it can be placed anywhere. In other words, the vertical partition line for matrix B can be placed between any columns. Its location is not controlled by the location of the partition lines in the A matrix. So let us choose to place the vertical partition line between the first and second columns. This gives

$$B = \begin{bmatrix} -1 & 0 & 7 \\ 3 & 8 & -2 \\ \hline 0 & 1 & 4 \end{bmatrix}$$

Carrying out the indicated multiplications in terms of the submatrices gives the following result that, as we know, must be the same as previously obtained.

$$AB = \begin{bmatrix} 2 & -1 & 0 \\ 4 & 2 & -3 \\ 1 & 6 & 5 \end{bmatrix} \begin{bmatrix} -1 & 0 & 7 \\ 3 & 8 & -2 \\ 0 & 1 & 4 \end{bmatrix}$$

$$= \begin{bmatrix} [2\;-1]\begin{bmatrix}-1\\3\end{bmatrix} + [0][0] & [2\;-1]\begin{bmatrix}0 & 7\\8 & -2\end{bmatrix} + [0]\,[1\;\;4] \\ \hline \begin{bmatrix}4 & 2\\1 & 6\end{bmatrix}\begin{bmatrix}-1\\3\end{bmatrix} + \begin{bmatrix}-3\\5\end{bmatrix}[0] & \begin{bmatrix}4 & 2\\1 & 6\end{bmatrix}\begin{bmatrix}0 & 7\\8 & -2\end{bmatrix} + \begin{bmatrix}-3\\5\end{bmatrix}[1\;\;4] \end{bmatrix}$$

$$= \begin{bmatrix} -5 & -8 & 16 \\ 2 & 13 & 12 \\ 17 & 53 & 15 \end{bmatrix}$$

**

This example illustrates a basic requirement for partitioning a matrix product that follows from the rule for multiplying matrices. This requirement is that the placement of partitions between columns in the first or left matrix requires the same placement of partitions between rows of the second or right matrix. In addition, the placement of partitions between rows in the first matrix and columns of the second is completely arbitrary and unrelated. However, these partitions will be the partitions remaining in the resulting product matrix.

Inversion in Terms of Partitions

The use of partitioning to invert large matrices is also of importance in some special circumstances such as efficient use of memory and storage in a computer as well as improving efficiency of computational schemes.

To illustrate the inversion of a matrix by the use of partitions, consider the same example matrix A defined in (9.1) above, and assume we want to compute its inverse, A^{-1}. From the definition of the inverse, we know that $AA^{-1} = I_n$. For convenience, let the unknown inverse be partitioned in an appropriate manner so that the product AB is conformable as we have discussed above. Let

$$B = A^{-1} = \left[\begin{array}{c|c} P_1 & Q_1 \\ \hline R_1 & S_1 \end{array} \right].$$

With A and B defined as in (9.1) and (9.3), the product is given by (9.5). Since the product is the identity matrix, we may now write

$$(9.6) \quad AA^{-1} = AB = \left[\begin{array}{cc} (PP_1+QR_1) & (PQ_1+QS_1) \\ (RP_1+SR_1) & (RQ_1+SS_1) \end{array} \right] = \left[\begin{array}{c|c} I_2' & 0' \\ \hline 0'' & I_1'' \end{array} \right]$$

Here, I_2' and I_1'' are subunit matrices of I_3 that, along with the two null matrices 0' and 0'', have been partitioned in an appropriate manner. For our particular third-order example, the resulting unit matrix would have to be

$$(9.7) \qquad\qquad AA^{-1} = AB = \left[\begin{array}{cc|c} 1 & 0 & 0 \\ 0 & 1 & 0 \\ \hline 0 & 0 & 1 \end{array} \right].$$

By expanding (9.6), we obtain the following four matrix equations.

(9.8)

$$
\begin{aligned}
\text{(a)} \quad & PP_1 + QR_1 = I' \\
\text{(b)} \quad & PQ_1 + QS_1 = 0' \\
\text{(c)} \quad & RP_1 + SR_1 = 0'' \\
\text{(d)} \quad & RQ_1 + SS_1 = I''
\end{aligned}
$$

Let us recall that since the matrix A is known, its submatrices P, Q, R, and S are known, and we are seeking to find the submatrices P_1, Q_1, R_1, and S_1 of A^{-1}. In terms of the known submatrices of A, these equations become

$$
\text{(a)} \quad \begin{vmatrix} 2 & -1 \\ 4 & 2 \end{vmatrix} P_1 + \begin{vmatrix} 0 \\ -3 \end{vmatrix} R_1 = \begin{vmatrix} 1 & 0 \\ 0 & 1 \end{vmatrix}
$$

$$
\text{(b)} \quad \begin{bmatrix} 2 & -1 \\ 4 & 2 \end{bmatrix} Q_1 + \begin{bmatrix} 0 \\ -3 \end{bmatrix} S_1 = \begin{bmatrix} 0 \\ 0 \end{bmatrix}
$$

$$
\text{(c)} \quad [1 \ 6] P_1 + 5 R_1 = [0 \ 0]
$$

$$
\text{(d)} \quad [1 \ 6] Q_1 + 5 S_1 = 1
$$

The solution for the submatrices proceeds as follows. From (9.8b) we find

(9.9) $\qquad Q_1 = -P^{-1}QS_1.$

Substituting this result into (9.8d) gives

$$
-RP^{-1}QS_1 + SS_1 = I''
$$

or $\qquad (S - RP^{-1}Q)S_1 = I'',$

from which we find a solution for S_1.

$$
S_1 = (S - RP^{-1}Q)^{-1}I'' = (S - RP^{-1}Q)^{-1}
$$

For the given values of the submatrices, we have

$$
S_1 = \left(5 - [1 \ 6] \begin{bmatrix} 2 & -1 \\ 4 & 2 \end{bmatrix}^{-1} \begin{bmatrix} 0 \\ -3 \end{bmatrix} \right)^{-1} = \frac{8}{79}.
$$

With S_1 known, equation (9.9) may be used to evaluate Q_1.

$$
Q_1 = -P^{-1}Q(S - RP^{-1}Q)^{-1}
$$

Or, for the given matrices,

$$Q_1 = -\begin{bmatrix} 2 & -1 \\ 4 & 2 \end{bmatrix}^{-1} \begin{bmatrix} 0 \\ -3 \end{bmatrix} \left(\tfrac{1}{8}\right) = \frac{1}{79} \begin{bmatrix} 3 \\ 6 \end{bmatrix}$$.

In a similar manner, equations (9.8c) and (9.8a) are used to evaluate the solutions for P_1 and R_1.

$$P_1 = (P - QS^{-1}R)^{-1}$$
$$R_1 = -S^{-1}R(P - QS^{-1}R)^{-1}$$

Again, using the given matrices,

$$P_1 = \left(\begin{bmatrix} 2 & -1 \\ 4 & 2 \end{bmatrix} - \begin{bmatrix} 0 \\ -3 \end{bmatrix} \left(\tfrac{1}{5}\right)[1 \ 6]\right)^{-1} = \frac{1}{79} \begin{bmatrix} 28 & 5 \\ -23 & 10 \end{bmatrix}$$

$$R_1 = -\frac{1}{5}[1 \ 6]\frac{1}{79}\begin{bmatrix} 28 & 5 \\ -23 & 10 \end{bmatrix} = \frac{1}{79}[22 \ -13]$$.

Finally, the inverse of A may be written as

$$B = A^{-1} = \begin{bmatrix} (P - QS^{-1}R)^{-1} & -P^{-1}Q(S - RP^{-1}Q)^{-1} \\ -S^{-1}R(P - QS^{-1}R)^{-1} & (S - RP^{-1}Q)^{-1} \end{bmatrix}$$.

Using the above specific values for the submatrices, the inverse of the given matrix A is

$$A^{-1} = \frac{1}{79}\begin{bmatrix} 28 & 5 & 3 \\ -23 & 10 & 6 \\ 22 & -13 & 8 \end{bmatrix}$$.

The above general result may appear quite unwieldy, but in reality, it can be time and space saving in that the inversion of a large-order matrix A has been reduced to the inversion of four lower-order matrices. Actually, partitioning may be applied with any number of submatrices, and the partitions may be done in any manner as long as the expansions satisfy the rules of matrix algebra.

EXERCISES

Use partitioning to evaluate the inverses of the following matrices. You may choose the partitions in any consistent manner you wish.

1.
$$\begin{bmatrix} 1 & 3 & -1 \\ 3 & 2 & 1 \\ -1 & 1 & 4 \end{bmatrix}$$

2.
$$\begin{bmatrix} 1 & -2 & 1 \\ 3 & 0 & 2 \\ -1 & 2 & 3 \end{bmatrix}$$

3.
$$\begin{bmatrix} 5 & -1 & 2 & 3 \\ -1 & 0 & 4 & -3 \\ 2 & 4 & -3 & 1 \\ 3 & -3 & 1 & 2 \end{bmatrix}$$

4.   Verify the result given by equation (10.5).

5.   Compute the inverse for the same example used in this section by using the following partitions.

$$A = \begin{bmatrix} 2 & -1 & 0 \\ 4 & 2 & -3 \\ \hline 1 & 6 & 5 \end{bmatrix} = \begin{bmatrix} P & Q \\ \hline R & S \end{bmatrix}$$

ADDITIONAL EXERCISES

1.*   Consider the following data.

x	0.00	0.25	0.50	0.75	1.00	1.25	1.50
y	0.010	0.046	0.200	0.444	0.790	1.220	1.750

For the least squares fit method, one measure of the degree of fit is given by the standard error of estimate[30], which is defined as

$$\text{Standard error} = \sqrt{\frac{\text{Total square error}}{n-2}},$$

where n is the number of data points and the total square error is defined in section 2. Using this as a measure of the degree of fit, determine whether a quadratic or cubic curve fit the given data better.

2.*   The least squares method can be applied to determining the coefficients k and m for the exponential relation $y = ke^{mt}$ in two different ways. One way is to convert the exponential relation to a linear equation as follows. First, the natural logarithm is taken to give

$$ln(y) = ln(k) + mt.$$

A new dependent variable z and two constants, a_0 and a_1, are defined as $z = ln(y)$, $a_0 = ln(k)$, and $a_1 = m$. With these terms, the above relation now has the form of a standard first-order polynomial

$$z = a_0 + a_1 t.$$

Apply this to the problem of determining the half-life of a radioactive substance for which the following data is known.

t(hours)	0	40	80	120	160	200
Radioactivity (milliCurie)	890	674	511	387	294	223

---

[30]   Hugh W. Coleman and W. Glenn Steele Jr., *Experimentation and Uncertainty Analysis for Engineers*, John Wiley & Sons, 1989, p. 173.

Half-life is defined as the time required for the radioactivity of the material to decay to half its original value. As we already stated above, radioactive decay obeys the relation $y = ke^{mt}$, where y is the radioactivity and k the level of y at $t = 0$.

3.*   What voltage will appear on the voltmeter in the circuit below? Hint: You will have to solve for all the mesh currents and note that the voltmeter indicates the voltage (current times resistance) drop across the resistance it is connected across. Also, an ideal voltmeter does not draw any current from the circuit it is measuring. All resistances have the same value of 20 ohm.

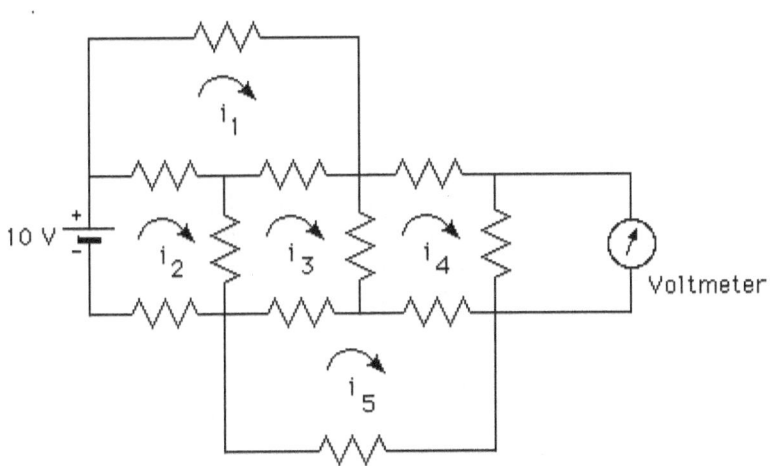

4.*   Write the matrix equation for the two circulating mesh-alternating currents in the circuit below, and solve for them.

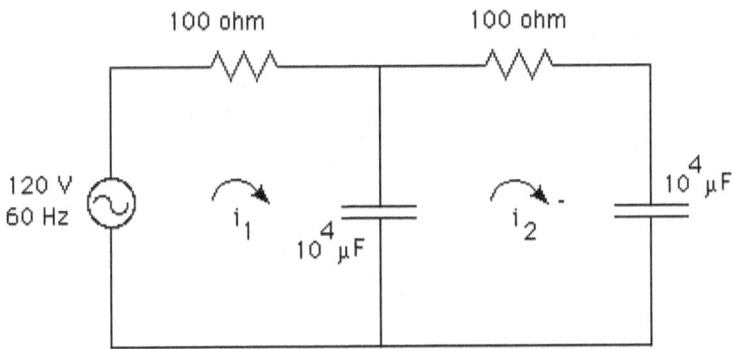

5.*   Write the matrix equation for the three circulating mesh-alternating currents in the circuit below, and solve for them.

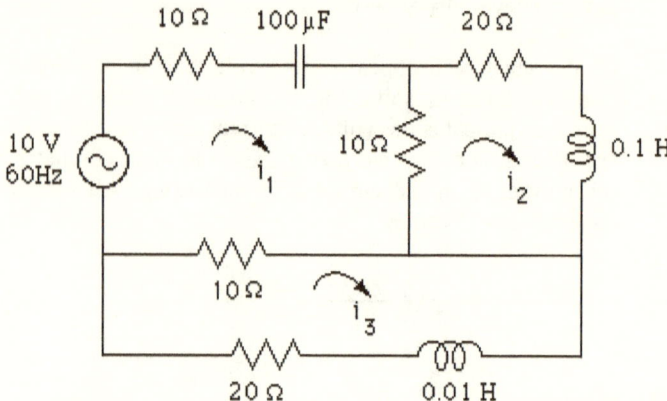

6.*   The example economy used in section 4 has been expanded by the inclusion of two additional segments, services and communications. The input-output table is given below. Construct the appropriate coefficient matrices, and solve for the production vector. Also determine the total output of the economy.

Produced For → / Consumed From ↓	Agriculture	Mining	Manufacturing	Transportation	Services	Communications	External Demand	Total Production
Agriculture	14	2	3	11	10	2	17	59
Mining	0	3	12	5	0	2	4	26
Manufacturing	11	6	48	12	31	10	88	206
Transportation	4	3	37	1	2	0	3	50
Services	1	2	22	2	7	4	30	68
Communications	1	0	8	3	6	1	6	25

Input-Output Table for Sample Economy
[Billions (10^9) Dollars]

7.*   If the external demand for services increases by 100% for the economy in additional exercise 6 above, what will the new production vector be? How does it compare with the original production vector? Assume the all characteristic constants defining the system do not change.

8.*   Set up the following heat-conduction problem for the steady state temperatures at the indicated nodes, and solve by either Gaussian elimination or Jacobi iteration.

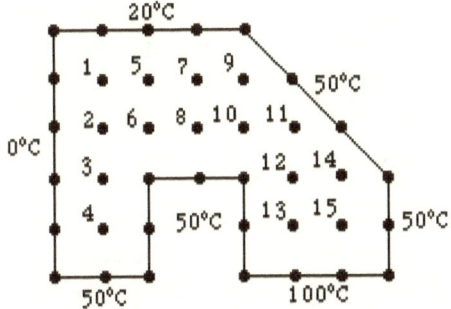

9.*   All of the example potential problems presented in this chapter have been those for which the boundary values of the solution are specified. Another very important type of boundary condition for potential problems is that of an insulating boundary or one across which no flow can occur. This type of boundary condition is sometimes called a reflecting boundary condition. How it is treated is illustrated in the figure below. The insulating or reflecting boundary is designated by the hatched border.

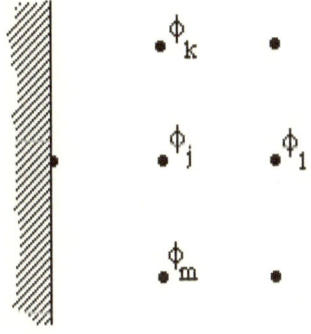

From the theory presented, the temperature at the jth node is given by

$$\phi_j = \frac{1}{4}(\phi_k + \phi_l + \phi_m + \phi_{boundary})$$

In the example problems in this chapter and the above exercises, the known boundary temperatures were substituted for $\phi_{boundary}$, with the resulting values included in the constant B vector. In the case of an insulating or reflecting boundary, the boundary value is reflected back to the neighboring node and given the same value. This means that $\phi_{boundary}$ is set equal to ϕ_j, resulting in the following equation.

$$\phi_j = \frac{1}{4}(\phi_k + \phi_l + \phi_m + \phi_j)$$

With this information, we derive the transfer matrix for the tee section below, which is identical to the one in exercise 2, section 6, except the sides are insulated as shown. Use either Gaussian elimination or Jacobi iteration to solve the internal node temperature vector.

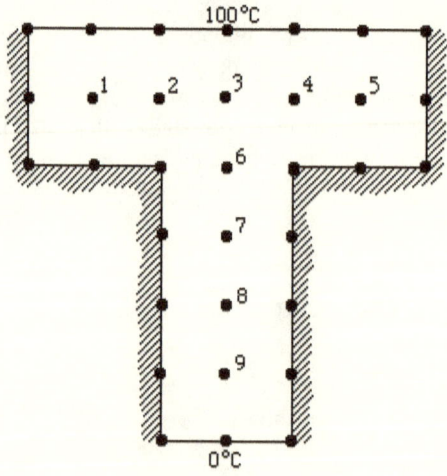

10.*   Solve for the temperature distribution (node temperatures) for the following geometry. Use the information about insulated boundaries given in additional exercise 9 above.

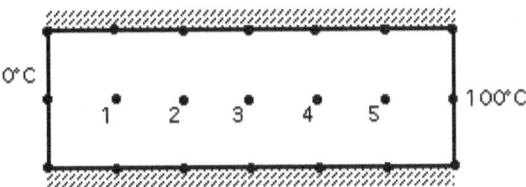

11.*   Solve for the temperature distribution (node temperatures) for the following geometry. Use the information about insulated boundaries given in problem additional exercise 9 above.

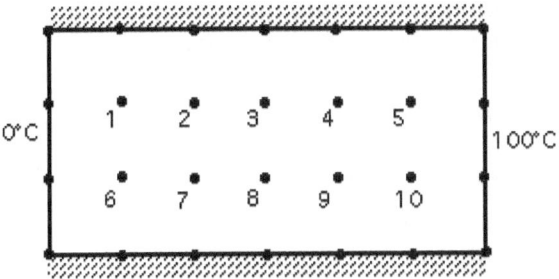

12.*   A police patrol car is assigned to patrol a section of town shown in the map below.

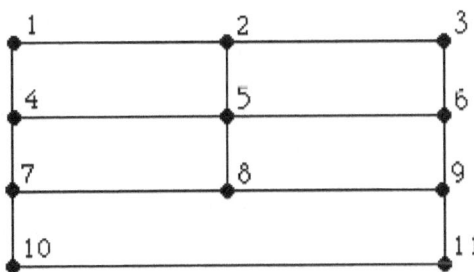

The patrol is required to move randomly (all paths from an intersection are equally probable) after stopping for ten minutes at an intersection.

a. Define the state vector, X_t for this problem as the vector of probabilities of being at each intersection at any time, t. Construct the Markov matrix that describes this situation.

b. Find the state vector expressing the likelihood the patrol car would be found at any intersection after an eight-hour shift if it began at the first intersection. Assume that a normal shift includes 1 1/2 hours for breaks.

c. What is the steady state vector?

13.* After an extensive study of several large corporations, you have made the following observations about the intercompany mobility of employees during their entire careers. 76% of staff people remain at the staff level while 21% move up to middle management and 3% are promoted directly to top management. Of those in middle management positions, 15% move back to staff positions while 25% move up to top management positions with the balance not changing level. Of those people in top management positions, 82% continue in their positions, 17% move back to middle management positions, and 1% drop directly back to staff positions.

a. Determine the Markov matrix for this problem.

b. A large corporation forms a new division with 10% of the employees in top management positions and the rest divided between staff and middle management. Based on your observations, what would you predict the steady state distribution to be between these three levels of employment?

14.* In the bacteria growth problem of example 6, set the survival rate of each of the first two stages to 50%. Study and discuss what happens to the total population over a large number of iterations. If the matrix computational software you have available has plotting capabilities, it is of interest to plot the total population versus iteration number (hours).

15.* In the bacteria growth problem of example 6, set the survival rate of each of the first stage to 49% and the second stage to 50%. Study and discuss what happens to the total population over a large number of iterations. If the matrix computational software you have available has plotting capabilities, it is of interest to plot the total population versus iteration number (hours).

16.* In the bacteria growth problem of example 6, set the survival rate of each of the first two stages to 50%. Also change the reproduction pattern so that offspring are produced in either of the first two stages. Each individual surviving to the third stage produces two new bacteria via cell division. Study and discuss what happens to the total population over a large number of iterations. If the matrix computational software you have available has plotting capabilities, it is of interest to plot the total population versus iteration number (hours).

17.* It is desired to design a tripod structure whose geometry is given below. A five-kilonewton (5 kN) gravitational load is to be supported at the apex. Determine the forces in each leg required to support the given load. Note that this is a three-dimensional problem, but equilibrium equations need to be written only for one joint.

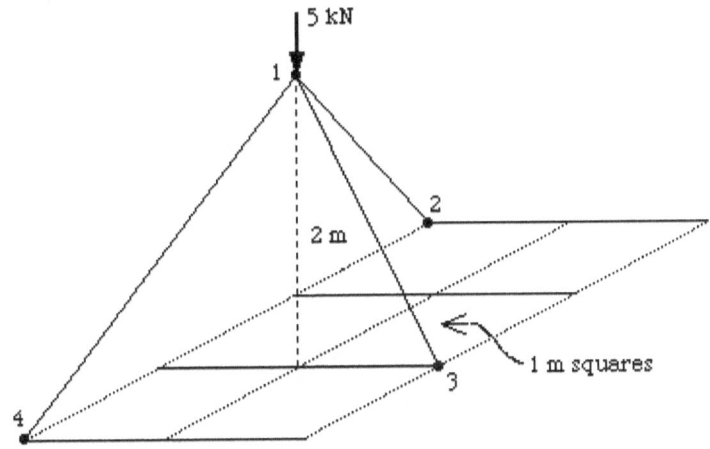

18.* Use partitioning to invert the following fourth-order matrix.

$$\begin{bmatrix} 2 & 1 & 4 & 3 \\ -1 & 1 & 0 & 2 \\ 3 & -2 & 2 & 1 \\ 0 & 1 & -3 & 3 \end{bmatrix}$$

IV

DETERMINANTS AND THE INVERSE MATRIX

1. INTRODUCTION

The theory of determinants is usually ascribed to G. W. Leibniz, who stated their basic properties in a letter written to the mathematician L'Hospital in 1693. However, a Japanese mathematician, Seki Kowa, had arrived at nearly the same development at least as early as 1683. The work of Leibniz seems to have been forgotten, and it was G. Cramer who rediscovered determinants and published a statement of their properties in 1750. The modern notation was introduced by A. Cayley in 1841. An excellent discussion of determinants and the historical development of their theory is found in the work of Sir Thomas Muir [920].[31]

The theory of determinants is intimately connected with that of matrices. The more important definitions and properties of determinants that are essential for the understanding of matrix algebra are presented in this chapter.

2. FUNDAMENTAL DEFINITIONS AND NOTATION

For any square matrix, its determinant is a scalar function that is evaluated in terms of its elements. If A is an nth-order matrix whose elements are a_{ij}, then its determinant is designated as Det A and symbolized as follows.

$$\text{Det } A = |A| = \begin{vmatrix} a_{11} & a_{12} & \cdots & a_{1n} \\ a_{21} & a_{22} & \cdots & a_{2n} \\ \vdots & \vdots & \ddots & \vdots \\ a_{n1} & a_{n2} & \cdots & a_{nn} \end{vmatrix}$$

Vertical bars around a matrix are used to designate the determinant of the matrix. Since the matrix A is n^{th} order, its determinant is also referred to as an nth-order determinant.

Before we develop the general rule for the expansion of an nth-order determinant, it is necessary to define a minor and a cofactor of an element of a determinant.

---

[31] This notation indicates a bibliographic entry contained in the references found in the appendix.

DEFINITION

Minor

The minor m_{ij} of the element a_{ij} of a determinant of order n is the determinant of order $(n-1)$ obtained by removing the ith row and the jth column of the original determinant.

EXAMPLE 1

Find the minor m_{23} of the determinant of a third-order general matrix A.

$$\text{Det } A = |A| = \begin{vmatrix} a_{11} & a_{12} & a_{13} \\ a_{21} & a_{22} & a_{23} \\ a_{31} & a_{32} & a_{33} \end{vmatrix}$$

SOLUTION

The minor of the element a_{23} is obtained by deleting the second row and the third column of $|A|$ and forming the determinant m_{23} from the remaining elements. Carrying out this procedure, the following determinant is obtained:

$$m_{23} = \begin{vmatrix} a_{11} & a_{12} \\ a_{31} & a_{32} \end{vmatrix} = \text{minor of } a_{23} \text{ of } |A|$$

DEFINITION

Cofactor

By definition, the cofactor of a given element of a determinant is the minor of the element with either a plus or a minus sign attached to it. The sign is obtained from the following formula.

$$\text{cofactor of } a_{ij} = c_{ij} = (-1)^{i+j}\, m_{ij},$$

where m_{ij} is the minor of a_{ij}.

EXAMPLE 2

For the general third-order determinant of example 1 above, evaluate the cofactor c_{23}.

SOLUTION

The cofactor of the element a_{23} for the third-order determinant above is given by

$$c_{23} = (-1)^{2+3} \, m_{23} = -m_{23}$$

DEFINITION

Determinant of general second-order matrix.

Let A be a general second-order matrix.

$$A = \begin{bmatrix} a_{11} & a_{12} \\ a_{21} & a_{22} \end{bmatrix}$$

The determinant, Det A, of this matrix is defined as

(2.1) $\qquad \text{Det } A = |A| = \begin{vmatrix} a_{11} & a_{12} \\ a_{21} & a_{22} \end{vmatrix} = a_{11}a_{22} - a_{21}a_{12}$.

EXAMPLE 3

Evaluate the determinant of the following matrix.

$$A = \begin{bmatrix} 1 & 2 \\ -1 & 3 \end{bmatrix}$$

SOLUTION

Using equation (2.1), the Det A is

$$\text{Det } A = (1)(3) - (-1)(2) = 5.$$

$$***$$

To illustrate the historical source of determinants and the above expansion of second-order determinants, consider the following general system of two linear algebraic equations in the two unknowns x_1 and x_2.

$$a_{11}x_1 + a_{12}x_2 = b_1$$

$$a_{21}x_1 + a_{22}x_2 = b_2$$

Let us solve for the unknown x_1 by solving the second equation for x_2 and substituting the result into the first equation. From the second equation we have

$$x_2 = \frac{1}{a_{22}}(b_2 - a_{21}x_1).$$

Substituting this result into the first equation and solving for x_1 gives the desired solution.

$$x_1 = \frac{(b_1a_{22} - b_2a_{12})}{(a_{11}a_{22} - a_{12}a_{21})}$$

In the above definition of a second-order determinant, the denominator of the solution is the determinant of the second-order matrix of coefficients. The numerator is also the same determinant of coefficients with the exception that the first column has been replaced with the column of constants, $[b_1 \; b_2]^T$. In other words,

$$x_1 = \frac{\begin{vmatrix} b_1 & a_{12} \\ b_2 & a_{22} \end{vmatrix}}{\begin{vmatrix} a_{11} & a_{12} \\ a_{21} & a_{22} \end{vmatrix}}.$$

This form of the result is an expression of the classical solution known as Cramer's rule that is defined later in section 6.

As a consequence of the definitions of minors and cofactors given above, it is seen that for Det A

$$c_{11} = a_{22}, \; c_{12} = -a_{21}, \; c_{21} = -a_{12}, \; c_{22} = a_{11}.$$

From these, the expansion of Det A may be written in the following four equivalent forms.

$$\text{Det } A = a_{11}c_{11} + a_{12}c_{12} = a_{11}c_{11} + a_{21}c_{21}$$

$$= a_{21}c_{21} + a_{22}c_{22} = a_{12}c_{12} + a_{22}c_{22}$$

Examining these relations carefully, we see that the expansion of Det A may be expressed as follows. The value of the Det A is the sum of the products of the elements of any single row or any single column of Det A times their corresponding cofactors. This result is true for determinants of any order and is known as Laplace's expansion.

3. THE LAPLACE EXPANSION OF A DETERMINANT

We have just shown how the determinant of a second-order matrix is evaluated by the equations

$$\text{Det } A = \sum_{j=1}^{2} a_{ij}c_{ij} \quad \text{or} \quad \text{Det } A = \sum_{i=1}^{2} a_{ij}c_{ij}$$

$$i = 1,2, \text{ or } 3 \qquad\qquad j = 1,2, \text{ or } 3$$
$$\text{(expansion by rows)} \qquad\qquad \text{(expansion by columns)}$$

where c_{ij} is the cofactor of a_{ij}. As stated, these types of expressions are applicable to determinants of any order.

DEFINITION

Laplace's Expansion of an nth-Order Determinant

If A is a matrix of nth order, the value of Det A is given by

$$\text{Det } A = \sum_{j=1}^{n} a_{ij}c_{ij}$$

$$i = \text{any}$$
$$\text{(expansion by rows)}$$

(3.1)

$$\text{or Det } A = \sum_{i=1}^{n} a_{ij}c_{ij}$$

$$j = \text{any}$$
$$\text{(expansion by columns)}$$

EXAMPLE 4

Evaluate the third-order determinant of a matrix A.

$$\text{Det } A = \begin{vmatrix} 1 & 2 & -1 \\ 3 & -5 & -2 \\ 0 & 4 & 3 \end{vmatrix}$$

SOLUTION

The value of Det A is uniquely defined by any one of the six equivalent expressions given by (3.1) with $n = 3$. For example, consider expanding about the second row, $i = 2$.

$$\text{Det } A = a_{21}C_{21} + a_{22}C_{22} + a_{23}C_{23} =$$

$$(3)(-1)^3 \begin{vmatrix} 2 & -1 \\ 4 & 3 \end{vmatrix} + (-5)(-1)^4 \begin{vmatrix} 1 & -1 \\ 0 & 3 \end{vmatrix} + (-2)(-1)^5 \begin{vmatrix} 1 & 2 \\ 0 & 4 \end{vmatrix}$$

Using the definition for the value of a second-order determinant given by equation (2.1), this expression is expanded again to determine the final result.

$$\text{Det } A = (3)(-1)(6 + 4) + (-5)(1)(3 - 0) + (-2)(-1)(4 - 0) = -37$$

EXAMPLE 5

It was stated above that the determinant may be evaluated by applying Laplace's expansion to any row or column. The important point of this is that all possible expansions of a determinant always gives the same result. To illustrate this, determine the value again of the third-order Det A given in example 5 above by expanding about the first column.

SOLUTION

Applying Laplace's expansion to the first column gives the following result.

$$\text{Det } A = a_{11}C_{11} + a_{21}C_{21} + a_{31}C_{31} =$$

$$(1)(-1)^2 \begin{vmatrix} -5 & -2 \\ 4 & 3 \end{vmatrix} + (3)(-1)^3 \begin{vmatrix} 2 & -1 \\ 4 & 3 \end{vmatrix} + (0)(-1)^4 \begin{vmatrix} 2 & -1 \\ -5 & -2 \end{vmatrix} =$$

$$(-7) + (-30) + (0) = -37$$

Not only has the same result occurred as it should, but we can also see the advantage of expanding about rows or columns that contain one or more zeros. The presence of zeros reduces the number of lower-order determinants that have to be evaluated.

***

As stated above, equation (3.1) is the Laplace expansion of the general n^{th}-order determinant. Equation (3.1) demonstrates that the determinant may be expanded by taking the sums of continued products of the elements and their corresponding cofactors along any row or along any column of the determinant. In this case, the cofactors c_{ij} are determinants of the $(n-1)^{th}$ order, but they in turn may be expanded by the rule (3.1) and so on until the determinant is finally evaluated.

A final important property of determinants is that the determinant of a product of two or more matrices is the product of their determinants. In other words,

$$\text{Det } (AB) = (\text{Det } A)(\text{Det } B)$$

EXERCISES

Evaluate the following determinants.

1.* $\begin{vmatrix} 7 & 2 & 4 \\ 3 & -4 & 5 \\ 1 & 3 & -2 \end{vmatrix}$
2.* $\begin{vmatrix} 3 & 6 & -1 \\ 1 & 9 & 2 \\ -4 & -2 & 1 \end{vmatrix}$
3.* $\begin{vmatrix} 2 & 1 & 0 \\ 1 & -1 & 1 \\ 1 & 5 & -1 \end{vmatrix}$

4.* $\begin{vmatrix} 2 & 5 & 6 \\ 3 & 7 & 9 \\ 2 & 4 & 0 \end{vmatrix}$
5.* $\begin{vmatrix} 0.15 & 0.35 & -0.44 \\ -0.70 & 0.04 & 0.68 \\ -0.93 & 0.23 & 0.89 \end{vmatrix}$
6.* $\begin{vmatrix} 5 & 2 & 2 \\ 3 & 6 & 3 \\ 6 & 6 & 9 \end{vmatrix}$

7.* $\begin{vmatrix} 1 & -4 & 2 & -2 \\ 4 & 7 & -3 & 5 \\ 3 & 0 & 8 & 0 \\ -5 & -1 & 6 & 9 \end{vmatrix}$
8.* $\begin{vmatrix} 2 & 2 & 5 & 3 \\ 2 & 3 & 3 & 2 \\ 3 & 3 & 7 & 4 \\ 3 & 2 & 9 & 7 \end{vmatrix}$

9.*   Select any pair of equal-order determinants above, and verify the relation that

$$\text{Det } (AB) = (\text{Det } A)(\text{Det } B).$$

***

4. FUNDAMENTAL PROPERTIES OF DETERMINANTS

From the basic definition (3.1), the following properties of determinants are deduced:

Property 1

If all the elements in a row or in a column are zero, the determinant is equal to zero.

EXAMPLE 6

Prove property 1 for the determinant of the matrix A below.

$$A = \begin{bmatrix} 2 & -1 & 3 \\ 0 & 0 & 0 \\ 4 & -2 & 1 \end{bmatrix}$$

SOLUTION

Applying Laplace's expansion to the second row, we find

$$\text{Det } A = -(0)\begin{vmatrix} -1 & 3 \\ -2 & 1 \end{vmatrix} + (0)\begin{vmatrix} 2 & 3 \\ 4 & 1 \end{vmatrix} - (0)\begin{vmatrix} 2 & -1 \\ 4 & -2 \end{vmatrix} = 0$$

Property 2

If all elements but one in a row or column are zero, the determinant is equal to the product of that element and its cofactor.

EXAMPLE 7

Develop a general proof of property 2.

SOLUTION

Consider a general matrix A for which all the elements of the ith row are zero except the rth. This means $a_{ij} = 0$ except for a_{ir}. Applying Laplace's expansion as given in equation (3.1) to determine Det A by expanding about the ith, row we find

$$\text{Det } A = \sum_{j=1}^{n} a_{ij}c_{ij} = \text{expansion about } i^{th} \text{ row}$$

.

Now as stated, all $a_{ij} = 0$ except a_{ir}. We thus have Det $A = a_{ir}c_{ir}$, which is the statement of property 2.

Property 3

The determinant of the transpose of a matrix is equal to the determinant of the matrix. Since Laplace's expansion may be applied to any row or column to obtain the value of a determinant, this property is easy to prove by applying it to the same column number of the transpose as row number of the given determinant.

Property 4

The interchange of any two columns or two rows of a determinant changes the sign of the determinant.

EXAMPLE 8

Verify property 4 by exchanging the second and third rows of the determinant given in example 4 and recomputing its value.

SOLUTION

From example 4 we have

$$
\text{Det A} = \begin{vmatrix} 1 & 2 & -1 \\ 3 & -5 & -2 \\ 0 & 4 & 3 \end{vmatrix} = -37
$$

Let us exchange the second and third rows and evaluate the resulting determinant.

$$
\begin{vmatrix} 1 & 2 & -1 \\ 0 & 4 & 3 \\ 3 & -5 & -2 \end{vmatrix} = (1)(-1)^2 \begin{vmatrix} 4 & 3 \\ -5 & -2 \end{vmatrix} + (0)(-1)^3 \begin{vmatrix} 2 & -1 \\ -5 & -2 \end{vmatrix} +
$$

$$
(3)(-1)^4 \begin{vmatrix} 2 & -1 \\ 4 & 3 \end{vmatrix} = (1)(-8+15) + (0) + (3)(6+4)
$$

$$
= +37 = -\text{Det A}
$$

Property 5

If two columns or two rows of a determinant are identical, the determinant is equal to zero.

EXAMPLE 9

Verify property 5 by evaluating the determinant of the following matrix in which the second and third rows are the same.

$$
A = \begin{bmatrix} 3 & -2 & 2 \\ 1 & 3 & 2 \\ 1 & 3 & 2 \end{bmatrix}
$$

SOLUTION

Applying Laplace's expansion to the third row yields the following.

$$\text{Det } A = (1)\begin{vmatrix} -2 & 2 \\ 3 & 2 \end{vmatrix} - (3)\begin{vmatrix} 3 & 2 \\ 1 & 2 \end{vmatrix} + (2)\begin{vmatrix} 3 & -2 \\ 1 & 3 \end{vmatrix}$$

$$= (1)(-4 - 6) - (3)(6 - 2) + (2)(9 + 2)$$

$$= -10 - 12 + 22 = 0$$

EXAMPLE 10

As a result of property 5, show that if Laplace's expansion is modified such that one sums the products of the elements of one row or column times the cofactors of any other row or column respectively, the result is identically zero.

SOLUTION

As given in equation (3.1), Laplace's expansion for the determinant of a general matrix A is

$$\text{Det } A = \sum_{k=1}^{n} a_{ik} c_{ik} \quad .$$

Here, the expansion has been taken about the ith row. If the expansion is taken about the jth row, one has

$$\text{Det } A = \sum_{k=1}^{n} a_{jk} c_{jk} \quad .$$

Clearly, both sums must be equal since they yield the same result.

Now consider the series that represents an expansion about the ith row with the cofactors replaced by those of the jth row.

$$s = \sum_{k=1}^{n} a_{ik} c_{jk}$$

This expansion is the same as an expansion about the jth row of a determinant in which the elements of the jth row are replaced with those of the ith row. In other words, this latter expansion is for the determinant of a matrix that appears as follows.

$$
A = \begin{bmatrix}
a_{11} & a_{12} & \cdots & a_{1n} \\
\vdots & \vdots & \vdots & \vdots \\
a_{i1} & a_{i2} & \cdots & a_{in} \\
\vdots & \vdots & \vdots & \vdots \\
a_{i1} & a_{i2} & \cdots & a_{in} \\
\vdots & \vdots & \vdots & \vdots
\end{bmatrix}
\begin{array}{l}
\\
\\
\leftarrow i^{th} \text{ row} \\
\\
\leftarrow j^{th} \text{ row} \\
\end{array}
$$

Hence, by property 5, the value of s must be zero.

$$***$$

The result of this example is stated in the following property.

Property 6

The expansion of any row or column of a determinant times the cofactors of any other row or column respectively always vanish identically.

Combining this property with Laplace's expansion gives the following general result for any n^{th}-order matrix A.

$$
\sum_{k=1}^{n} a_{ik}\, c_{jk} = \begin{cases} 0 \text{ if } i \neq j \\ \\ \text{Det } A \text{ if } i = j \end{cases},
$$

where c_{jk} is the cofactor of element a_{jk}.

Property 7

If a row or column is added to or subtracted from another row or column, the value of the determinant is unchanged.

EXAMPLE 11

Develop a general proof of property 7.

SOLUTION

Consider a general matrix A in which the ith row is replaced by the sum of the ith and k^{th} rows. That is, if $A = [a_{ij}]$, then the altered matrix, say B, is

$$B = \begin{bmatrix} a_{11} & a_{12} & \cdots & a_{1n} \\ \vdots & \vdots & \vdots & \vdots \\ a_{i1} + a_{k1} & a_{i2} + a_{k2} & \cdots & a_{in} + a_{kn} \\ \vdots & \vdots & \vdots & \vdots \\ a_{n1} & a_{n2} & \cdots & a_{nn} \end{bmatrix}.$$

Applying Laplace's expansion to the ith row of B yields

$$B = \sum_{r=1}^{n} (a_{ir} + a_{kr}) c_{ir} = \sum_{r=1}^{n} a_{ir} c_{ir} + \sum_{r=1}^{n} a_{kr} c_{ir}$$

$$= \text{Det } A + 0 = \text{Det } A \qquad .$$

Now the first of the latter two sums is Det A from Laplace's expansion. The second is an expansion about the kth row times the cofactors of the ith row. By example 10, this expansion must vanish, thus proving property 7.

Property 8

If all the elements in any row or column are multiplied by a scalar, the determinant is multiplied by that scalar.

This property is proved by application of Laplace's expansion (3.1). If d represents the determinant to be expanded and we consider an expansion about the ith row, we would have

$$d = \sum_{j=1}^{n} a_{ij} c_{ij} \qquad .$$

Now assume that a new determinant d_3 is formed by multiplying every element in the ith row of d by the scalar k. Now d_3 is identical to d except the ith row has the elements $ka_{ij}, j = 1, 2, \ldots, n$. If Laplace's expansion is applied to the ith row of d_3, we have

$$d_3 = \sum_{j=1}^{n} (ka_{ij})\, c_{ij} = k \sum_{j=1}^{n} a_{ij} c_{ij} = k\, d \quad ,$$

which proves the property.

Property 9

Multiplying the elements of any row or column by the same constant and adding the result to any other row or column does not change the value of the determinant.

EXAMPLE 12

Verify property 9 for the following general third-order determinant:

$$d = \begin{vmatrix} a_{11} & a_{12} & a_{13} \\ a_{21} & a_{22} & a_{23} \\ a_{31} & a_{32} & a_{33} \end{vmatrix}$$

SOLUTION

Let the third column be multiplied by the scalar m and added to the first column. The resulting determinant is

$$\begin{vmatrix} (a_{11}+ma_{13}) & a_{12} & a_{13} \\ (a_{21}+ma_{23}) & a_{22} & a_{23} \\ (a_{31}+ma_{33}) & a_{32} & a_{33} \end{vmatrix} .$$

Now expand this determinant about the first column by Laplace's expansion.

$$\sum_{i=1}^{n} (a_{i1} + ma_{i3})\, c_{i1}$$

This summation can be separated into two sums as follows.

$$\sum_{i=1}^{n} (a_{i1} + ma_{i3})c_{i1} = \sum_{i=1}^{n} a_{i1}c_{i1} + m\sum_{i=1}^{n} a_{i3}c_{i1}$$

The first summation is the expansion of d about its first column. The second summation must vanish because it represents the expansion of a determinant about its first column with the cofactors being those of the third column. As in example 10 above, the expansion about any row or column with the cofactors from any other row or column is always identically zero. With this observation, the property is verified.

***

In conclusion, it should be noted that all the above properties involve elementary operations on rows or columns that can be performed on matrices as well as used for evaluating determinants.

5.   THE EXPANSION OF DETERMINANTS

The use of Laplace's expansion and the properties of determinants given in section 4 will now be illustrated by expanding determinants that have numerical elements.

EXAMPLE 13

Expand the following determinant.

$$d = \begin{vmatrix} 2 & 0 & 3 & 3 \\ 0 & -1 & 2 & 1 \\ 1 & -1 & 3 & 1 \\ 0 & 3 & 1 & 2 \end{vmatrix}$$

SOLUTION

Without the use of any of the properties of section 5, this determinant can be expanded directly by applying Laplace's expansion to the first column. We have selected this column because it contains two zeros, which simplifies the expansion. If we now expand this determinant in terms of the first column, we find

$$d = 2 \begin{vmatrix} -1 & 2 & 1 \\ -1 & 3 & 1 \\ 3 & 1 & 2 \end{vmatrix} + 1 \begin{vmatrix} 0 & 3 & 3 \\ -1 & 2 & 1 \\ 3 & 1 & 2 \end{vmatrix} .$$

These two third-order determinants must now be expanded. For convenience, let us expand each about their first rows. The result is as follows.

$$d = 2 \left\{ (-1) \begin{vmatrix} 3 & 1 \\ 1 & 2 \end{vmatrix} - 2 \begin{vmatrix} -1 & 1 \\ 3 & 2 \end{vmatrix} + \begin{vmatrix} -1 & 3 \\ 3 & 1 \end{vmatrix} \right\} +$$
$$\left\{ -3 \begin{vmatrix} -1 & 1 \\ 3 & 2 \end{vmatrix} + 3 \begin{vmatrix} -1 & 2 \\ 3 & 1 \end{vmatrix} \right\}$$

After evaluating each second-order determinant, we find that the value of the determinant is -16.

$$***$$

The determinant d may also be expanded by the properties of section 4 in the following manner. Leaving the last three rows unchanged, form a new first row by multiplying the third row by -2 and adding it to the first row (property 9). The resulting determinant may then be expanded about its first column to yield

$$d = \begin{vmatrix} 0 & 2 & -3 & 1 \\ 0 & -1 & 2 & 1 \\ 1 & -1 & 3 & 1 \\ 0 & 3 & 1 & 2 \end{vmatrix} = \begin{vmatrix} 2 & -3 & 1 \\ -1 & 2 & 1 \\ 3 & 1 & 2 \end{vmatrix} .$$

Now the third-order determinant may be further reduced by adding twice the second row to the first and three times the second row to the third row. These operations give the following result.

$$d = \begin{vmatrix} 0 & 1 & 3 \\ -1 & 2 & 1 \\ 0 & 7 & 5 \end{vmatrix} .$$

Applying Laplace's expansion to the first column and evaluating the single second-order determinant brings us to the final answer.

$$d = -(-1) \begin{vmatrix} 1 & 3 \\ 7 & 5 \end{vmatrix} = (5 - 21) = -16$$

This is the same answer as before. By application of elementary operations, all but a single element in a row or column may be reduced to zero without changing the value of the determinant. Laplace's expansion then reduces the original determinant to one of one order lower. By continuing this reduction process, only one determinant of each lower order must be evaluated. In the example above, there were two third-order and five second-order determinants that had to be evaluated to find the value of d. By using the elementary operations, this was reduced to evaluating a single third-order and a single second-order determinant. In the worst case of a fourth-order determinant, there could have been four third-order and therefore twelve second-order determinants to be evaluated to find d.

At this point, it should be noted that the process of reducing a matrix to upper triangular form that was presented in the previous chapter utilized elementary operations that are included in the above properties. This provides a very convenient way of evaluating determinants as it may be shown that the determinant of an upper triangular matrix is simply the product of all the diagonal elements. The verification of this is left to the reader as an exercise (AE.9).

EXAMPLE 14

Determine the value of the determinant of the following matrix by transforming it to upper triangular form.

$$A = \begin{bmatrix} 2 & -1 & 5 & 1 \\ 1 & 4 & 6 & 3 \\ 4 & 2 & 7 & 4 \\ 3 & 1 & 2 & 5 \end{bmatrix}$$

SOLUTION

From section 9 of chapter I, the elementary matrix that will reduce all the first-column elements below the first to zero is

$$
\begin{bmatrix}
1 & 0 & 0 & 0 \\
-\dfrac{1}{2} & 1 & 0 & 0 \\
-2 & 0 & 1 & 0 \\
-\dfrac{3}{2} & 0 & 0 & 1
\end{bmatrix}.
$$

Premultiplying the given matrix by this elementary matrix yields

$$
\begin{bmatrix}
2 & -1 & 5 & 1 \\
0 & \dfrac{9}{2} & \dfrac{7}{2} & \dfrac{5}{2} \\
0 & 4 & -3 & 2 \\
0 & \dfrac{5}{2} & -\dfrac{11}{2} & \dfrac{7}{2}
\end{bmatrix}.
$$

Repeating this process twice more reduces this matrix to upper triangular form. The result is

$$
A_1 =
\begin{bmatrix}
2 & -1 & 5 & 1 \\
0 & 9/2 & 7/2 & 5/2 \\
0 & 0 & -55/9 & -2/9 \\
0 & 0 & 0 & 131/55
\end{bmatrix}.
$$

As may be shown, (AE.9), the value of the determinant is the product of the diagonal elements that gives Det $A = -131$. Note that the transformation of the given matrix to upper triangular form does not alter the value of the determinant because every operation falls under property 9.

EXERCISES

Evaluate the following determinants by application of the above rules to simplify the numerical calculations.

1. $\begin{vmatrix} 7 & 2 & 4 \\ 3 & -4 & 5 \\ 1 & 3 & -2 \end{vmatrix}$
2. $\begin{vmatrix} 3 & 6 & -1 \\ 1 & 9 & 2 \\ -4 & -2 & 1 \end{vmatrix}$
3. $\begin{vmatrix} 2 & 1 & 0 \\ 1 & -1 & 1 \\ 1 & 5 & -1 \end{vmatrix}$

4. $\begin{vmatrix} 2 & 5 & 6 \\ 3 & 7 & 9 \\ 2 & 4 & 0 \end{vmatrix}$
5. $\begin{vmatrix} 0.15 & 0.35 & -0.44 \\ -0.70 & 0.04 & 0.68 \\ -0.93 & 0.23 & 0.89 \end{vmatrix}$
6. $\begin{vmatrix} 2 & 1 & 1 & 1 \\ 0 & 1 & 2 & -3 \\ 1 & -2 & 1 & 6 \\ 1 & 0 & 2 & 1 \end{vmatrix}$

7. $\begin{vmatrix} -1 & -2 & 1 & 2 \\ 1 & 1 & 0 & 1 \\ 0 & 1 & -1 & 2 \\ 2 & 3 & 2 & -1 \end{vmatrix}$
8. $\begin{vmatrix} -1 & -2 & 1 & 2 \\ 0 & 3 & 0 & 1 \\ 0 & 0 & -1 & 2 \\ 0 & 0 & 0 & -4 \end{vmatrix}$
9. $\begin{vmatrix} 2 & 6 & -2 & 4 \\ -1 & -2 & 1 & 0 \\ 7 & 13 & 5 & 9 \\ 4 & 8 & -6 & 11 \end{vmatrix}$

***

6.   THE SOLUTION OF LINEAR EQUATIONS: CRAMER'S RULE

Many problems that arise in applied mathematics require the solution of a system of linear algebraic equations of the form

(6.1)
$$a_{11}x_1 + a_{12}x_2 + \ldots + a_{1n}x_n = b_1$$
$$a_{21}x_1 + a_{22}x_2 + \ldots + a_{2n}x_n = b_2$$
$$\ldots$$
$$a_{n1}x_1 + a_{n2}x_2 + \ldots + a_{nn}x_n = b_n$$

As discussed in chapter I, these equations are easily represented in matrix form as follows.

$$AX = B,$$

where X is the column vector of unknowns x_i, B the column vector of constants b_i, and A the matrix of coefficients a_{ij}. Solutions for the unknowns may be obtained by taking ratios of determinants, which is expressed as Cramer's rule.

DEFINITION

Cramer's Rule

   The general set of equations (6.1) are solved in a systematic manner by determinants that are known as Cramer's rule. The solution to the system of equations, (6.1), by Cramer's rule is expressed by the following relation,

(6.2) $$x_r = \frac{\text{Det } A_r}{\text{Det } A} = \frac{|A_r|}{|A|}, \quad r = 1, 2, 3, \ldots n$$

provided Det $A \neq 0$. In this expression, A is the matrix of coefficients, and A_r is a matrix formed by replacing the the rth column of A by the column vector B.

EXAMPLE 15

   Apply Cramer's rule, equation (6.2), to solve for the circulating current i_1 in the example circuit problem introduced in chapter II. The circuit considered is

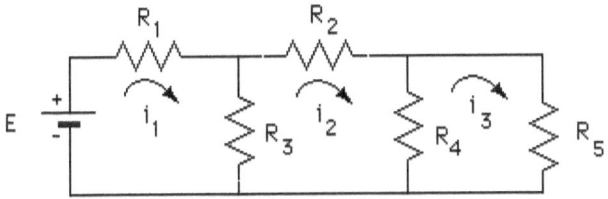

Fig. 4.1

It has already been shown that the equations for the circulating currents are

$$11i_1 - 6i_2 \qquad = 12$$

(6.3) $$-6i_1 + 24i_2 - 10i_3 = 0$$

$$-10i_2 + 20i_3 = 0.$$

Clearly, this is a specific third-order example of the general case given above in equation (6.1). In matrix notation, the coefficient matrix, unknown, and constant vectors are the following.

$$A = \begin{bmatrix} 11 & -6 & 0 \\ -6 & 24 & -10 \\ 0 & -10 & 20 \end{bmatrix}, \quad X = \begin{bmatrix} i_1 \\ i_2 \\ i_3 \end{bmatrix}, \quad B = \begin{bmatrix} 12 \\ 0 \\ 0 \end{bmatrix}$$

SOLUTION

From equation (6.2), we have

$$i_1 = \frac{\begin{vmatrix} 12 & -6 & 0 \\ 0 & 24 & -10 \\ 0 & -10 & 20 \end{vmatrix}}{\begin{vmatrix} 11 & -6 & 0 \\ -6 & 24 & -10 \\ 0 & -10 & 20 \end{vmatrix}} = \frac{4,560}{3,460} = 1.328 \text{ Amps}$$

***

From equation (6.2), it is seen that if the determinant of the matrix A of coefficients of the system of equations, Det A, is not zero, this equation gives the complete solution to the problem of solving the system (8.1). However, the use of (6.3) is not practical for the solution of more than three simultaneous equations. The reason for this lies in the fact that to evaluate a determinant of order n by Laplace's expansion requires $(n!)\,(n-1)$ multiplications. It therefore follows that to evaluate x_r by (6.2) for a system of n equations would require $(n+1)\,(n!)\,(n-1)$ multiplications and n divisions. For a system of four equations and four unknowns, this would involve a total of 292 multiplications and divisions. More efficient methods for the solution of systems of equations will be presented later.

EXERCISES

1.   Solve equation (6.3) for the remaining unknown loop currents i_2, i_3 by Cramer's rule.

Solve the following systems of equations by Cramer's rule.

2. $\begin{aligned} 2x - y &= 5 \\ 4x + 3y &= -1 \end{aligned}$

3. $\begin{aligned} x + y &= 3 \\ x - 2y &= 1 \end{aligned}$

4.
$$\begin{aligned} 2x - y &= 5 \\ -x + y &= 1 \end{aligned}$$

5.*
$$\begin{aligned} x + y + z &= 0 \\ x \quad\quad + 2z &= 3 \\ 2y - z &= 1 \end{aligned}$$

6.
$$\begin{aligned} x + y + z &= 1 \\ 2x + y - z &= 2 \\ x - 2y - z &= 1 \end{aligned}$$

7.*
$$\begin{aligned} x + y + z &= 2 \\ 2x - y - z &= 1 \\ x + 2y - z &= 3 \end{aligned}$$

8.*
$$\begin{aligned} 2x \quad\quad + 3z &= 7 \\ -x \quad\quad + 2z &= -14 \\ y + z &= 21 \end{aligned}$$

9.*
$$\begin{aligned} 3x + 2y + z &= 1 \\ 2x - y - z &= 1 \\ x + 4y \quad\quad &= 1 \end{aligned}$$

***

7. THE INVERSE MATRIX

We have now reached a point at which we may develop an operational definition for the inverse of a matrix that provides a means for its direct computation. To begin with, the definition of an adjoint matrix needs to be established. Consider an nth-order matrix A, and replace each element a_{ij} with its cofactor c_{ij}. The transpose of this matrix is called the adjoint matrix of A. It is represented as follows.

$$\text{Adj } A = [\, c_{ij}\,]^T = [\, c_{ji}\,]$$

EXAMPLE 16

Determine the adjoint Adj A of the matrix A below, and evaluate the product A (Adj A).

$$A = \begin{bmatrix} 1 & -2 & 3 \\ 3 & 5 & 1 \\ 6 & 4 & 2 \end{bmatrix}$$

SOLUTION

Replacing each element of A by its cofactor gives

$$\begin{bmatrix} 6 & 0 & -18 \\ 16 & -16 & -16 \\ -17 & 8 & 11 \end{bmatrix}.$$

The transpose of this matrix gives the desired adjoint matrix.

$$\text{Adj } A = \begin{bmatrix} 6 & 16 & -17 \\ 0 & -16 & 8 \\ -18 & -16 & 11 \end{bmatrix}$$

We now may compute the product A (Adj A).

$$A \text{ Adj } A = \begin{bmatrix} 1 & -2 & 3 \\ 3 & 5 & 1 \\ 6 & 4 & 2 \end{bmatrix} \begin{bmatrix} 6 & 16 & -17 \\ 0 & -16 & 8 \\ -18 & -16 & 11 \end{bmatrix}$$

$$= \begin{bmatrix} -48 & 0 & 0 \\ 0 & -48 & 0 \\ 0 & 0 & -48 \end{bmatrix} = (-48) \, I_3$$

It is also of importance to note that -48 is the determinant of A.

***

Let us now examine in more general terms the product obtained by postmultiplying the matrix A by its adjoint. In other words, let the matrix B be defined as

$$B = A \, (\text{Adj } A).$$

From the definition of matrix multiplication each element, b_{ij}, of B is the product of the ith row vector of A times the jth column of Adj A. Since Adj A is the transpose of the matrix of cofactors of A, then each main diagonal element b_{ii} of B must have the value of Det A from the application of Laplace's expansion. The elements of B not on the main diagonal, b_{ij}, must vanish as they are the sum of products of the ith row of A times the cofactors of the jth row of A. This is the result of property 6 above. Thus, the matrix B is simply a diagonal matrix whose diagonal elements are all equal to Det A. We therefore have the following useful relation.

$$A \, (\text{Adj } A) = (\text{Det } A) \, I_n = |\, A \,| \, I_n$$

It is left as an exercise for the reader to verify the result that a matrix and its adjoint commute, that is, $A \, (\text{Adj } A) = (\text{Adj } A) \, A$ by use of the previous relation. Based on the adjoint matrix, we may now introduce the following computational rule for the inverse of an nth-order matrix A.

If A is an nth-order matrix for which Det $A \neq 0$, then

(7.1)
$$A^{-1} = \frac{\text{Adj } A}{|\, A \,|} = \frac{\text{Adj } A}{\text{Det } A}.$$

From the previous relation, we can see that the ratio Adj A / Det A yields the inverse of A since

$$AA^{-1} = A^{-1}A = I_n.$$

DEFINITION

Singular and Nonsingular Matrices

By definition, if the determinant of a square matrix A vanishes so that $|\, A \,| = 0$, A is a singular matrix. If the determinant does not vanish, that is, if $|\, A \,| \neq 0$, then the matrix A is said to be nonsingular. Expressing this another way, the inverse of a square matrix exists only if the matrix is nonsingular (its determinant does not vanish).

EXAMPLE 17

Apply the above definition to determine the inverse of the following matrix:

$$A = \begin{bmatrix} 2 & 2 & 0 \\ 0 & 3 & 1 \\ 1 & 0 & 1 \end{bmatrix}$$

SOLUTION

The transposed matrix of A is

$$A^T = \begin{bmatrix} 2 & 0 & 1 \\ 2 & 3 & 0 \\ 0 & 1 & 1 \end{bmatrix}.$$

The adjoint matrix of A is now obtained by replacing each element of A^T by its corresponding cofactor. Doing this, the following matrix is obtained:

$$\text{Adj } A = \begin{bmatrix} 3 & -2 & 2 \\ 1 & 2 & -2 \\ -3 & 2 & 6 \end{bmatrix}.$$

The determinant $|A|$ is now easily computed by applying Laplace's expansion to the first row of A^T, which is the first column of A. Note that the cofactors of the elements of the first row of A^T are the first row of Adj A. The value of the determinant is

$$|A| = (2)(3) + (0)(-2) + (1)(2) = 8.$$

The inverse of A is now obtained by dividing the adjoint matrix by the determinant $|A|$; it is

$$A^{-1} = \frac{1}{8} \begin{bmatrix} 3 & -2 & 2 \\ 1 & 2 & -2 \\ -3 & 2 & 6 \end{bmatrix}.$$

***

Although equation (7.1) is formally complete for the evaluation of the inverse of a matrix, it is not useful for the computation of the inverse of large matrices. The reason for this lies in the fact that to evaluate a determinant of order n from its algebraic definition requires $(n!)(n-1)$ multiplications. The computation of the adjoint matrix requires the computation of n^2 determinants of the $(n-1)^{th}$ order, or $n^2(n-1)!(n-2)$ multiplications. When n is large, this value is approximated as $(n^2)n!$.

From this we have the result that the inversion of an nth-order matrix by equation (7.1) would require $(n!)(n^2-n-1)$ multiplications and n^2 divisions. This is a prohibitive amount of labor. Much more efficient methods for the inversion of large matrices are discussed in later chapters.

EXERCISES

Compute the adjoints of the following matrices and verify $A(\text{Adj } A) = |A|I$.

1. $\begin{bmatrix} 3 & 5 \\ 5 & 12 \end{bmatrix}$
2. $\begin{bmatrix} 1 & -1 & 0 \\ 0 & 2 & 1 \\ 3 & -2 & 1 \end{bmatrix}$
3.* $\begin{bmatrix} 2 & 1 & 0 \\ 1 & -1 & 1 \\ 1 & 5 & 1 \end{bmatrix}$

4.* $\begin{bmatrix} 1 & -1 & 1 \\ 4 & -3 & 0 \\ -2 & 2 & -2 \end{bmatrix}$
5.* $\begin{bmatrix} 1 & -1 & 1 & -2 \\ 5 & -4 & 1 & 0 \\ 4 & -3 & 0 & 1 \\ 2 & 0 & 3 & 2 \end{bmatrix}$

6.*   Compute the inverses for the matrices given in exercises 2 through 5 above and verify that $AA^{-1} = A^{-1}A = I$.

7.*   Compute $A^{-1}B$ and AB^{-1}, where

$$A = \begin{bmatrix} 1 & 2 & 3 \\ 1 & 3 & 5 \\ 1 & 5 & 12 \end{bmatrix}, \quad B = \begin{bmatrix} 1 & 1 & 1 \\ 1 & 2 & 3 \\ 1 & 4 & 9 \end{bmatrix}.$$

ADDITIONAL EXERCISES

1.   Without making use of Laplace's expansion, evaluate the following determinant.

$$\begin{vmatrix} 1 & x & x^2 \\ 1 & y & y^2 \\ 1 & z & z^2 \end{vmatrix}$$

2.   Without making use of Laplace's expansion, evaluate the following determinant.

$$\begin{vmatrix} a & 1 & 1 & 1 \\ 1 & a & 1 & 1 \\ 1 & 1 & a & 1 \\ 1 & 1 & 1 & a \end{vmatrix}$$

3.   If the vertices of a triangle have the rectangular coordinates (x_1, y_1), (x_2, y_2), (x_3, y_3), show that the area of this triangle is expressed by the following determinant:

$$\frac{1}{2} \begin{vmatrix} 1 & x_1 & y_1 \\ 1 & x_2 & y_2 \\ 1 & x_3 & y_3 \end{vmatrix}$$

4.   As a consequence of the result of exercise (AE.2), show that the equation of the straight line $ax + by + c = 0$ that passes through the points (x_1, y_1) and (x_2, y_2) can be written in the form

$$\begin{vmatrix} 1 & x & y \\ 1 & x_1 & y_1 \\ 1 & x_2 & y_2 \end{vmatrix} = 0$$

Hint: Consider the point (x, y) as being an arbitrary point on the line.

5.   Apply Laplace's expansion as an alternate means of verifying exercise (AE.3) above.

6.   Show that the application of Laplace's expansion in evaluating a determinant requires $n!(n-1)$ multiplications as stated in section 6.

7.   Consider the two homogeneous linear equations

$$a_{11}x_1 \quad + \quad a_{12}x_2 \quad = \quad 0$$

$$a_{21}x_1 \quad + \quad a_{22}x_2 \quad = \quad 0$$

or in matrix form, $AX = [0]$. Show by Cramer's rule that $a_1 = 0$, $a_2 = 0$, and therefore, for a nontrivial solution—that is, $x_1 \neq 0$, $x_2 \neq 0$—to possibly exist, the determinant of the coefficients Det A must vanish. Show also that if Det $A = 0$ the ratio of x_1 to x_2 is given by

$$\frac{x_1}{x_2} = -\frac{a_{12}}{a_{11}} = -\frac{a_{22}}{a_{21}}.$$

8.   Show that, in the cases of n homogeneous linear equations in n unknowns,

$$a_{11}x_1 + a_{12}x_2 + \ldots + a_{1n}x_n = 0$$
$$a_{21}x_1 + a_{22}x_2 + \ldots + a_{2n}x_n = 0$$
$$\ldots$$
$$a_{n1}x_1 + a_{n2}x_2 + \ldots + a_{nn}x_n = 0,$$

or $AX = [0]$ in matrix form, a necessary condition for this system of equations to have a solution other than the trivial solution $x_i = 0$ for all i is that the determinant of the coefficients A must vanish.

9.   Apply elementary operations to reduce the example fourth-order matrix A of section 5 to upper triangular form.

10.   Prove that the value of an upper or lower triangular determinant is the product of the diagonal elements.

11.   Show that the inverse of a nonsingular diagonal matrix is diagonal. If you have access to a matrix processing program, you should try this with several matrices of various orders.

12.*   Given the matrix

$$A = \begin{bmatrix} 8 & 4 & 3 \\ 2 & 1 & 1 \\ 1 & 2 & 1 \end{bmatrix},$$

compute A^{-2} by two methods.

13.   Show that if $C = [c_{ij}]$ is the matrix of cofactors of the nth-order matrix A, the following result is valid.

$$| AC^T | = (| A |)^n$$

14.   Given an nth-order square matrix, prove that its inverse exists only if its rank is also n.

15.*   Evaluate the determinants of several elementary matrices (producing different operations), and summarize your results.

16.*   Repeat exercise (AE.14*) for the inverses of several elementary matrices.

17.*   Prove that if any two rows or columns of an nth-order determinant are linearly dependent, then the value of the determinant must be identically zero. Note that if R_1 and R_2 are two rows or columns (vectors) of the nth order, they are linearly dependent if $R_1 = kR_2$, where k is a nonzero scalar.

V

EIGENVALUES AND EIGENVECTORS

1. INTRODUCTION

The study of many physical systems of practical importance—such as the investigation of the elastic vibrations of a bridge or any other solid structure, the flutter vibrations of an airplane wing, the transient oscillations of an electric network, or the buckling of an elastic structure—leads to the solution of what is known in the literature as an eigenvalue problem. In this chapter, a general mathematical discussion of the eigenvalue problem of a matrix will be given.

2. THE EIGENVALUES OF A MATRIX

Let A be a square matrix of order n and X be a column matrix or vector of the nth order. By the rule of the multiplication of matrices discussed in chapter I, multiplication of the vector X by the matrix A generates a new vector Y so that

$$(2.1) \qquad AX = Y.$$

The vector Y can be conceived as a transformation of the original vector X. We may now be asked whether or not it may happen that the vector Y has the same direction as the vector X. If they have the same direction, Y is simply proportional to X or $Y = \mu X$, where μ is a scalar multiplier and (2.1) becomes

$$(2.2) \qquad AX = \mu X.$$

The scalar multiplier μ must now be determined. To do so, equation (2.2) is rearranged as a set of n homogeneous algebraic equations as follows,

$$(2.3) \qquad \mu X - AX = (\mu I_n - A)X = 0_n,$$

where I_n is an nth-order identity matrix and 0_n is an nth-order zero column vector.

Now as discussed in chapter II, section 6, n homogeneous equations (2.3) in the n unknowns X_i have no solution other than the trivial solution unless the determinant of the coefficient matrix is zero. Hence, for nontrivial solutions to exist,

(2.4) $| \mu I - A | = 0.$

In matrix algebra, the matrix

(2.5) $K = \mu I_n - A$

is called the characteristic matrix of the matrix A. The determinant $| \mu I_n - A |$ is called the characteristic function of A. Since $| \mu I_n - A | = p(\mu)$ is a polynomial in μ, it is also called the characteristic polynomial of A. When set equal to zero, this is called the characteristic equation of A and is denoted by $p(\mu) = 0$.

In equation (2.4), $p(\mu)$ is a polynomial in μ of degree n since the order of the matrix A is n and has the form

(2.6) $p(\mu) = \mu^n + c_{n-1}\mu_n^{-1} + c_{n-2}\mu^{n-2} + \ldots + c_1\mu + c_0 = 0.$

It is thus apparent that the scalar multiplier μ that we are seeking may be any of the roots of the algebraic equation (2.6). From the theory of equations, it is known that an algebraic equation of the nth degree always has n roots. In general, the roots are complex numbers. It is also possible that some of the roots may be multiple or repeated roots. Each multiple root must be counted according to its degree of multiplicity.

The Eigenvalues of A

The roots $\mu_1, \mu_2, \mu_3, \ldots, \mu_n$ of the characteristic equation $p(\mu) = 0$ are called the eigenvalues of the matrix A. In the mathematical literature, the eigenvalues are also called the characteristic roots, secular values, proper values, or the latent roots of the matrix A.

EXAMPLE 1

Determine the eigenvalues for the matrix below.

$$A = \begin{bmatrix} 21 & 3 \\ 3 & 15 \end{bmatrix}$$

SOLUTION

The characteristic matrix is

$$K = \begin{bmatrix} \mu\text{-}21 & -3 \\ -3 & \mu\text{-}15 \end{bmatrix}.$$

The characteristic polynomial is

$$p(\mu) = \begin{vmatrix} \mu\text{-}21 & -3 \\ -3 & \mu\text{-}15 \end{vmatrix} = (\mu\text{-}21)(\mu\text{-}15) - 9$$

$$= \mu^2 - 36\mu + 306 = 0 \qquad .$$

The two eigenvalues are $m_1 = 18 + 3\sqrt{2}$ and $m_2 = 18 - 3\sqrt{2}$.

***

The Eigenvectors of A

For every value of $\mu = \mu_i$, $i = 1, 2, \ldots, n$, a nontrivial solution of the homogeneous equation (2.3) can be found (see chapter II, section 6). In the case of multiple roots, the situation can become a little more complicated. There is always at least one solution, but there may not be as many linearly independent solution vectors as the degree of multiplicity of the root.

Let $X = X_i$ be the vector associated with $\mu = \mu_i$ in (2.3); we may then write

(2.7) $(\mu_i I_n - A)X_i = 0, i = 1, 2, 3, \ldots, n$

provided the characteristic equation $p(\mu) = 0$ has n distinct roots. The vectors X_i are called the eigenvectors or principal vectors of the matrix A. Since the eigenvectors are the solutions of a homogeneous set of equations, the solution is determined only up to a constant factor, and only the ratios of the elements in the columns X_i are uniquely determined. The geometrical interpretation of this is that the eigenvectors are uniquely determined only in their direction, but their length or absolute value is arbitrary.

EXAMPLE 2

Determine the eigenvectors for the matrix of the previous example.

SOLUTION

For the given matrix, the characteristic matrix is

$$K = mI_2 - A = \begin{bmatrix} \mu-21 & -3 \\ -3 & \mu-15 \end{bmatrix}.$$

We write the eigenvalue problem as a system of homogeneous equations as given in equation (2.7) or $KX = 0_2$.

(2.8)
$$\begin{bmatrix} \mu_i-21 & -3 \\ -3 & \mu_i-15 \end{bmatrix}\begin{bmatrix} x_1^i \\ x_2^i \end{bmatrix} = \begin{bmatrix} 0 \\ 0 \end{bmatrix}$$

We have introduced a superscript i on the elements of the unknown eigenvector to designate that they are elements of the ith eigenvector X_i. Similarly, the subscript i has been placed on the eigenvalue to indicate that there are two different values for this problem. Substituting the first eigenvalue from example 1 gives the system

$$\begin{bmatrix} \sqrt{2}-1 & -1 \\ -1 & \sqrt{2}+1 \end{bmatrix}\begin{bmatrix} x_1^1 \\ x_2^1 \end{bmatrix} = \begin{bmatrix} 0 \\ 0 \end{bmatrix}$$

This is a homogeneous system of two equations and two unknowns, and we know the rank is one as the determinant of the coefficient matrix was set to zero in order to find the eigenvalues. This means that the solution is not unique, so let us arbitrarily set $x_2^1 = 1$ and solve for x_1^1. This gives $x_1^1 = 1/(\sqrt{2} - 1)$. The first eigenvector is

$$X_1 = \begin{bmatrix} \dfrac{1}{\sqrt{2}-1} \\ 1 \end{bmatrix}.$$

Substituting the second eigenvalue into (2.8) gives

$$\begin{bmatrix} -\sqrt{2}-1 & -1 \\ -1 & -\sqrt{2}+1 \end{bmatrix}\begin{bmatrix} x_1^2 \\ x_2^2 \end{bmatrix} = \begin{bmatrix} 0 \\ 0 \end{bmatrix}.$$

Following the same procedure gives the second eigenvector to be

$$X_2 = \begin{bmatrix} \dfrac{-1}{\sqrt{2}+1} \\ 1 \end{bmatrix}.$$

The reader may verify these results by substituting the eigenvectors into the left-hand side of equation (2.2) and seeing that the result agrees with the right-hand side.

EXERCISES

1.   If μ_i are the eigenvalues of A, show that the eigenvalues of kA, $k =$ constant, are $k\mu_i$.

Determine the eigenvalues and eigenvectors each of the following matrices.

1. $\begin{bmatrix} 3 & -1 \\ -1 & 3 \end{bmatrix}$ 2. $\begin{bmatrix} 0 & 1 \\ 1 & 0 \end{bmatrix}$ 3. $\begin{bmatrix} 1 & -1 \\ -1 & 1 \end{bmatrix}$

4. $\begin{bmatrix} 7 & -2 \\ -2 & 7 \end{bmatrix}$ 5. $\begin{bmatrix} 1 & -2 \\ -4 & -1 \end{bmatrix}$ 6. $\begin{bmatrix} -5 & 3 \\ 3 & -5 \end{bmatrix}$

7*. $\begin{bmatrix} 1 & 1 & -1 \\ 2 & 0 & 2 \\ 3 & -1 & 5 \end{bmatrix}$ 8*. $\begin{bmatrix} 1 & 2 & 2 \\ 2 & 3 & 3 \\ 2 & 3 & 1 \end{bmatrix}$

9*. $\begin{bmatrix} 1 & 2 & 2 & 1 \\ 2 & 3 & 3 & 2 \\ 2 & 3 & 2 & 1 \\ 1 & 2 & 1 & 0 \end{bmatrix}$ 10*. $\begin{bmatrix} -13 & 16 & -27 & -19 \\ 10 & -6 & 18 & 10 \\ 7 & -4 & 13 & 7 \\ 10 & -12 & 18 & 16 \end{bmatrix}$

11*. $\begin{bmatrix} 4 & 2 & 4 & 2 & 10 \\ 2 & -16 & 6 & 12 & 4 \\ 7 & -4 & 7 & 9 & 10 \\ 10 & -8 & 14 & 12 & 16 \\ -15 & 16 & -19 & -23 & -30 \end{bmatrix}$

***

3. GEOMETRICAL INTERPRETATION OF THE EIGENVALUE PROBLEM

Before we consider the general treatment of the eigenvalue problem, the geometrical significance of the eigenvalues and eigenvectors of certain special matrices will be discussed. By this approach, the operations involved in determining the eigenvalues and eigenvectors of a matrix may be seen to have a very close relationship to the analytical geometry of second-order or quadratic surfaces. The result is that a more intuitive feeling for the purely algebraic operations of the process may be attained.

In order not to obscure, the main ideas with unnecessary geometrical and algebraic detail, two-dimensional problems will be considered first; the treatment will later be extended to spaces of any number of dimensions.

The Equation of a Central Ellipse

Let x_1 and x_2 be two Cartesian coordinates of a plane as shown in figure 5.1. From the analytic geometry of conic sections, the equation of a centrally located ellipse with reference to this Cartesian coordinate system is

(3.1)
$$\frac{x_1^2}{a^2} + \frac{x_2^2}{b^2} = 1$$
.

To write equation (3.1) in matrix notation, it is necessary to introduce the following matrices:

$$D = \begin{bmatrix} \frac{1}{a^2} & 0 \\ 0 & \frac{1}{b^2} \end{bmatrix}, \quad X = \begin{bmatrix} x_1 \\ x_2 \end{bmatrix}$$

D is a square diagonal matrix of the second order, and X is the coordinate vector. In terms of these matrices, it is easy to see that as a consequence of the rule for the multiplication of matrices, equation (3.1) may be written in the following form:

$$X^TDX = 1$$

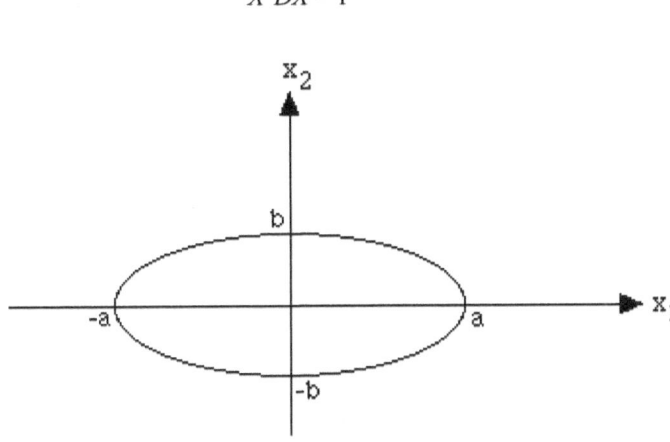

Fig. 5.1 Central Ellipse

From vector analysis, it is known that a vector normal to the ellipse at every point on its boundary is given by computing the gradient of the function $\Phi(x_1, x_2)$ where Φ is the equation of the quadratic surface or

(3.2) $\Phi(x_1, x_2) = X^TDX = 1$

In the notation of matrix algebra, the gradient or grad Φ may be written in the following form.

$$N = \text{grad } \Phi = \begin{bmatrix} \dfrac{\partial \Phi}{\partial x_1} \\[2ex] \dfrac{\partial \Phi}{\partial x_2} \end{bmatrix} = \begin{bmatrix} \dfrac{\partial}{\partial x_1}\left(\dfrac{x_1^2}{a^2} + \dfrac{x_2^2}{b^2}\right) \\[2ex] \dfrac{\partial}{\partial x_2}\left(\dfrac{x_1^2}{a^2} + \dfrac{x_2^2}{b^2}\right) \end{bmatrix}$$

(3.3)

$$= \begin{bmatrix} \dfrac{2x_1}{a^2} \\[2ex] \dfrac{2x_2}{b^2} \end{bmatrix} = 2\begin{bmatrix} \dfrac{1}{a^2} & 0 \\[2ex] 0 & \dfrac{1}{b^2} \end{bmatrix}\begin{bmatrix} x_1 \\ x_2 \end{bmatrix} = 2DX$$

Let us suppose now that the directions of the principal axes of the ellipse $\Phi(x_1, x_2) = 1$ are not known[32] and that it is required to determine these directions. The vector X given by (3.2) is a vector from the origin of the coordinate system to the surface of the ellipse and is, in general, not parallel to the vector N given by (3.3). The vectors X and N are parallel to each other only when X has the direction of the principal axes. Now if N is to be parallel to X, we must have

(3.4) $N = \mu X$

for some scalar μ. Since $N = 2DX$ from (3.3), we must have $DX = (\mu/2)X$. If we now set $\mu/2 = \mu$, we have the standard form of the eigenvalue problem:

$$DX = \mu X$$

Hence, the possible values of μ are the roots of the characteristic polynomial $p(\mu)$ of the matrix D given by

$$p(\mu) = |\mu I - D| = \begin{bmatrix} \left(\mu - \dfrac{1}{a^2}\right) & 0 \\ 0 & \left(\mu - \dfrac{1}{b^2}\right) \end{bmatrix} = 0$$

or

$$p(\mu) = \left(\mu - \frac{1}{a^2}\right)\left(\mu - \frac{1}{b^2}\right) = 0 .$$

The two eigenvalues of D are

(3.5) $\mu_1 = \dfrac{1}{a^2}, \quad \mu_2 = \dfrac{1}{b^2} .$

It is thus evident that equation (3.1) of the centrally located ellipse may be written in terms of the eigenvalues of D in the form

$$\mu_1 x_1^2 + \mu_2 x_2^2 = 1 .$$

---

[32] Principal axes of any quadratic surface are the axes of symmetry. In figure 5.1 above, the axes x_1, x_2 are the axes of symmetry and thus the principal axes of the ellipse. By an arbitrary set of axes, we are referring to any other set with the same origin but rotated by some angle to the x_1 axis.

The Eigenvectors of D

It remains now to find the eigenvectors of the matrix D. These eigenvectors will give the directions of the principal axes of the ellipse. To find them, we use equation (2.2), which may be written in the form

$$DX_i = \mu_i X_i, \; i = 1, 2.$$

Now

$$D = \begin{bmatrix} \mu_1 & 0 \\ 0 & \mu_2 \end{bmatrix},$$

where equation (3.5) has been employed to redefine the values of the elements of D. Expanding the eigenvalue problem above gives

(3.6) $$\mu_1 x'_1 = \mu_i x'_1, \; \mu_2 x'_2 = \mu_i x'_2, \; i = 1, 2.$$

The superscripts have been introduced to designate that the element in question is an element of the ith eigenvector. Equations (3.6) may be solved to give the eigenvectors

$$X_1 = \begin{bmatrix} r \\ 0 \end{bmatrix}, \quad X_2 = \begin{bmatrix} 0 \\ s \end{bmatrix},$$

where r and s are arbitrary constants. These results indicate that the line $x_2 = 0$ is the principal axis associated with the eigenvalue μ_1, and the line $x_1 = 0$ is the principal axis associated with the eigenvalue μ_2.

In this particular example, the equation of the ellipse had the very simple form (3.1) because its principal axes coincided with the coordinate directions. In order to extend the geometrical interpretation to a more complicated case, the equation of the ellipse relative to a rotated system of axes will be considered.

Rotation of the Coordinate Axes in the Plane

Let (x_1, x_2) be a set of Cartesian coordinates in a plane as shown in figure 5.2, and let (y_1, y_2) be another set of Cartesian coordinates that have the same origin as the first set of coordinates but which are inclined at an angle θ with respect to the first set as shown.

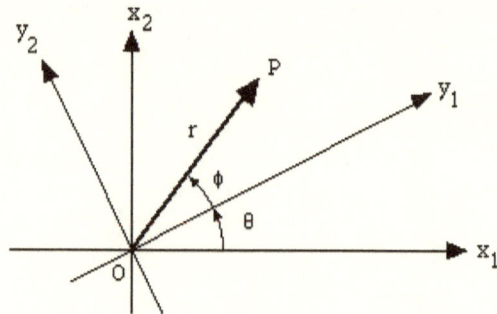

Fig. 5.2. Rotation of Axes

Let OP be a vector of length r drawn from the origin, and suppose OP makes an angle of ϕ with respect to y_1. The vector OP therefore makes an angle $\theta + \phi$ with respect to x_1. The coordinates of P relative to x_1 and x_2 are

(3.7)

$$x_1 = r \cos(\theta + \phi) = r \cos\theta\cos\phi - r \sin\theta\sin\phi$$

$$x_2 = r \sin(\theta + \phi) = r \sin\theta\cos\phi + r \cos\theta\sin\phi$$

The coordinates of P with respect to y_1 and y_2 are

$$y_1 = r \cos\phi, \quad y_2 = r \sin\phi.$$

If these expressions for y_1 and y_2 are used to replace the terms $r \cos\phi$ and $r \sin\phi$ in (3.7), the results are

(3.8)

$$x_1 = y_1 \cos\theta - y_2 \sin\theta$$

$$x_2 = y_1 \sin\theta + y_2 \cos\theta$$

In order to express the effect of the rotation of coordinates in matrix form, it is convenient to introduce the following matrices:

$$X = \begin{bmatrix} x_1 \\ x_2 \end{bmatrix}, \quad Y = \begin{bmatrix} y_1 \\ y_2 \end{bmatrix}, \quad R(\theta) = \begin{bmatrix} \cos\theta & \sin\theta \\ -\sin\theta & \cos\theta \end{bmatrix}$$

In terms of these matrices, equation (3.8) may be written in the form

$$X = R(-\theta)Y = R(\theta)^{-1}Y$$

This expresses the original coordinates X in terms of the rotated coordinates Y. If the last equation is premultiplied by $R(\theta)$, the resulting equation is

(3.9) $Y = R(\theta)X.$

The Rotation Matrix $R(\theta)$

The matrix $R(\theta)$ is generally called a rotation matrix. As can be seen from (3.9), premultiplication of the coordinate matrix X by the rotation matrix yields another coordinate matrix Y that is inclined at an angle θ with respect to X. The matrix $R(\theta)$ has the following important properties:

$$R(-\theta) = R(\theta)^T = R(\theta)^{-1} , \; | R(\theta) | = 1$$

Since $R(\theta)^T = R(\theta)^{-1}$, $R(\theta)$ is seen to be an orthogonal matrix .

Returning to the example, the sum of the squares of the coordinates X and Y are invariant to the transformation (3.9). This is true because the length of the vector OP, r, is constant. That is, letting $R = R(\theta)$,

$$y_1^2 + y_2^2 = Y^TY = X^TR^TRX = X^TX = x_1^2 + x_2^2 \quad .$$

The above result follows as a consequence of the reversal law of transposed products and the fact that $R^TR = R^{-1}R = I_2$.

The rotation matrix may be used to determine the principal axes of an ellipse when its principal axes do not coincide with the coordinates x_1 and x_2 as given by equation (3.1). In general, the equation of an ellipse whose center is at the origin of the coordinates x_1 and x_2 has the form

$$ax_1^2 + 2bx_1x_2 + cx_2^2 = 1.$$

In matrix notation, this equation takes the form

$$X^T \begin{bmatrix} a & b \\ b & c \end{bmatrix} X = 1$$

or

(3.10) $$X^T AX = 1, \quad A = \begin{bmatrix} a & b \\ b & c \end{bmatrix}.$$

Let the axes now be rotated by the transformation $X = R^T Y$. If this transformation is applied to (3.10), the result is

(3.11) $$Y^T R A R^T Y = 1.$$

Now if the new coordinates Y coincide with the principal axes of the ellipse (3.10), equation (3.11) does not contain the term $y_1 y_2$ and the matrix RAR^T must be a diagonal matrix. If the product (3.11) is carried out directly, the result will be seen to have the form

$$a'y_1{}^2 + 2b'y_1 y_2 + c'y_2{}^2 = 1.$$

The coefficient of the $y_1 y_2$ term is

(3.12) $$b' = (c - a)\sin(2\theta) + b\cos(2\theta).$$

If b' is to be made to vanish, θ must be chosen so that

$$\tan(2\theta) = \frac{b}{a - c}.$$

Let us now denote the product RAR^T as the rotated matrix M_r. By expanding this product, we see that it has the general form as shown below:

(3.13)

$$M_r = \begin{bmatrix} (\frac{a+c}{2}) + (\frac{a-c}{2})\cos(2\theta) + b\sin(2\theta) & (c-a)\sin(2\theta) + b\cos(2\theta) \\ (c-a)\sin(2\theta) + b\cos(2\theta) & (\frac{a+c}{2}) - (\frac{a-c}{2})\cos(2\theta) - b\sin(2\theta) \end{bmatrix}$$

In this result and (3.12) above, we have made use of the standard trigonometric double-angle relations $\sin(2\theta) = 2\sin\theta\cos\theta$, $\sin^2\theta = [1 - \cos(2\theta)]/2$, $\cos^2\theta = [1 + \cos(2\theta)]/2$.

Invariants of the Rotated Matrix Mr

There are certain properties of the matrix A that are also properties of the matrix M_r and hence are not changed by the rotation transformation $M_r = RAR^T$. For example, the determinant of A is equal to the determinant of M_r. This is a consequence of the rule of multiplication of determinants and the fact that Det $R =$ Det $R^T = 1$ so that

(3.14) $\qquad$ Det $M_r = ($Det $R)($Det $A)($Det $R^T) = $ Det A.

At this point, it is convenient to introduce the trace of a matrix that is defined to be the sum of the elements along the principal diagonal. The trace of A is designated as Tr A. From (3.13) it is seen that

$$\text{Tr } M_r = \text{Tr } A = a + c.$$

It will now be shown that the characteristic polynomial of A is the same as the characteristic polynomial of M_r and that hence, A and M_r have the same characteristic equation and eigenvalues. By definition, the characteristic polynomial of M_r, $p(\mu)$ is the determinant of the characteristic matrix K of M_r given by

(3.15) $\qquad$ $K = \mu I_2 - M_r, \ p(\mu) = |K|$.

As we have defined above, $M_r = RAR^T$. Since R is an orthogonal matrix, $R^{-1} = R^T$. This means that we may write $RR^{-1} = RR^T = I_2$. Substituting into equation (3.15) gives the result

$$\text{Det } (\mu I_2 - M_r) = | \ \mu RR^T - RAR^T \ |$$

$$= | \ R(\mu I_2 - A)R^T \ |$$

$$= | \ R \ || \ \mu I_2 - A \ || \ R^T \ |$$

This last relation results from the property of determinants that the determinant of a product of two or more matrices (all square and of equal order) is the product of the determinants of the individual matrices.

Now since the determinants of R and its transpose are equal to one, the last relation above gives the desired result.

$$\text{Det } (\mu I_2 - M_r) = \text{Det } (\mu I_2 - A)$$

This equation establishes that the characteristic polynomials of A and M_r are the same. Thus, A and M_r must have the same eigenvalues and eigenvectors.

Hence, we have the final result that the eigenvalues of A are invariant under the rotation R. Geometrically, this is evident from the fact that, as seen by (3.5), the eigenvalues are the reciprocals of the squares of the semimajor and semiminor axes of the ellipse and do not depend on the orientation of the ellipse with respect to the coordinate axes.

EXAMPLE 3

The equation of a central ellipse is

$$29x_1^2 + 24x_1x_2 + 36x_2^2 = 180.$$

Express this in matrix form; determine the magnitude (semimajor and semiminor) and location of the principle axes.

SOLUTION

First, this equation must be divided through by the factor 180 to place it in standard form. The result is

$$(29/180)x_1^2 + (1/15)x_1x_2 + (1/5)x_2^2 = 1$$

Introducing matrix form gives

$$\begin{bmatrix} x_1 \\ x_2 \end{bmatrix}^T \begin{bmatrix} \dfrac{29}{180} & \dfrac{1}{15} \\ \dfrac{1}{15} & \dfrac{1}{5} \end{bmatrix} \begin{bmatrix} x_1 \\ x_2 \end{bmatrix} = 1$$

The two pair of eigenvalues and associated eigenvectors are $1/9$, $[0.8 \ {-0.6}]^T$ and $1/4$, $[0.6 \ 0.8]^T$. From the above statement, the semimajor axis is $1/\sqrt{1/9} = 3$. Similarly, the semiminor axis is $1/\sqrt{1/4} = 2$. We can now sketch the ellipse in the x_1x_2 coordinate system as shown below.

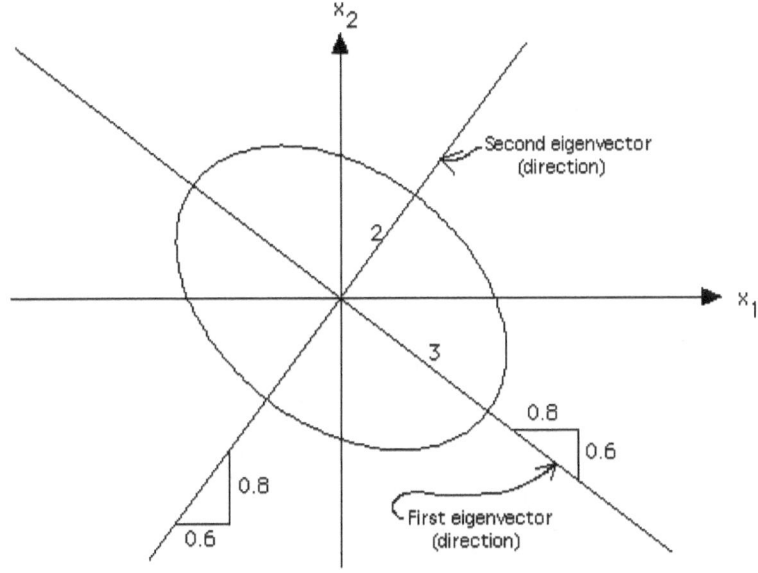

Fig. 5.3

EXERCISES

1.   The equation of a central ellipse is $5x_1^2 + 6x_1x_2 + 5x_2^2 = 8$. Write the equation in matrix (quadratic) form, and locate the principal axes.

2.   The equation of a central ellipse is $0.2153 \ x_1^2 - 0.1202 \ x_1x_2 + 0.1458 \ x_2^2 = 1$. Write the equation in matrix (quadratic) form, and locate the principal axes.

3.   Find the rotation matrix R that will reduce the following second-order matrix A to diagonal form by means of the transformation RAR^T.

$$A = \begin{bmatrix} 4 & -1 \\ -1 & 2 \end{bmatrix}$$

4.   Find a rotation matrix R that will reduce the second-order matrix A below to upper triangular form by means of the transformation RAR^T.

$$A = \begin{bmatrix} 1 & 2 \\ 3 & 4 \end{bmatrix}$$

***

4. ALGEBRAIC DISCUSSION OF THE EIGENVECTORS OF A MATRIX

Before we extend the geometrical interpretation of the eigenvalue problem to a quadratic surface of n dimensions, an algebraic discussion of the eigenvalue problem will be given.

Consider a square matrix A of order n whose elements are real numbers. It is required to determine the eigenvalues and the eigenvectors of A. From equation (2.4), we recall that the eigenvalues of A are the roots of the characteristic equation

(4.1) $p(\mu) = |\mu I_n - A| = 0.$

Where $p(\mu)$ is the characteristic polynomial of A and is in general a polynomial of the nth degree in μ. In general, the characteristic equation (4.1) will have n roots $\mu_1, \mu_2, \mu_3, \ldots, \mu_n$. In some cases, the roots may be multiple roots and must be counted according to their degree of multiplicity.

The Eigenvectors of A

By the defining equation (2.7), the eigenvector associated with the eigenvalue μ_i is a nonzero column vector X_i of the nth order, which satisfies the homogeneous equation

(4.2) $(\mu_i I_n - A)X_i = 0_n,$ or $AX_i = \mu_i X_i.$

Let it be assumed for the present that the characteristic equation (4.1) does not have multiple roots, so the eigenvalues μ_i are a set of n distinct numbers. In such a case, there are n distinct eigenvectors X_i that satisfy equation (4.2) for the index $i = 1, 2, \ldots, n$. It must be remembered that since the elements of X_i are solutions of the homogeneous equation (4.2), they are not uniquely determined, but only the ratios of the elements are determined. If X_i is regarded as a vector in n-dimensional space, this means that its length is not determined, only its direction. The eigenvectors for each individual eigenvalue may be determined by the method illustrated in chapter II, section 10.

The Modal Matrix of A

Let a square matrix M be constructed from the eigenvector columns X_i in the following manner:

$$M = [X_1, X_2, \ldots, X_n]$$

That is, the columns of the matrix M are the eigenvectors of A. The square matrix M is called the modal matrix of A. It is easy to show that the set of n equations, equations (4.2), may be written as one matrix equation in terms of the modal matrix. This equation has the form

(4.3) $AM = MD,$

where D is the nth-order diagonal matrix of eigenvalues.

(4.4)
$$D = \begin{bmatrix} \mu_1 & 0 & \cdots & 0 \\ 0 & \mu_2 & \cdots & 0 \\ \cdot & \cdot & \cdots & \cdot \\ 0 & 0 & \cdots & \mu_n \end{bmatrix}$$

Let us assume in general that the given matrix A is not a symmetric matrix so that $A^T \neq A$, then it is possible to obtain the eigenvectors of A^T in a similar manner. The eigenvalues of A^T are the roots of the characteristic equation

(4.5) $\text{Det} (\mu I_n - A^T) = \text{Det} (\mu I_n - A)^T = \text{Det} (\mu I_n - A) = p(\mu) = 0.$

Equation (4.5) follows from the fact that the value of a determinant is unaltered by interchanging its rows and its columns. From (4.5), it is seen that A^T has the same characteristic equation as A and, hence, that the eigenvalues of A^T are the same as those of A.

The Eigenvectors of AT

Let the nth-order column vector Y_i be the eigenvector of A^T associated with the eigenvalue μ_i. This eigenvector satisfies the equation

(4.6) $(\mu_i I_n - A^T)Y_i = 0_n,$ or $A^T Y_i = \mu_i Y_i,$

where $i = 1, 2, \ldots, n.$

The modal matrix of A^T, N, may now be constructed from the eigenvectors Y_i in the following manner:

$$N = [Y_1, Y_2, \ldots, Y_n]$$

The set of equations (4.6) may be written in terms of the diagonal matrix (4.4) in the form

(4.7) $$A^T N = ND.$$

Equations (4.3) and (4.7) are similar to each other and will now be used to obtain a relation between the modal matrices M and N. The columns of the matrix M are called the principal axes of A, and the columns of the matrix N are called the adjoint axes of A.

If the eigenvalues of A are distinct so that the characteristic equation of A does not have multiple roots, it can be shown that the modal matrices M and N are nonsingular matrices and therefore have inverses M^{-1} and N^{-1}. In this case, let (4.3) be premultiplied by M^{-1}, and let (4.7) be premultiplied by N^{-1}. The results of these operations are

(4.8) $$M^{-1}AM = D$$

and

(4.9) $$N^{-1}A^T N = D.$$

Both of these equations equal the same diagonal matrix D because A and A^T have the same eigenvalues. Also, the eigenvectors are taken in the same order so that X^i and Y^i are both associated with μ_i, etc.

Now we take the transpose of equation (4.9) and obtain

(4.10) $$N^T A (N^T)^{-1} = D^T = D.$$

If (4.10) is compared with (4.8), it can be seen that N^T and M^{-1} must be related to each other by the equation

(4.11) $$N^T = WM^{-1} \text{ or } W = N^T M,$$

where W must be a diagonal matrix. As has been mentioned above, the lengths of the vectors X_i and Y_i of the modal matrices M and N are not determined by equations (4.2) and (4.6). By multiplication with suitable constants, however, these vectors may be adjusted to satisfy the relation

$$Y_i^T X_i = 1$$

Then the vectors X_i and Y_i are said to be normalized. It now may be shown that the diagonal matrix W becomes the identity matrix I_n so (4.11) becomes

$$N^T = M^{-1} \text{ or } N^T M = M^T N = I_n.$$

It is thus seen that the principal axes of the matrix A, X_i, are orthogonal to its adjoint axes Y_i. If the matrix A is a symmetric matrix so that $A = A^T$, then $M = N$, and in this case, the previous result becomes

$$M^T = M^{-1} \text{ or } M^T M = I_n.$$

This equation shows that the eigenvectors of a symmetric matrix are orthogonal. This also means that the modal matrix A of a symmetric matrix A is an orthogonal matrix if the eigenvectors of A have been normalized to have unit length.

The Inverse of A

If the modal matrices M and N of a matrix A and its eigenvalues are known, then a simple relation may be obtained for A^{-1}. To obtain this expression, we write (4.8) in the form

$$A = MDM^{-1} = MDN^T.$$

If the inverse is now taken, the result is

$$A^{-1} = (N^T)^{-1} D^{-1} M^{-1} = MD^{-1} N^T$$

It should be noted that the inverse of a diagonal matrix is simply a diagonal matrix whose diagonal elements are the reciprocals of the original matrix. It is also important to remember that all of the above steps were developed under the assumption that the eigenvalues of A were distinct. In a practical sense, this last equation is not very useful for obtaining the inverse of a matrix unless a complete eigenvalue analysis of the matrix A has been performed.

5. EIGENVALUES OF A REAL SYMMETRIC MATRIX ARE REAL

Since the eigenvalues of a matrix are the roots of its characteristic equation, which is, in general, a high-degree algebraic equation, it might be supposed that some of the eigenvalues might be complex numbers. It will now be proved that if the matrix A is real and symmetric so that $A^T = A$, then the eigenvalues μ_i are real numbers.

To see this, let X_i be an eigenvector of the matrix A associated with the eigenvalue μ_i. By definition, we have

(5.1) $$AX_i = \mu_i X_i$$

If μ_i is a complex number, and if the elements of the matrix A are real numbers, then the elements of X_i must be complex numbers. Now let all the complex quantities of (5.1) be replaced by their complex conjugates. If this is done, (5.1) becomes

(5.2) $$A\bar{X}_i = \bar{\mu}_i \bar{X}_i,$$

where the bar is used to denote the complex conjugate.

We now premultiply (5.1) by $\bar{X}_i^T$ and (5.2) by X_i^T and thus obtain

(5.3) $$\bar{X}_i^T A X_i = \mu_i \bar{X}_i^T X_i$$

and

(5.4) $$X_i^T A \bar{X}_i = \bar{\mu}_i X_i^T \bar{X}_i.$$

Taking the transpose of Equation (5.4) gives the result

(5.5) $$\bar{X}_i^T A X_i = \bar{\mu}_i \bar{X}_i^T X_i,$$

where use has been made of the fact that A is symmetric; hence, $A^T = A$. Now subtract equation (5.3) from equation (5.5) and obtain

(5.6) $$0 = (\bar{\mu}_i - \mu_i)\bar{X}_i^T X_i.$$

Now, $\overline{X}_i^T X_i = |X_1|^2 + |X_2|^2 + \cdots + |X_n|^2$ is the sum of positive quantities and cannot be zero since the eigenvectors themselves cannot be zero; hence, (5.6) is satisfied only if

(5.7) $$\mu_i = \overline{\mu_i}$$

Since μ_i is equal to its complex conjugate, it must be real. If the eigenvalue μ_i is real, then it follows that in order for (5.1) and (5.2) to be satisfied, we must have

(5.8) $$X_i = \overline{X_i},$$

and thus the elements of the eigenvector X_i must be real numbers. Since the above argument holds for any eigenvalue and its associated eigenvector of the matrix A, it therefore follows that all the eigenvalues and eigenvectors of the real symmetric matrix A are real. This is not generally true if the matrix A were real but not symmetric.

EXAMPLE 4

Verify equation (4.8) for the matrix of examples 1 and 2 above.

SOLUTION

From examples 1 and 2, the matrix and its pair of associated eigenvalues and eigenvectors are

$$A = \begin{bmatrix} 21 & 3 \\ 3 & 15 \end{bmatrix}$$

$$\mu_1 = 18 + 3\sqrt{2}, \quad X_1 = \begin{bmatrix} \dfrac{1}{\sqrt{2} - 1} \\ 1 \end{bmatrix}$$

$$\mu_2 = 18 - 3\sqrt{2}, \quad X_2 = \begin{bmatrix} \dfrac{-1}{\sqrt{2} + 1} \\ 1 \end{bmatrix}.$$

The modal matrix of A is

$$M = \begin{bmatrix} \dfrac{1}{\sqrt{2} - 1} & \dfrac{-1}{\sqrt{2} + 1} \\ 1 & 1 \end{bmatrix}.$$

If we now compute the expression $M^{-1}AM$, we find

$$M^{-1}AM = \begin{bmatrix} 18 + 3\sqrt{2} & 0 \\ 0 & 18 - 3\sqrt{2} \end{bmatrix},$$

which agrees with equation (4.8).

Finally, if we normalize[33] the eigenvectors, the new vectors will be

$$Y_1 = \frac{X_1}{||X_1||} = \begin{bmatrix} \dfrac{1}{\sqrt{2}\sqrt{2} - \sqrt{2}} \\ \dfrac{\sqrt{2} - 1}{\sqrt{2}\sqrt{2} - \sqrt{2}} \end{bmatrix}$$

$$Y_2 = \frac{X_2}{||X_2||} = \begin{bmatrix} \dfrac{-1}{\sqrt{2}\sqrt{2} + \sqrt{2}} \\ \dfrac{\sqrt{2} + 1}{\sqrt{2}\sqrt{2} + \sqrt{2}} \end{bmatrix}.$$

The modal matrix constructed from these normalized eigenvectors will be an orthogonal matrix and have the property that $M^{-1} = M^T$.

EXERCISES

1.   For the orthonormal eigenvectors of the matrix A determined in example 5, verify that the modal matrix constructed from these vectors is an orthogonal matrix.

---

[33]   As previously defined, the L_2 norm must be used.

Find the eigenvalues and eigenvectors for the following matrices.

2*. $\begin{bmatrix} 5 & -3 \\ -3 & 2 \end{bmatrix}$ 3*. $\begin{bmatrix} -5 & 2 & 1 \\ 2 & -4 & -2 \\ 1 & -2 & -3 \end{bmatrix}$ 4*. $\begin{bmatrix} 4 & -2 & 2 \\ -2 & -5 & 1 \\ 2 & 1 & 3 \end{bmatrix}$ 5*. $\begin{bmatrix} -6 & 3 & 2 \\ 3 & 0 & -2 \\ 2 & -2 & 4 \end{bmatrix}$

6*. $\begin{bmatrix} 1 & -1 & 0 & 3 \\ -1 & 2 & -2 & 1 \\ 0 & -2 & 0 & 5 \\ 3 & 1 & 5 & -1 \end{bmatrix}$

***

6. HERMITIAN MATRICES

Sometimes it is necessary to determine the eigenvalues and eigenvectors of matrices whose elements are complex numbers. In quantum theory, for example, one frequently encounters matrices whose elements are complex numbers but which have the property that

(6.1) $$A^T = \overline{A}.$$

That is, the elements of A are such that

$$a_{ji} = \overline{a}_{ij}.$$

Matrices which satisfy this condition are called Hermitian matrices as was already defined in chapter I, section 11. By an argument similar to that of section 4 for symmetric matrices, it can be shown that the eigenvalues of a Hermitian matrix are real numbers. In this case, if X_i is an eigenvector of A associated with an eigenvalue μ_i, we have

(6.2) $$AX_i = \mu_i X_i.$$

If the complex conjugate of equation (6.2) is taken, the result is

(6.3) $$\overline{A}\,\overline{X}_i = \overline{\mu}_i \overline{X}_i.$$

We now premultiply (6.2) by $\overline{X}_i^T$ and (6.3) by X_i^T and thus obtain

(6.4) $$\overline{X}_i^T A X_i = \mu_i \overline{X}_i^T X_i$$

and

(6.5) $$X_i^T \overline{A} \overline{X_i} = \overline{\mu}_i X_i^T \overline{X_i} .$$

Now take the transpose of (6.5). This gives

(6.6) $$X_i^T \overline{A}^T X_i = \overline{\mu}_i \overline{X_i}^T X_i = \overline{X_i}^T A X_i ,$$

where use has been made of the property (6.1). If now (6.4) is subtracted from (6.6), the result is

$$0 = (\overline{\mu}_i - \mu_i) \overline{X_i}^T X_i .$$

Since $\overline{X_i}^T X_i \neq 0,$ it follows that

$$(\overline{\mu}_i - \mu_i) = 0,$$

and therefore, the eigenvalues of the Hermitian matrix are real. The eigenvectors X_i are complex vectors. Since $\overline{A} = A^T$, it follows that

$$\text{Det } \overline{A} = \text{Det } A^T = \text{Det } A.$$

Hence, since the determinant of $\overline{A}$ is equal to the determinant of A, it follows that the determinant of A must be a real number. Hermitian matrices correspond in the complex domain to symmetric matrices in the real domain. If A is Hermitian, it follows from the fact that $A^T = \overline{A}$ that if M is the modal matrix of A and N is the modal matrix of A^T, then

$$N = \overline{M}$$

and

$$M^T \overline{M} = I_n.$$

EXAMPLE 5

Show that the following matrix is Hermitian, and find its eigenvalues and eigenvectors.

$$A = \begin{bmatrix} -2 & 2-j \\ 2+j & 2 \end{bmatrix}$$

SOLUTION

Taking the complex conjugate of A gives

$$\overline{A} = \begin{bmatrix} -2 & 2+j \\ 2-j & 2 \end{bmatrix}.$$

Since this is also the transpose of A, the given matrix is Hermitian.

Now the characteristic matrix of A is

$$K = \mu I_2 - A = \begin{bmatrix} \mu + 2 & -2+j \\ -2-j & \mu - 2 \end{bmatrix}.$$

The characteristic equation is

$$p(\mu) = \text{Det } K = \begin{vmatrix} \mu+2 & -2+j \\ -2-j & \mu-2 \end{vmatrix} = \mu^2 - 9 = 0.$$

The roots of this equation are the eigenvalues $\mu_1 = 3$, $\mu_2 = -3$.

To find the eigenvectors, we must find nontrivial solutions for the homogeneous equation $KX_i = 0$. Using the notation of section 2, these equations may be written as follows.

$$\begin{bmatrix} \mu_i + 2 & -2+j \\ -2-j & \mu_i - 2 \end{bmatrix} \begin{bmatrix} x_1^i \\ x_2^i \end{bmatrix} = \begin{bmatrix} 0 \\ 0 \end{bmatrix} \text{ where } i = 1, 2.$$

For the first eigenvalue, $\mu_1 = 3$, we have

$$\begin{bmatrix} 3+2 & -2+j \\ -2-j & 3-2 \end{bmatrix} \begin{bmatrix} x_1^i \\ x_2^i \end{bmatrix} = \begin{bmatrix} 5 & -2+j \\ -2-j & 1 \end{bmatrix} \begin{bmatrix} x_1^i \\ x_2^i \end{bmatrix} = \begin{bmatrix} 0 \\ 0 \end{bmatrix}.$$

For convenience, set $x_2' = 1$ and solve for x_1' from the first equation. This yields $x_1' = (1/5)(2-j)$. The first eigenvector is now

$$X_1 = \begin{bmatrix} \dfrac{2-j}{5} \\ 1 \end{bmatrix}.$$

In the same manner, we find the eigenvector associated with the second eigenvalue, $\mu_2 = -3$. The result is

$$X_2 = \begin{bmatrix} -2+j \\ 1 \end{bmatrix}.$$

EXERCISES

1.* Verify equation (4.8) for the matrix of example 6.

2.* Show that the matrix below is Hermitian, and determine the eigenvalues and eigenvectors.

$$M = \begin{bmatrix} 3 & 1+j \\ 1-j & 2 \end{bmatrix}$$

3.* Show that the following matrix is Hermitian, and find the characteristic equation.

$$M = \begin{bmatrix} 2 & -2j & -1+j \\ 2j & -3 & 2j \\ -1-j & -2j & -1 \end{bmatrix}$$

4.* Find the eigenvalues and eigenvectors for the matrix of exercise 3 above.

5.* Find the eigenvalues and eigenvectors for the matrix below.

$$\begin{bmatrix} 1 & 1+j & 2-j & 1+2j \\ 1-j & 0 & 1-2j & 3-4j \\ 2+j & 1+2j & -1 & 4+3j \\ 1-2j & 3+4j & 4-3j & 2 \end{bmatrix}$$

***

7. LOCATION OF THE EIGENVALUES OF A MATRIX

At various times in analysis, it is useful to know estimates or bounds for the eigenvalues of a matrix. One possible application would be in providing fairly accurate initial estimates of eigenvalues for an iterative procedure to guarantee or increase the rate of convergence.

One common method[34] of obtaining estimates or bounds of eigenvalues is to begin with the eigenvalue equation

$$AX = \mu X.$$

When expanded, this system can be written in the form

(7.1)
$$
\begin{aligned}
(\mu - a_{11})x_1 - a_{12}x_2 - \ldots - a_{1n}x_n &= 0 \\
-a_{21}x_1 + (\mu - a_{22})x_2 - \ldots - a_{2n}x_n &= 0 \\
&\ldots \\
-a_{n1}x_1 - a_{n2}x_2 - \ldots + (\mu - a_{nn})x_n &= 0
\end{aligned}
$$

Let us now consider a given eigenvector X associated with a particular eigenvalue. Also suppose that the kth element, x_k, of this eigenvector has the largest magnitude, i.e.,

(7.2)
$$\left|\frac{x_m}{x_k}\right| \le 1, \quad m = 1, 2, \ldots, n .$$

The kth equation of (7.1) is

$$-a_{k1}x_1 - \ldots - a_{k,k-1}x_{k-1} + (\mu - a_{kk})x_k - a_{k,k+1}x_{k+1} - \ldots - a_{kn}x_n = 0$$

Rearranging terms, we may write

(7.3) $(\mu - a_{kk}) = a_{k1}\dfrac{x_1}{x_k} + \cdots + a_{k,k-1}\dfrac{x_{k-1}}{x_k} + a_{k,k+1}\dfrac{x_{k+1}}{x_k} + \cdots + a_{kn}\dfrac{x_n}{x_k}$.

Since in general we may be dealing with complex quantities, we take the magnitude of (7.3). After applying the triangle inequality, i.e., $|z_1 + z_2| \le |z_1| + |z_2|$, equation (7.3) becomes

(7.4) $|\mu - a_{kk}| \le |a_{k1}|\left|\dfrac{x_1}{x_k}\right| + \cdots + |a_{k,k-1}|\left|\dfrac{x_{k-1}}{x_k}\right| + |a_{k,k+1}|\left|\dfrac{x_{k+1}}{x_k}\right| + \cdots + |a_{kn}|\left|\dfrac{x_n}{x_k}\right|$.

---

[34]   The material in this section is based on the Gerschgorin Circle Theorem, which may be found in Golub and Van Loan, 1989, p. 341.

By virtue of (7.2), equation (7.4) may be written

(7.5)
$$|\mu - a_{kk}| \le |a_{k1}| + \cdots + |a_{k,k-1}| + |a_{k,k+1}| + \cdots + |a_{kn}|$$

$$\le \sum_{\substack{i=1 \\ (i \ne k)}}^{n} |a_{ki}|$$

Equation (7.5) states that an eigenvalue of A will lie within a circular disk centered at the point a_{kk} with a radius of r_k.

$$r_k = \sum_{\substack{i=1 \\ (i \ne k)}}^{n} |a_{ki}|$$

From this we may make the following assertion. For any matrix of order n, the n eigenvalues lie within the n disks whose centers are given by the elements of the main diagonal of the matrix. The radius of each disk is given by the sum of the magnitudes of all the off-diagonal elements in each respective row. This result is illustrated in figure 5.4 below. The reader should note that more than one eigenvalue can lie inside a given disk.

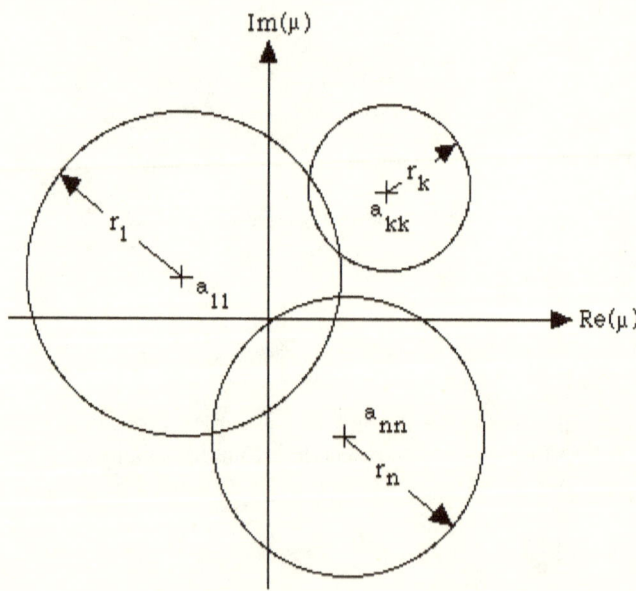

Fig. 5.4

If the matrix is real and symmetric, then the eigenvalues are real, and the above statement may be altered as follows.

Let

(7.6)
$$c_k = a_{kk} \quad \text{and} \quad l_k = \sum_{\substack{i=1 \\ (i \neq k)}}^{n} |a_{ki}|$$

The real eigenvalues will lie in the closed intervals

$$c_k - l_k \leq \mu \leq c_k + l_k, \ k = 1, 2, \ldots, n.$$

EXAMPLE 6

Determine estimates for the values of the eigenvalues for the following matrix.

$$A = \begin{bmatrix} 21 & 3 & 1 \\ 3 & 15 & 4 \\ 1 & 4 & 10 \end{bmatrix}$$

SOLUTION

From equation (7.6), we have

$$c_1 = |a_{11}| = 21, \ l_1 = |3| + |1| = 4$$
$$c_2 = 15, \ l_2 = 7$$
$$c_3 = 10, \ l_3 = 5$$

The bounding intervals for the eigenvalues are

$$17 \leq \mu \leq 25, \ 8 \leq \mu \leq 22, \ 5 \leq \mu \leq 15.$$

To one decimal place of accuracy, the eigenvalues are 22.9, 14.3, and 6.8, which can be seen to fall within the bounding intervals.

EXERCISES

Apply the method of this section to estimate the location (bounds) for the eigenvalues of the following matrices. Determine the actual eigenvalues to evaluate your results.

1. $\begin{bmatrix} 3 & -1 \\ -1 & 3 \end{bmatrix}$
2. $\begin{bmatrix} -5 & 3 \\ 3 & -5 \end{bmatrix}$
3. $\begin{bmatrix} 3 & 1+j \\ 1-j & 2 \end{bmatrix}$
4*. $\begin{bmatrix} 1 & 1 & 1 \\ 1 & 2 & 2 \\ 1 & 2 & 3 \end{bmatrix}$

5*. $\begin{bmatrix} 2 & 1 & 1 \\ 1 & 2 & 1 \\ 1 & 1 & 2 \end{bmatrix}$
6*. $\begin{bmatrix} -2 & 1 & 1 \\ 2 & -3 & 1 \\ 2 & -1 & -1 \end{bmatrix}$
7*. $\begin{bmatrix} 5 & 2 & 1 \\ 1 & 4 & -1 \\ -1 & -2 & 3 \end{bmatrix}$

8.* $\begin{bmatrix} 1 & 1-2j & 2+j \\ 1+2j & -2 & 3-2j \\ 2-j & 3+2j & 1 \end{bmatrix}$

***

8. ORTHOGONAL TRANSFORMATIONS

A particular kind of transformation that plays a prominent role in applied mathematics is the orthogonal transformation. Geometrically, this transformation in the plane corresponds to a rigid rotation about the origin or reflection through a line of a Cartesian coordinate system. An example of an orthogonal transformation in the plane has already been mentioned. The three-dimensional case will now be considered, and the treatment will later be extended to the general case of n dimensions.

Let it be supposed that all points P in three-dimensional space are subjected to the same rotation about the origin of a Cartesian coordinate system. Each point P will be displaced to a corresponding point P', and the origin alone will remain fixed. If x_1, x_2, x_3 are the coordinates of the point P, and y_1, y_2, y_3 the coordinates of the point P', we may construct a column matrix or vector X, whose elements are the coordinates of P, and similarly another vector Y. The vector X and the vector Y will be related in a definite way in any given rotation. The relationship expresses the orthogonal transformation.

$$Y = TX$$

Instead of supposing that the points are rotated about the origin O, we may equivalently imagine that the points of space remain fixed but that we rotate the coordinate axes rigidly around the origin to some new position. If we call Ox_1, Ox_2, Ox_3 the original axes, the rotated axes will be represented by Oy_1, Oy_2, Oy_3. If this rotation of axes is assumed, a point P whose coordinates are x_1, x_2, x_3 in the original system of axes will have coordinates y_1, y_2, y_3 in the rotated system. For the applications we have in mind, it will be convenient to interpret an orthogonal transformation as a rotation of axes.

In order to understand the analytical properties of orthogonal transformation and of their matrices, consider first the general linear transformation between the variables x and the variables y. This transformation may be written in the form

(8.1)
$$\begin{aligned}
x_1 &= s_{11}y_1 + s_{12}y_2 + s_{13}y_3 \\
x_2 &= s_{21}y_1 + s_{22}y_2 + s_{23}y_3 \\
x_3 &= s_{31}y_1 + s_{32}y_2 + s_{33}y_3,
\end{aligned}$$

where the elements s_{ik} are constants. This transformation relates the original coordinates x with the rotated coordinates y. The transformation matrix, (8.1), is

$$S = \begin{vmatrix} s_{11} & s_{12} & s_{13} \\ s_{21} & s_{22} & s_{23} \\ s_{31} & s_{32} & s_{33} \end{vmatrix}.$$

Now since the transformation (8.1) represents a rigid rotation from the coordinates x to the coordinates y, the square of the distance from the origin O to the point P must be the same when expressed in terms of either coordinate system. This is

(8.2) $$X^TX = x_1^2 + x_2^2 + x_3^2 = y_1^2 + y_2^2 + y_3^2 = Y^TY$$

as seen in figure 5.5 below.

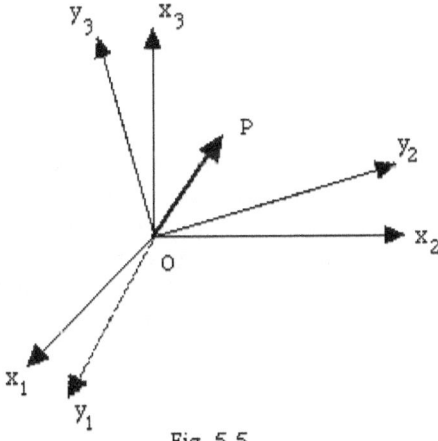

Fig. 5.5

In terms of the transformation matrix S, the linear transformation (8.1) may be expressed in the compact form

(8.3) $$X = SY.$$

Hence, (8.2) can be expressed in terms of the transformation matrix S by the equation

$$X^T X = Y^T S^T S Y = Y^T Y.$$

This equation is true for all Y, and hence, the transformation matrix S satisfies the relation

(8.4) $$S^T S = I, \text{ or } S^{-1} = S^T.$$

That is, if S is to represent a rotation from the original coordinates x to the new coordinates y, it must satisfy the relation $S^{-1} = S^T$.

The transformation (8.3) is then said to be an orthogonal transformation, and the transformation matrix S is called an orthogonal matrix. The numerical values of the elements of S will of course depend on the particular rotation performed, but in all cases, provided we are dealing with a rotation of the coordinates and hence with orthogonal matrices, the general relations (8.4) are satisfied. As a consequence of (8.4), we see that

$$(\text{Det } S^T)(\text{Det } S) = (\text{Det } S)^2 = 1.$$

Hence,
$$\text{Det } S = \pm 1.$$

The possibility of the determinant of S having the value -1 implies that an orthogonal transformation may also represent a rotation followed by a reflection since this transformation still keeps the distance from the origin to a given point invariant to the transformation.

If (8.1) is an orthogonal transformation, the geometrical significance of the elements s_{ik} is easily obtained. The elements of the first row of S, s_{11}, s_{12}, s_{13} are the cosines of the angles that the original axis Ox_1 makes with the three new axes Oy_1, Oy_2, Oy_3. Those elements in the first column, namely s_{11}, s_{21}, s_{31}, are the cosines of the angles that the new axis Oy_1 makes with the three original axes Ox_1, Ox_2, Ox_3. The other elements may be obtained in a similar manner.

The properties of orthogonal matrices may be extended to matrices having any number of rows and columns. Thus, if we have a linear transformation connecting a set of n variables $x_1, x_2, \ldots, x_n$ with a set of n variables $y_1, y_2, \ldots, y_n$, the geometrical representation of the transformation will be a rotation of the axes in a space of n dimensions. In this case, the transformation matrix S will still satisfy the relation $S^{-1} = S^T$ but will contain n rows and n columns.

EXERCISES

1.   Consider three sets of Cartesian coordinates (x_1, x_2, x_3), (y_1, y_2, y_3), and (z_1, z_2, z_3). Let (y_1, y_2, y_3) be obtained from (x_1, x_2, x_3) by a rotation through an angle θ about the x_3, y_3 axis as shown in figure 5.6 below.

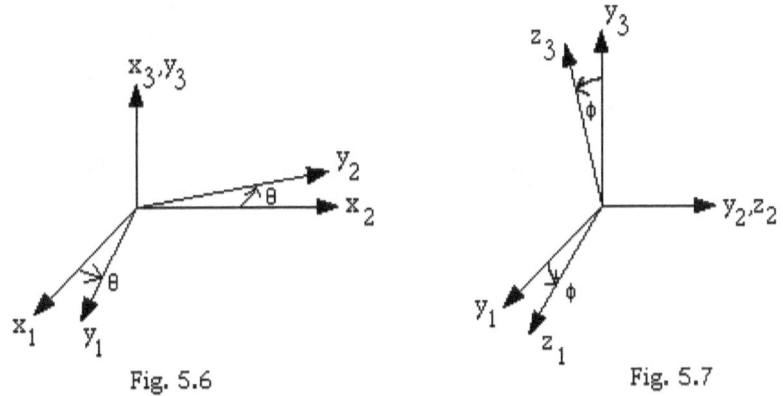

Fig. 5.6   Fig. 5.7

Also let (z_1, z_2, z_3) be generated form (y_1, y_2, y_3) by a rotation through an angle ϕ about the y_2, z_2 axis as shown in figure 5.7. Derive the rotation matrices $R(\theta)$ and $R(\phi)$ such that

$$Y = R(\theta)X \text{ and } z = R(\phi)Y.$$

2.   For the same situation as given in exercise 1 above, demonstrate geometrically that the transformation from X to Z is given by the product $R(\phi)R(\theta)$, that is, if $Z = MX$, then $M = R(\phi)R(\theta)$.

***

9.   SYMMETRIC MATRICES AND QUADRATIC FORMS

Let the following general symmetric matrix with real elements be given:

$$A = \begin{bmatrix} a_{11} & a_{12} & a_{13} \\ a_{12} & a_{22} & a_{23} \\ a_{13} & a_{23} & a_{33} \end{bmatrix}$$

This symmetric matrix may be associated with the following quadratic form in the three variables x_1, x_2, x_3,

$$X^T A X = a_{11}x_1^2 + a_{22}x_2^2 + a_{33}x_3^2 + 2a_{12}x_1x_2 + 2a_{13}x_1x_3 + 2a_{23}x_2x_3,$$

where X is a column vector whose elements are the three variables x_1, x_2, x_3. From the relation above, we conclude that an appropriate symmetric matrix may be connected with each quadratic form. If the matrix is given, the quadratic form is determined and vice versa.

It is advantageous to give a geometric interpretation of a quadratic form. Let the variable x_1, x_2, x_3 be the Cartesian coordinates of a point in space, and let us consider the equation obtained by equating the quadratic form to the number one. This equation is

(9.1) $X^T A X = 1.$

It can be shown that the points P, whose coordinates x_1, x_2, x_3 satisfy (9.1), lie on a quadratic surface having the origin as its center. The precise nature of the quadratic surface depends on the values of the elements a_{ik}.

If the determinant of A does not vanish, the quadratic surface is an ellipsoid or a hyperboloid situated in ordinary three-dimensional space. The determinant of the matrix A is called the discriminant of the quadratic form. If Det $A = 0$, the quadratic surface (9.1) degenerates into a cylinder of the elliptic or hyperbolic type or else into two parallel planes symmetrically situated with respect to the origin.

Let it be supposed that the quadratic surface (9.1) is the equation of an ellipsoid as shown in figure 5.8. This ellipsoid having the point O as center will, in general, have three principal axes. The principal axes of the ellipsoid are the three mutually perpendicular axes of symmetry that pass through the center O. It will be assumed that the principal axes differ in length. In this case, their orientation is uniquely determined.

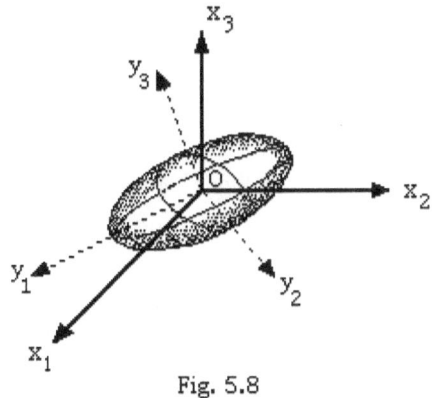

Fig. 5.8

The Transformation of a Quadratic Form into a Sum of Squares by Means of an Orthogonal Transformation

Let it be supposed that we wish to determine the lengths and orientation of the principal axes of the ellipsoid defined by equation (9.1). To do so, we shall introduce a new set of axes y by an orthogonal transformation (8.4) so that the axes Oy_1, Oy_2, Oy_3 coincide with the principal axes of the ellipsoid. Now if the new axes coincide with the principal axes of the ellipsoid, its surface will be symmetrically situated with respect to them. This symmetry requires that the quadratic form, which enters into the equation of the ellipsoid in terms of these new axes, should degenerate into a sum of squares. Hence, the equation of the ellipsoid in the new axes should have the form

(9.2) $\qquad b_1 y_1{}^2 + b_2 y_2{}^2 + b_3 y_3{}^2 - 1 = Y^T B Y,$

where

$$B = \begin{bmatrix} b_1 & 0 & 0 \\ 0 & b_2 & 0 \\ 0 & 0 & b_3 \end{bmatrix}.$$

The quantities b_1, b_2, b_3 are positive constants. They will be shown to be the eigenvalues of the matrix A of (9.1). The lengths of the principal axes of the ellipsoid d_i are given by

(9.3) $$d_1 = \frac{2}{\sqrt{b_1}}, \quad d_2 = \frac{2}{\sqrt{b_2}}, \quad d_3 = \frac{2}{\sqrt{b_3}}.$$

In order to obtain b_1, b_2, b_3 we introduce the rotated axes by means of the orthogonal transformation

$$X = SY.$$

We now substitute this relation into (9.1) and thus obtain

$$Y^T S^T A S Y = 1.$$

If the axes have been properly rotated so that they coincide with the principal axes of the ellipsoid, then this relation should have the same form as (9.2) so that

(9.4) $$S^T A S = S^{-1} A S = B,$$

where use has been made of the orthogonal relation $S^T = S^{-1}$. If equation (9.4) is premultiplied by S, the resulting equation is

(9.5) $$AS = SB.$$

Now let the transformation matrix S be partitioned into three columns so that

$$S = [S_1, S_2, S_3],$$

where S_1, S_2, and S_3 are the three columns of the square matrix S. Since B is a diagonal matrix, it can be seen that equation (9.5) is equivalent to the three equations,

(9.6)
$$\begin{aligned}
AS_1 &= b_1 S_1, \\
AS_2 &= b_2 S_2, \\
AS_3 &= b_3 S_3.
\end{aligned}$$

The form of these equations indicates that the vectors S_1, S_2, and S_3 are eigenvectors of the matrix A and that the numbers b_1, b_2 and b_3 are the eigenvalues of A and are the three roots of the characteristic equation of A, which may be written in the form

(9.7) $p(b) = \text{Det}\,(bI - A) = 0.$

The characteristic equation of A, in general, is a cubic equation. If equation (9.7) has three distinct roots b_1, b_2, and b_3, then the quadratic surface (9.2) has three principal axes whose lengths are given by (9.3). The matrix A and equation (9.6) fix the directions of the vectors S_i, and the equation $S^T S = I$ or

(9.8) $S_i^T S_i = 1,\, i = 1, 2, 3$

determines the length of the vectors S_1, S_2, and S_3 . The directions of the y_1, y_2, and y_3 axes may be determined with respect to the coordinates x_1, x_2, and x_3 by means of the equation

$$X = SY.$$

For example, the line for the y_1 axis for which $y_2 = 0$ and $y_3 = 0$ may be expressed in terms of the x coordinates by writing equation (9.8) in the form

$$\begin{bmatrix} x_1 \\ x_2 \\ x_3 \end{bmatrix} = \begin{bmatrix} s_{11} & s_{12} & s_{13} \\ s_{21} & s_{22} & s_{23} \\ s_{31} & s_{32} & s_{33} \end{bmatrix} \begin{bmatrix} y_1 \\ 0 \\ 0 \end{bmatrix}.$$

If this expression is expanded, the result may be written in the form

(9.9) $X_1 = y_1 S_1.$

In this equation, we have used the subscript $_1$ on the X vector to denote that it is associated with the y_1 axis. Equation (9.7) shows that the axis y_1 has the direction of the first column vector S_1 of the transformation matrix S. It may be shown similarly that the axis y_2 has the direction of the vector S_2 and the axis y_3, the direction of the vector S_3. If (9.9) is substituted into the equation of the quadratic surface (9.1), we obtain

$$X_1^T A X_1 = y_1 S_1^T A S_1 y_1 = y_1^2 S_1^T A S_1 = 1.$$

But from (9.6), $A S_1 = b_1 S_1$, and hence the previous result can be written as

(9.10) $b_1 y_1^2 S_1^T S_1 = b_1 y_1^2 = 1$

Equation (9.10) shows that the axis y_1 intersects the quadratic surface at the distance $1/R$—b_1 from the origin.

Thus far, we have assumed that the principal axes of the ellipsoid are of unequal length. When two or all three of the principal axes of the ellipsoid have the same length (that is, when two or all three of the eigenvalues b_1 are equal), the orthogonal matrix S becomes indeterminate.

The geometrical picture enables us to understand the reason for this indeterminateness. If, for example, the ellipsoid is a spheroid and the axis Ox_1 is rotated to bring it into position Oy_1, which coincides with the axis of symmetry of the spheroid, then the axes Oy_2 and Oy_3 will always coincide with two other principal axes no matter how much the coordinate axes are pivoted about Oy_1.

In the case that the ellipsoid has all three eigenvalues equal to each other, it degenerates into a sphere. In this case, any three mutually perpendicular diameters may serve as the principal axes of that sphere.

In the special example that has been discussed, the analysis was applied to three-dimensional ellipsoids and rotations in three-dimensional space. Exactly the same problem can be considered in a space of any number n of dimensions, although the geometrical representation now fails us unless we are able to visualize ellipsoids and rotations in a space on n dimensions. From the standpoint of matrix algebra, the problem consists in transforming a symmetric matrix of order n into a diagonal one by means of an orthogonal transformation, $S^T A S = B$, where B is an nth-order diagonal matrix whose elements are the eigenvalues of A. In this case, the eigenvalues are the roots of an algebraic equation of the nth degree and are real numbers.

The existence of multiple roots does not invalidate the existence of n distinct and mutually perpendicular axes but only indicates that some of these axes are no longer uniquely determined but can be replaced by other equally valid axes. The multiplicity of the roots of the characteristic equation of the matrix A introduces certain difficulties in the numerical calculations of the problem. If certain eigenvalues are almost equal to each other without collapsing into one, the associated principal axes or eigenvectors are still theoretically determined. However, to find these axes with any degree of accuracy becomes increasingly difficult as the corresponding eigenvalues approach equality.

Tangent Lines and Tangent Planes

Consider a quadratic curve or surface given by the following standard equation.

$$X^T A X = 1$$

Let a point on the curve or surface be denoted by the vector X_p. It is of interest to note that if the quadratic form is a curve, then the equation of the line tangent to the curve at the point X_p is given by

$$X^T A X_p = 1.$$

If the quadratic form is a surface, then the above relation gives the equation of the tangent plane at the given point.

EXAMPLE 7

Write the equation of a circle whose radius is on in the matrix quadratic form. Also find the equation of the tangent line at the point where the line $x_1 = x_2$ crosses the circle.

SOLUTION

In $x_1 x_2$ coordinates, the equation of a circle whose radius is one is

$$x_1^2 + x_2^2 = 1.$$

The corresponding matrix quadratic equation for this curve is

$$x_1^2 + x_2^2 = 1 = X^T \begin{bmatrix} 1 & 0 \\ 0 & 1 \end{bmatrix} X.$$

The line $x_1 = x_2$ intersects the circle where $x_1 = x_2$. The coordinates of this point may be found from the equation of the circle by placing $x_1 = x_2$. It should be noted that there are two symmetric intersection points.

$$x_1^2 + x_1^2 = 2 x_1^2 = 1$$

From this we find that $x_1 = \pm 1/\sqrt{2}$. For convenience we will take the positive root to locate the point in the first quadrant. This result means that the coordinates of the intersection point are $x_1 = 1/\sqrt{2}, x_2 = 1/\sqrt{2}$ or

$$X_p = \frac{1}{\sqrt{2}} \begin{bmatrix} 1 \\ 1 \end{bmatrix}.$$

As stated above, the equation of the tangent line at this point is given by

$$X^T \begin{bmatrix} 1 & 0 \\ 0 & 1 \end{bmatrix} X_p = X^T \begin{bmatrix} 1 & 0 \\ 0 & 1 \end{bmatrix} \frac{1}{\sqrt{2}} \begin{bmatrix} 1 \\ 1 \end{bmatrix} = 1$$

$$\text{or} \quad x_1 + x_2 = \sqrt{2}$$

We can apply a little geometric reasoning to verify this answer. The slope of the desired line must be -1, and its intercept on the x_2 axis is $\sqrt{2}$. Hence, the equation of the line should be $x_2 = \sqrt{2} - x_1$. This agrees exactly with the above result.

EXERCISES

1.*   Given the quadratic surface

$$4x_1{}^2 + 4x_2{}^2 + 2x_3{}^2 - 4x_1x_2 + 4x_1x_3 + 4x_2x_3 = 1,$$

determine the associated matrix M of this quadratic surface. By introducing an orthogonal transformation $Y = SX$, determine the directions and lengths of the principal axes of the quadratic surface.

2.   A two-dimensional ellipse whose semimajor and semiminor axes are 2 and 1 respectively is defined by the following quadratic form.

$$X^T \begin{bmatrix} \frac{1}{4} & 0 \\ 0 & 1 \end{bmatrix} X = 1$$

Verify that the line $x_2 = x_1$ intersects the ellipse at the point $\left(2/\sqrt{5}, 2/\sqrt{5}\right)$ and that the equation of the tangent line is given by

$$X^T \begin{bmatrix} \frac{1}{4} & 0 \\ 0 & 1 \end{bmatrix} X_p = 1 \text{ , where } X_p = \frac{2}{\sqrt{5}} \begin{bmatrix} 1 \\ 1 \end{bmatrix}$$

3.*   For the quadratic surface given in exercise 1 above, determine where the line $x_1 = x_2 = x_3$ pierces the surface, and find the equation of the plane tangent to the surface at this point.

***

10. COMMUTATIVE MATRICES

It has been seen that, in general, the multiplication of matrices is not a commutative operation, but it has been mentioned that in special cases, matrices commute; for example, two diagonal matrices of the same order do. The general rule that determines under what conditions matrices will commute may be given a simple geometric interpretation when the matrices considered are real and symmetric.

Thus, let A and B be two symmetric matrices that we may suppose to be of order 3. In this case, we may associate with each matrix a corresponding quadratic surface in a space of three dimensions. Also assume that the quadratic surfaces are ellipsoids. The geometric rule for the possibility of commutation of two matrices in multiplication states the following:

Two symmetric matrices of the same order will always commute in multiplication when the principal axes of their associated ellipsoids coincide. The matrices will not commute when these axes do not coincide.

To see this, let the quadratic surfaces associated with the matrices A and B be written in the form

$$X^T A X = 1$$

and

$$X^T B X = 1.$$

If the principal axes of these two quadratic surfaces coincide, there must exist an orthogonal matrix S that transforms both A and B to the diagonal forms

$$S^{-1} A S = D_1$$

and

$$S^{-1} B S = D_2,$$

where D_1 and D_2 are diagonal matrices whose elements are the eigenvalues of A and B [see equation (9.4)]. By premultiplication of these equations by S and postmultiplication by S^{-1}, they can be written in the form

$$A = S D_1 S^{-1}$$

and

$$B = SD_2S^{-1}.$$

We now form the matrix product

$$AB = (SD_1S^{-1})(SD_2S^{-1})$$
$$= SD_1D_2S^{-1}$$

On the other hand, the product $[B][A]$ is

$$BA = SD_2D_1S^{-1}$$

Since D_1 and D_2 are diagonal matrices, they commute in multiplication so that $D_1D_2 = D_2D_1$, and the right-hand members of both the above equations are equal to each other; hence,

$$AB = BA.$$

This proves the rule. It is interesting to see how the rule of commutation applies in special cases.

Diagonal Matrices

Diagonal matrices may be proved to commutate by direct multiplication; however, it is interesting to apply the commutation rule to them. It has been seen that a diagonal matrix can be associated with a quadratic surface whose center is at the origin and whose principal axes coincide with the axes of coordinates. Then we consider two diagonal matrices of the same order. Since the principal axes of their associated quadratic surfaces coincide with the coordinate axes, these principal axes necessarily coincide, and our geometric rule indicates that the two diagonal matrices commute.

The fact that a diagonal matrix cannot commute with an arbitrary matrix is evident since the principal axes of an arbitrary matrix do not coincide with the coordinates axes, and hence, its axes do not coincide with those of the diagonal matrix. The sole exception to this is the unit matrix I_2.

A scalar matrix is a diagonal matrix whose elements are identical and which has consequently three identical eigenvalues. A scalar matrix is therefore associated with a sphere having the origin as its center. Since the principal diagonals of a sphere are defined by any three mutually perpendicular diameters of the sphere, it follows that a certain triad of these mutually perpendicular diameters will always coincide with the principal axes of the arbitrary ellipsoid. Since the principal axes of the two surfaces always coincide, we conclude—in accordance with the geometric rule—that a scalar matrix commutes with any arbitrary matrix.

If two symmetric matrices do not commute, they cannot be simultaneously transformed into the diagonal form by means of the same orthogonal transformation.

11. FUNDAMENTAL PROPERTIES OF THE CHARACTERISTIC DETERMINANT AND THE CHARACTERISTIC EQUATION OF A MATRIX

The determination of the eigenvalues and eigenvectors of a given square matrix is a problem of great importance in applied mathematics. In this section, several of the more important properties of the characteristic equation and the characteristic determinant will be discussed. These properties are very useful in the solution of matrix eigenvalue problems. The characteristic equation of a square matrix A was defined in section 2 to be

(11.1) $$\mathrm{Det}\,(\mu I_n - A) = p(\mu) = 0.$$

The polynomial $p(\mu)$ is the characteristic polynomial of A. If A is an nth-order matrix, the polynomial $p(\mu)$ is of the nth degree and has the general form

(11.2) $$p(\mu) = \mu^n + a_{n-1}\mu^{n-1} + a_{n-2}\mu^{n-2} + \ldots + a_1\mu + a_0 = 0.$$

If the parameter μ is placed equal to zero in (11.2), the result is

$$p(0) = a_0.$$

However, as a consequence of (11.1), we have

$$p(0) = \mathrm{Det}\,(-A) = (-1)^n\,\mathrm{Det}\,A.$$

Hence, it is apparent that the coefficient a_0 of the characteristic equation is given by

(11.3) $$a_0 = (-1)^n\,\mathrm{Det}\,A.$$

Now let it be supposed that the matrix A has n distinct eigenvalues $\mu_1, \mu_2, \ldots, \mu_n$. In such a case, the characteristic equation (11.2) may be written in the factored form

(11.4) $(\mu - \mu_1)(\mu - \mu_2) \ldots (\mu - \mu_n) = p(\mu) = 0.$

Now if μ is placed equal to zero, the result is

$$p(0) = (-1)^n \mu_1 \mu_2 \ldots \mu_n = a_0 = (-1)^n \text{ Det } A.$$

Hence, we obtain the important relation that

$$\text{Det } A = \mu_1 \mu_2 \ldots \mu_n.$$

Since the determinant of the matrix A is the product of its eigenvalues, it follows that if one of the eigenvalues of A is zero, then $\text{Det } A = 0$ and the matrix A is a singular matrix. The above expression for a_0 also holds for repeated and complex eigenvalues as well.

If equation (11.4) is multiplied out in order to obtain the coefficients of the various powers of μ, it will be found that the coefficient of μ^{n-1}, which is written as a_{n-1} in (11.2), is given by

(11.5) $a_{n-1} = - (\mu_1 + \mu_2 + \ldots + \mu_n).$

If, on the other hand, the determinant $\text{Det } (\mu I_n - A) = p(\mu)$ is expanded in order to obtain the coefficient of μ^{n-1} in the polynomial $p(\mu)$, the result is

(11.6) $a_{n-1} = - (m_{11} + m_{22} + \ldots + m_{nn}) = - \text{Tr } A.$

By definition, the trace of the matrix A, $\text{Tr } A$, is the sum of the principal diagonal terms of A. If (11.5) and (11.6) are compared, it is seen that we have another important relation.

(11.7) $\text{Tr } A = (\mu_1 + \mu_2 + \ldots + \mu_n)$

If the modal matrix M of the matrix A is known, it can be seen from equation (5.4) that the matrix A can be written as

$$A = MDM^{-1},$$

where D is the diagonal matrix whose elements are the eigenvalues μ_i of the matrix A. If A is multiplied by itself p times, the result is

$$A^p = MD^p M^{-1}$$

The diagonal matrix D^p has as elements the quantities $(\mu_i)^p$, which are the eigenvalues of A raised to the power p. It therefore follows from this relation that the eigenvalues of A^p are the numbers $(\mu_1)^p$, $(\mu_2)^p$, ... , $(\mu_n)^p$, and hence, as a consequence of equation (11.7), we have

$$\mathrm{Tr}\ M^p = (\mu_1^p + \mu_2^p + ... + \mu_n^p).$$

Bocher's Formulas for the Coefficients of the Characteristic Polynomial

The mathematician M. Bocher has given some very useful relations for obtaining the coefficients a_i of the characteristic polynomial $p(\mu)$ of the matrix A in terms of the quantities $S_p = \mathrm{Tr}\ A^p$. These relations make it possible to obtain the polynomial (11.2) in a much more efficient manner than by direct expansion of the determinant $\mathrm{Det}\ (\mu I_n - A) = p(\mu)$.

If $S_p = \mathrm{Tr}\ A^p$, Bocher has shown that

$$a_{n-1} = -S_1.$$

$$a_{n-2} = -(a_{n-1}S_1 + S_2)/2$$

(11.8) $$a^{n-3} = -(a_{n-2}S_1 + a_{n-1}S_2 + S_3)/3$$

$$. . .$$

$$a_0 = -(a_1S_1 + a_2S_2 + ... + a_{n-1}S_{n-1} + S_n)/n$$

EXAMPLE 8

As an illustration of the use of these formulas, apply them to determine the characteristic polynomial for the following matrix.

(11.9) $$A = \begin{bmatrix} 2 & -1 & 0 \\ 9 & 4 & 6 \\ -8 & 0 & -3 \end{bmatrix}$$

SOLUTION

From the given matrix, we find

$$S_1 = \mathrm{Tr}\ A = 3.$$

To find S_2 and S_3, we need the square and cube of the matrix A. These are

$$A^2 = \begin{bmatrix} -5 & -6 & -6 \\ 6 & 7 & 6 \\ 8 & 8 & 9 \end{bmatrix}$$

and

$$A^3 = \begin{bmatrix} -16 & -19 & -18 \\ 27 & 22 & 24 \\ 16 & 24 & 21 \end{bmatrix}.$$

From these we find

$$S_2 = \operatorname{Tr} A^2 = 11, \text{ and } S_3 = \operatorname{Tr} A^3 = 27.$$

Then by the Bocher formulas (11.8), we have

$$a_2 = -S_1 = -3$$

$$a_1 = -(a_2 S_1 + S_2)/2$$

$$= -[(-3)(3) + 11]/2 = -1$$

$$a_0 = -(a_1 S_1 + a_2 S_2 + S_3)/3$$

$$= -[(-1)(3) + (-3)(11) + 27)/3 = 3.$$

Hence, the characteristic polynomial of the matrix (11.9) is

$$p(\mu) = \mu^3 - 3\mu^2 - \mu + 3 = (\mu - 1)(\mu + 1)(\mu - 3).$$

In this case, the characteristic polynomial can be factored as shown and the eigenvalues of A are seen to be $\mu_1 = 1$, $\mu_2 = -1$, and $\mu_3 = 3$.

$$***$$

The determinant of the matrix A of this previous example can be obtained if the characteristic polynomial is known since equation (11.3) may be written in the form

$$\operatorname{Det} A = (-1)^n a_0 = -3.$$

It is instructive to compute the modal matrix M of the matrix given in (11.9). The eigenvectors of A, X_i, satisfy the equation

(11.10) $AX_i = \mu_i X_i$.

For this example case, (11.10) has the explicit form

$$
\begin{bmatrix}
(\mu_i-2) & 1 & 0 \\
-9 & (\mu_i-4) & -6 \\
8 & 0 & (\mu_i+3)
\end{bmatrix}
\begin{bmatrix}
x_1 \\ x_2 \\ x_3
\end{bmatrix}
=
\begin{bmatrix}
0 \\ 0 \\ 0
\end{bmatrix}.
$$

If μ_i is replaced by $\mu_1 = 1$, $\mu_2 = -1$, and $\mu_3 = 3$ respectively, the following columns are obtained.

$$
X_1 = m\begin{bmatrix} 1 \\ 1 \\ -2 \end{bmatrix}, \quad
X_2 = p\begin{bmatrix} 1 \\ 3 \\ -4 \end{bmatrix}, \quad
X_3 = q\begin{bmatrix} 1 \\ -1 \\ -4/3 \end{bmatrix},
$$

where the numbers m, p, and q are arbitrary since only the directions of the vectors X_i are specified by (11.10). If we set $m = p = 1$ and $q = 3$, the above eigenvectors can be assembled to form a modal matrix M.

$$
M = \begin{bmatrix}
1 & 1 & 3 \\
1 & 3 & -3 \\
-2 & -4 & -4
\end{bmatrix}
$$

The inverse is

$$
M^{-1} = \frac{1}{4}\begin{bmatrix}
12 & 4 & 6 \\
-5 & -1 & -3 \\
-1 & -1 & -1
\end{bmatrix}.
$$

Direct multiplication gives the result

$$
D = M^{-1}AM = \begin{bmatrix}
1 & 0 & 0 \\
0 & -1 & 0 \\
0 & 0 & 3
\end{bmatrix}.
$$

The transformation relating the matrices D and M is called a similarity transformation in the mathematical literature. It is easy to see that the matrices D and A related by the above transformation have the following properties:

$$\text{Det } D = \text{Det } A$$

$$\text{Tr } D = \text{Tr } A$$

Therefore, the eigenvalues of D = eigenvalues of A.

EXERCISES

Use Bocher's formulas to obtain the characteristic polynomials for each of the following matrices.

1. $\begin{bmatrix} -5 & 3 \\ 3 & -5 \end{bmatrix}$ 2. $\begin{bmatrix} 3 & 1+j \\ 1-j & 2 \end{bmatrix}$ 3*. $\begin{bmatrix} 1 & 1 & 1 \\ 1 & 2 & 2 \\ 1 & 2 & 3 \end{bmatrix}$

5*. $\begin{bmatrix} -2 & 1 & 1 \\ 2 & -3 & 1 \\ 2 & -1 & -1 \end{bmatrix}$ 6*. $\begin{bmatrix} 5 & 2 & 1 \\ 1 & 4 & -1 \\ -1 & -2 & 3 \end{bmatrix}$

***

12. CIRCULANT MATRICES

In the study of certain dynamical systems that have circular symmetry such as balanced multiphase electrical networks, mechanical vibrations, hydrocarbon molecule vibrations, and digital communication theory, we encounter dynamical matrices that are of the circulant type. In this section, the fundamental properties of circulant matrices will be presented.

A circulant matrix of order n is a matrix of the form

$$(12.1) \qquad C = \begin{bmatrix} c_1 & c_2 & c_3 & \cdots & c_{n-1} & c_n \\ c_n & c_1 & c_2 & \cdots & c_{n-2} & c_{n-1} \\ c_{n-1} & c_n & c_1 & \cdots & c_{n-3} & c_{n-2} \\ \cdot & \cdot & \cdot & \cdots & \cdot & \cdot \\ c_2 & c_3 & c_4 & \cdots & c_n & c_1 \end{bmatrix}.$$

It will be noticed that the elements are equal in lines parallel to the leading diagonal and circulate from the end back to the beginning. This circulation insures that C commutes with the permutation matrix P, which permutes the elements in cyclic order. The permutation matrix P has the form

(12.2)
$$P = \begin{bmatrix} 0 & 1 & 0 & \cdots & 0 & 0 \\ 0 & 0 & 1 & \cdots & 0 & 0 \\ 0 & 0 & 0 & \cdots & 0 & 0 \\ \cdot & \cdot & \cdot & \cdots & \cdot & \cdot \\ 1 & 0 & 0 & \cdots & 0 & 0 \end{bmatrix}.$$

The Third-Order Case

To fix the general ideas, the third-order case will be discussed in detail. The general case of order n follows the case of order 3 in the manner by which its eigenvalues and eigenvectors are obtained. The third order circulant C has the form

$$C = \begin{bmatrix} c_1 & c_2 & c_3 \\ c_3 & c_1 & c_2 \\ c_2 & c_3 & c_1 \end{bmatrix}.$$

The third-order permutation matrix has the form

$$P = \begin{bmatrix} 0 & 1 & 0 \\ 0 & 0 & 1 \\ 1 & 0 & 0 \end{bmatrix}.$$

It may be shown by direct multiplication that

$$P^2 = \begin{bmatrix} 0 & 0 & 1 \\ 1 & 0 & 0 \\ 0 & 1 & 0 \end{bmatrix}, \quad P^3 = \begin{bmatrix} 1 & 0 & 0 \\ 0 & 1 & 0 \\ 0 & 0 & 1 \end{bmatrix} = I$$

Thus, the cube of the third-order permutation matrix is the identity matrix. As a consequence of the forms of P, P^2, and P^3, it may be seen that C can be expressed in the form

(12.3) $C = c_1 I + c_2 P + c_3 P^2.$

This shows that the third-order circulant matrix may be expressed as a second-degree polynomial of the matrix P, the permutation matrix of order 3.

The Eigenvalues and Eigenvectors of P of Order 3

The eigenvalue equation for the matrix P has the form

$$PX = \mu X,$$

where X is a third-order column vector. If the above equation is expanded, the result is

$$x_2 = \mu x_1, \ x_3 = \mu x_2, \ x_1 = \mu x_3.$$

Starting with the last equation and working backward, we have

(12.4) $$x_1 = \mu x_3 = \mu \mu x_2 = \mu^2 x_2 = \mu^2 (\mu x_1) = \mu^3 x_1,$$

which leads to

$$\mu^3 = 1$$

since we must require x_1 to be nonzero to avoid a null eigenvector. The characteristic equation of P is

$$\mu^3 - 1 = 0.$$

Therefore, the three eigenvalues of P are the three cube roots of one

(12.5)
$$\mu_1 = 1$$
$$\mu_2 = -\frac{1}{2} + j\frac{\sqrt{3}}{2}, \quad j = \sqrt{-1}$$
$$\mu_3 = -\frac{1}{2} - j\frac{\sqrt{3}}{2}$$

It is convenient to introduce the notation

$$a = \exp\left(\frac{2\pi j}{3}\right) = -\frac{1}{2} + j\frac{\sqrt{3}}{2}$$

so that (12.5) may be written as

(12.6) $$\mu_1 = 1, \ \mu_2 = a, \ \mu_3 = a^2$$

since

$$a^2 = \left[\exp\left(\frac{2\pi j}{3}\right)\right]^2 = \exp\left(\frac{4\pi j}{3}\right)$$

$$= \cos\left(\frac{4\pi}{3}\right) + j\sin\left(\frac{4\pi}{3}\right) = -\frac{1}{2} - j\frac{\sqrt{3}}{2}.$$

The Eigenvectors of P

We now return to equations (12.4). Let x_1 be an arbitrary number. Then as can be seen from (12.4), we may construct the column

$$\begin{bmatrix} x_1 \\ x_2 \\ x_3 \end{bmatrix} = \begin{bmatrix} x_1 \\ \mu x_1 \\ \mu^2 x_1 \end{bmatrix} = x_1 \begin{bmatrix} 1 \\ \mu \\ \mu^2 \end{bmatrix}.$$

Since x_1 is arbitrary let $x_1 = 1$, then for each eigenvalue μ_i, $i = 1, 2, 3$, we may construct an eigenvector X_i of the form

$$X_i = \begin{bmatrix} 1 \\ \mu_i \\ \mu_i^2 \end{bmatrix}, \quad i = 1, 2, 3.$$

Since the eigenvalues are given by (12.6), we obtain the following three eigenvectors.

$$X_1 = \begin{bmatrix} 1 \\ 1 \\ 1 \end{bmatrix}, \quad X_2 = \begin{bmatrix} 1 \\ a \\ a^2 \end{bmatrix}, \quad X_3 = \begin{bmatrix} 1 \\ a^2 \\ a \end{bmatrix}$$

The Modal Matrix of P

We may now construct the modal matrix of P as

$$S = [\, X_1 \ \ X_2 \ \ X_3 \,] = \begin{bmatrix} 1 & 1 & 1 \\ 1 & a & a^2 \\ 1 & a^2 & a \end{bmatrix}$$

The three eigenvector equations,

$$PX_i = \mu_i X_i,$$

may be written as one matrix equation

$$PS = SD = S\begin{bmatrix} 1 & 0 & 0 \\ 0 & a & 0 \\ 0 & 0 & a^2 \end{bmatrix},$$

where D is the diagonal matrix[35] of the eigenvalues of P.

If we postmultiply the last result by S^{-1}, we obtain

$$P = SDS^{-1}.$$

If P is multiplied by itself, we have

$$P^2 = (SDS^{-1})(SDS^{-1}) = SD^2S^{-1}.$$

If the above forms for P and P^2 are substituted into (12.3), we obtain,

(12.7) $\qquad C = S(c_1 I + c_2 D + c_3 D^2)S^{-1}.$

Now let

$$D_0 = c_1 I + c_2 D + c_3 D^2 =$$

$$\begin{bmatrix} c_1 + c_2 + c_3 & 0 & 0 \\ 0 & c_1 + ac_2 + a^2 c_3 & 0 \\ 0 & 0 & c_1 + a^2 c_2 + a c_3 \end{bmatrix} =$$

$$\begin{bmatrix} \alpha_1 & 0 & 0 \\ 0 & \alpha_2 & 0 \\ 0 & 0 & \alpha_3 \end{bmatrix}$$

With this notation, equation (12.7) has the form

$$C = SD_0 S^{-1}.$$

---

[35] In some applications, the diagonal matrix of eigenvalues is referred to as the spectral matrix of C.

From this result, we may see that the eigenvalues of the circulant matrix C are

$$\mu_1 = c_1 + c_2 + c_3,$$

$$\mu_2 = c_1 + ac_2 + a^2 c_3,$$

$$\mu_3 = c_1 + a^2 c_2 + ac_3.$$

The matrix S is also the modal matrix of C.

The Inverse of S

The inversion of the modal matrix S of the circulant matrix C may be carried out by the following procedure. Let us write

$$R = SV,$$

where R and V are column matrices of third order. Let Y be an arbitrary column vector of third order. We now premultiply the above by Y^T to obtain

$$Y^T R = Y^T SV.$$

Expanding, we obtain,

$$y_1 r_1 + y_2 r_2 + y_3 r_3 = \begin{bmatrix} y_1 \\ y_2 \\ y_3 \end{bmatrix}^T \begin{bmatrix} 1 & 1 & 1 \\ 1 & a & a^2 \\ 1 & a^2 & a \end{bmatrix} \begin{bmatrix} v_1 \\ v_2 \\ v_3 \end{bmatrix} .$$

This equation may be written in the form

$$(12.8) \qquad y_1 r_1 + y_2 r_2 + y_3 r_3 = \begin{bmatrix} y_1 + y_2 + y_3 \\ y_1 + ay_2 + a^2 y_3 \\ y_1 + a^2 y_2 + ay_3 \end{bmatrix}^T \begin{bmatrix} v_1 \\ v_2 \\ v_3 \end{bmatrix} .$$

If we now let $y_1 = y_2 = y_3 = 1$, (12.8) reduces to

$$r_1 + r_2 + r_3 = \begin{bmatrix} 3 & 0 & 0 \end{bmatrix} \begin{bmatrix} v_1 \\ v_2 \\ v_3 \end{bmatrix} = 3v_1 ,$$

where we have used the relation $1 + a + a^2 = 0$.

If we let $y_1 = 1$, $y_2 = a^2$, $y_3 = a$, then (12.8) reduces to

(12.10) $r_1 + a^2 r_2 + a r_3 = 3v_2.$

Finally, if we let $y_1 = 1$, $y_2 = a$, $y_3 = a^2$, (12.8) gives

(12.11) $r_1 + a r_2 + a^2 r_3 = 3v_3.$

From equations (12.9), (12.10), and (12.11), we obtain v_1, v_2, v_3. The results may be written in the form

(12.12)
$$V = \begin{bmatrix} v_1 \\ v_2 \\ v_3 \end{bmatrix} = \frac{1}{3} \begin{bmatrix} 1 & 1 & 1 \\ 1 & a^2 & a \\ 1 & a & a^2 \end{bmatrix} \begin{bmatrix} r_1 \\ r_2 \\ r_3 \end{bmatrix}.$$

From the fact that $R = SV$, we have

$$V = S^{-1} R.$$

Hence comparing this relation, and (12.12) we see that the desired inverse of S is given by

(12.13)
$$S^{-1} = \frac{1}{3} \begin{bmatrix} 1 & 1 & 1 \\ 1 & a^2 & a \\ 1 & a & a^2 \end{bmatrix} = \frac{1}{3} \bar{S},$$

where $\bar{S}$ is the complex conjugate of the matrix S.

The Determinant of S

The determinant of the modal matrix S may be expanded in terms of its first column in the form,

$$\text{Det } S = \begin{vmatrix} 1 & 1 & 1 \\ 1 & a^2 & a \\ 1 & a & a^2 \end{vmatrix} = \begin{vmatrix} a & a^2 \\ a^2 & a \end{vmatrix} - \begin{vmatrix} 1 & 1 \\ a^2 & a \end{vmatrix} + \begin{vmatrix} 1 & 1 \\ a & a^2 \end{vmatrix}$$
$$= (a^2 - a^4) - (a - a^2) + (a^2 - a)$$
$$= 3a(a - 1) = -3j\sqrt{3}.$$

This result follows from $\bar{S} S = 3 S^{-1} S = 3I$, from which we can see that $\text{Det}(\bar{S}S) = 27$.

The Circulant of Order n

The general circulant of order n has the form (12.1). It may be expressed as a polynomial of degree $n-1$ in the permutation matrix P of order n as given by (12.2) in the form

(12.14) $$C = c_1 I + c_2 P + \ldots + c_n P^{n-1},$$

where I is the nth-order identity matrix and the c_i's are the elements of the circulant C. By the same method as that used in the third-order case, we obtain the characteristic equation

$$\mu^n = 1.$$

The eigenvalues of P are therefore

$$\mu_k = \exp(\frac{2\pi kj}{n}), k = 0,1,2,\square \ , n-1.$$

The eigenvectors of P may be taken to be

$$X = \begin{bmatrix} 1 \\ \mu_k \\ \vdots \\ \mu_k^{n-1} \end{bmatrix}, k = 0,1,2,\square \ , n-1.$$

The modal matrix of $[P]$ has the form,

$$S = \begin{bmatrix} X_0 & X_1 & \cdots & X_{n-1} \end{bmatrix}$$

$$= \begin{bmatrix} 1 & 1 & \cdots & 1 \\ 1 & \mu_1 & \cdots & \mu_{n-1} \\ \vdots & \vdots & \cdots & \vdots \\ 1 & \mu_1^{n-1} & \cdots & \mu_{n-1}^{n-1} \end{bmatrix} .$$

By the same procedure used to obtain (12.13), it may be shown that S has the inverse,

$$S^{-1} = \frac{1}{n} \bar{S},$$

where $\bar{S}$ is the complex conjugate of S.

The nth-order permutation matrix P may be reduced to the form

(12.15) $\qquad\qquad P = SDS^{-1},$

where D is the diagonal matrix

$$D = \begin{bmatrix} \mu_0 & 0 & \cdots & 0 \\ 0 & \mu_1 & \cdots & 0 \\ \vdots & \vdots & \cdots & \vdots \\ 0 & 0 & \cdots & \mu_{n-1} \end{bmatrix}.$$

Since the mth power of P as expressed in (12.15) may be written in the form

$$P^m = SD^m S^{-1},$$

The form (12.14) for the nth-order circulant matrix may be expressed as

$$C = S(c_1 I + c_2 D + \ldots + c_n D_{n-1}) S^{-1}$$

$$= SD_0 S^{-1},$$

where D_0 is the diagonal matrix of the eigenvalues of the nth-order circulant C. The diagonal matrix D_0 has the form

$$D_0 = \begin{bmatrix} \alpha_0 & 0 & \cdots & 0 \\ 0 & \alpha_1 & \cdots & 0 \\ \vdots & \vdots & \cdots & \vdots \\ 0 & 0 & \cdots & \alpha_{n-1} \end{bmatrix},$$

where $\alpha_0, \alpha_1, \ldots, \alpha_{n-1}$ are the n eigenvalues of C. The eigenvalues of C are therefore seen to be

$$(12.16) \qquad \alpha_k = (c_1 + c_2\mu_k + \ldots + c_n\mu_k^{n-1}), \ k = 0, 1, \ldots, n-1,$$

and the modal matrix of C is S.

A Special Case

In some special cases of practical importance, the circulant matrix appears in the form

$$(12.17) \qquad C = \begin{bmatrix} b & c & \cdots & c \\ c & b & \cdots & c \\ \cdot & \cdot & \cdots & \cdot \\ c & c & \cdots & b \end{bmatrix}.$$

This is a special case of the general circulant with $c_1 = b$, $c_2 = c_3 = \ldots = c_n = c$. In this case, the eigenvalues of C are given by (12.16) in the form

$$(12.18) \qquad \alpha_k = b + c(\mu_k + \mu_k^2 + \ldots + \mu_k^{n-1}) , \ k = 0, 1, \ldots, n-1.$$

If $k = 0$, we have $\mu_0 = 1$, and therefore, equation (12.18) gives the result

$$\alpha_0 = b + (n-1)c.$$

If $k \neq 0$, μ_k is given by (12.18); it involves the geometric progression

$$\mu_k + \mu_k^2 + \cdots + \mu_k^{n-1} = \frac{\mu_k^n - \mu_k}{\mu_k - 1}$$

$$= \frac{e^{2\pi jk} - \mu_k}{\mu_k - 1} = \frac{1 - \mu_k}{\mu_k - 1} = -1.$$

Hence, if we substitute these relations into (12.18), we obtain

$$\alpha_k = b - c, \ k \neq 0, \ k = 1, 2, \ldots, n-1$$

Therefore, the matrix C of (12.17) has one eigenvalue $\alpha_0 = b + (n-1)c$, and $(n-1)$ repeated eigenvalues of the form $\alpha_k = b - c$ with $k \neq 0$.

EXERCISES

Find the eigenvalues and eigenvectors for the following circulant matrices.

1. $\begin{bmatrix} 2 & -1 & 1 \\ 1 & 2 & -1 \\ -1 & 1 & 2 \end{bmatrix}$
2. $\begin{bmatrix} 2 & 1 & 1 \\ 1 & 2 & 1 \\ 1 & 1 & 2 \end{bmatrix}$
3. $\begin{bmatrix} 3 & 2 & 1 \\ 1 & 3 & 2 \\ 2 & 1 & 3 \end{bmatrix}$

***

13. SYMMETRIC MATRICES WITH MULTIPLE EIGENVALUES.

The eigenvectors of a symmetric matrix A whose eigenvalues μ_i are all distinct form an orthogonal set of vectors. However, from the geometrical discussion given in section 9, it is apparent that if A is a symmetric matrix of order n, we can always construct a set of n mutually orthogonal eigenvectors whether the eigenvalues of A are distinct or not. In this section, a method by which a set of mutually orthogonal eigenvectors may be constructed is described.

The method will be illustrated by two examples that illustrate the general procedure to be followed.

EXAMPLE 9

Find the eigenvalues and eigenvectors for the following matrix.

$$A = \begin{bmatrix} 0 & 1 & 1 \\ 1 & 0 & -1 \\ 1 & -1 & 0 \end{bmatrix}$$

SOLUTION

The characteristic equation for this matrix is

$$\text{Det } K = \text{Det } (\mu I_3 - A) = \begin{vmatrix} \mu & -1 & -1 \\ -1 & \mu & 1 \\ -1 & 1 & \mu \end{vmatrix}$$

$$= \mu^3 - 3\mu + 2 = 0 = (\mu + 2)(\mu - 1)^2.$$

This matrix has the following eigenvalues:

$$\mu_1 = -2, \ \mu_2 = \mu_3 = 1$$

The eigenvector belonging to $\mu_1 = -2$ may be shown to be

(13.1)
$$X_1 = \begin{bmatrix} -1 \\ 1 \\ 1 \end{bmatrix}.$$

For the repeated eigenvalue $\mu_2 = \mu_3 = 1$, the eigenvector equation $AX_2 = \mu_2 X_2$ may be put in terms of the following homogeneous algebraic equation.

(13.2)
$$\begin{bmatrix} 1 & -1 & -1 \\ -1 & 1 & 1 \\ -1 & 1 & 1 \end{bmatrix} \begin{bmatrix} x_1 \\ x_2 \\ x_3 \end{bmatrix}_2 = 0$$

It should be easy to see that the rank of the coefficient matrix is one indicating that there is only one independent equation. The fact that the rank is one also tells us that there will be $3 - 1 = 2$ (order minus rank) or two linearly independent eigenvectors associated with the repeated eigenvalue. From equation (13.2), it should be easy to see that the single independent equation may be written as

(13.3) $x_3 = x_1 - x_2.$

To obtain the required eigenvectors, two solutions of (13.3) must be found that are orthogonal to each other and to the first eigenvector. If we let $x_1 = x_2 = 1$, then from (13.3) we obtain $x_3 = 0$. Accordingly, we may take

$$X_2 = \begin{bmatrix} 1 \\ 1 \\ 0 \end{bmatrix}$$

as one eigenvector belonging to the eigenvalue $\mu_2 = 1$. We may also notice that X_2 is orthogonal to X_1 by evaluating the product $X_1^T X_2$ and seeing that it vanishes. To obtain the third eigenvector, let us choose $x_2 = h$ and $x_3 = k$ since they are arbitrary. Equation (13.3) then gives $x_1 = (h + k)$, so that the third eigenvector may be written as

(13.4)
$$X_3 = \begin{bmatrix} h+k \\ h \\ k \end{bmatrix}.$$

Again it may be shown that X_3 is orthogonal to X_1. We must now adjust X_3 so that is is orthogonal to X_2. We therefore must have

$$X_2{}^T X_3 = 0.$$

This relationship requires that $k = -2h$. Substituting this into equation (13.4) gives

$$X_3 = \begin{bmatrix} -h \\ h \\ -2h \end{bmatrix} = -h \begin{bmatrix} -1 \\ 1 \\ -2 \end{bmatrix}.$$

Since h is arbitrary, let us set $h = -1$. We finally obtain

$$(13.5) \qquad X_1 = \begin{bmatrix} -1 \\ 1 \\ 1 \end{bmatrix}, \; X_2 = \begin{bmatrix} 1 \\ 1 \\ 0 \end{bmatrix}, \; X_3 = \begin{bmatrix} -1 \\ 1 \\ -2 \end{bmatrix}$$

as a set of three orthogonal and thus linearly independent eigenvectors of A. Orthogonality was used to generate the third eigenvector as it is easier to utilize computationally than the less specific condition of linear independence in these situations.

***

In this example we have carried out the determination of the eigenvectors by direct algebraic evaluation. It is important for a clearer understanding of the general methods to review what occurs when the Gaussian elimination method is applied.

To begin, consider the general augmented matrix of the characteristic matrix of the homogeneous system

$$\text{Aug } K = \begin{bmatrix} \mu & -1 & -1 & 0 \\ -1 & \mu & 1 & 0 \\ -1 & 1 & \mu & 0 \end{bmatrix}.$$

For the first eigenvalue of -2, we have

$$\text{Aug } K = \begin{bmatrix} -2 & -1 & -1 & 0 \\ -1 & -2 & 1 & 0 \\ -1 & 1 & -2 & 0 \end{bmatrix}.$$

Applying Gaussian elimination to this matrix, it may be seen that it reduces to the following form.

$$\begin{bmatrix} -2 & -1 & -1 & 0 \\ 0 & -1 & 1 & 0 \\ 0 & 0 & 0 & 0 \end{bmatrix}$$

Since the rank is two, there will be only one nontrivial solution. Following the procedure of chapter II, section 11, we set the third element of the eigenvector to one, that is, $x_3 = 1$. Back substitution gives $x_2 = 1$ and $x_1 = -1$, which is the identical result for the first eigenvector as given in (13.1) above.

Now for the repeated eigenvalue of -1, the augmented matrix is

$$\begin{bmatrix} 1 & -1 & -1 & 0 \\ -1 & 1 & 1 & 0 \\ -1 & 1 & 1 & 0 \end{bmatrix}.$$

Gaussian elimination reduces this to

$$\begin{bmatrix} 1 & -1 & -1 & 0 \\ 0 & 0 & 0 & 0 \\ 0 & 0 & 0 & 0 \end{bmatrix}.$$

Since the rank is one, there will be two independent nontrivial eigenvectors. Again following the procedure of chapter II, section 11, we first set $x_3 = 1$, $x_2 = 0$ and solve for x_1. The result is $x_1 = 1$, which gives for the second eigenvector

$$X_2 = \begin{bmatrix} 1 \\ 0 \\ 1 \end{bmatrix}$$

If we now set $x_3 = 0$, $x_2 = 1$, the third eigenvector is found to be

$$X_3 = \begin{bmatrix} 1 \\ 1 \\ 0 \end{bmatrix}.$$

The complete set of eigenvectors is now

$$(13.6) \qquad X_1 = \begin{bmatrix} -1 \\ 1 \\ 1 \end{bmatrix}, \ X_2 = \begin{bmatrix} 1 \\ 0 \\ 1 \end{bmatrix}, \ X_3 = \begin{bmatrix} 1 \\ 1 \\ 0 \end{bmatrix}.$$

In comparing this set of eigenvectors with those of (13.5), we see that the first of each set are the same as they must be since the eigenvalue is distinct. It should also be remembered at this point that eigenvectors are always determined only up to an arbitrary multiplicative constant. In other words, the first eigenvector above could just as easily have been given as

$$X_1 = \begin{bmatrix} 3 \\ -3 \\ -3 \end{bmatrix}.$$

The latter two eigenvectors of the sets above are associated with the double eigenvalue. The second vector of (13.5) is the same as the third of (13.6), but it appears that X_3 of (13.5) does not correspond with X_2 of (13.6). In actuality, there is no difficulty in this if we recall that since the rank of the augmented matrix is one for the repeated eigenvalue, there are always two linearly independent eigenvectors associated with the repeated eigenvalue. Linear independence does not imply uniqueness, which is evidenced by this example. Recall that X_3 of (13.5) was generated by requiring that it be orthogonal to X_1 and X_3. Orthogonality was used in that it is a relatively easy way to generate an independent vector. The eigenvectors of (13.6) were simply generated by the Gaussian elimination process.

If we carefully inspect the third eigenvector of (13.5), we can see that it is linearly dependent with the second vector of (13.5) [or the third of (13.6)] and the second of (13.6). After some examination, it should be seen that

$$\begin{bmatrix} -1 \\ 1 \\ -2 \end{bmatrix} = \begin{bmatrix} 1 \\ 1 \\ 0 \end{bmatrix} - 2\begin{bmatrix} 1 \\ 0 \\ 1 \end{bmatrix}.$$

In other words, both sets of eigenvectors comprise valid sets. Both are linearly independent sets but (13.5) are also orthogonal. The point of this previous discussion was that Gaussian elimination can be useful in generating the eigenvectors.

EXAMPLE 10

Find the eigenvalues and eigenvectors for the upper triangular matrix below. This is an example of a nonsymmetric matrix but of a type frequently encountered in many matrix computational algorithms.

$$U = \begin{bmatrix} 2 & -1 & 2 \\ 0 & -4 & 6 \\ 0 & 0 & -4 \end{bmatrix}$$

SOLUTION

The eigenvalues of U are easily found to be

$$\mu_1 = 2, \; \mu_2 = \mu_3 = -4.$$

The general augmented matrix of the homogeneous eigenvector equation is

(13.7) $$\text{Aug } U = \begin{bmatrix} \mu-2 & 1 & -2 & 0 \\ 0 & \mu+4 & -6 & 0 \\ 0 & 0 & \mu+4 & 0 \end{bmatrix}.$$

For the distinct first eigenvector, $\mu_1 = 2$, the augmented matrix becomes

$$\begin{bmatrix} 0 & 1 & -2 & 0 \\ 0 & 6 & -6 & 0 \\ 0 & 0 & 6 & 0 \end{bmatrix}.$$

The rank of this matrix is two, indicating that there is only one independent eigenvector associated with the first distinct eigenvalue. By the back substitution process, we find

$$X_1 = \begin{bmatrix} 1 \\ 0 \\ 0 \end{bmatrix}.$$

When $\mu_2 = \mu_3 = -4$ for the repeated eigenvalue, equation (13.7) reduces to the following.

$$\begin{bmatrix} -6 & 1 & -2 & 0 \\ 0 & 0 & -6 & 0 \\ 0 & 0 & 0 & 0 \end{bmatrix}$$

The rank of this matrix is two, which means that there is only one independent eigenvector associated with the double eigenvalue. The homogeneous system of equations associated with the above augmented matrix is

$$\begin{bmatrix} -6 & 1 & -2 \\ 0 & 0 & -6 \\ 0 & 0 & 0 \end{bmatrix} \begin{bmatrix} x_1 \\ x_2 \\ x_3 \end{bmatrix} = \begin{bmatrix} 0 \\ 0 \\ 0 \end{bmatrix}.$$

Either by the back substitution process or examination of the above system indicates that x_3 must vanish, x_2 is arbitrary, and $6x_1 = x_2$. The second eigenvector may be written as

$$X_2 = \begin{bmatrix} 1 \\ 6 \\ 0 \end{bmatrix}.$$

We notice that X_2 is not orthogonal to X_1, but they are linearly independent. Because only a single independent eigenvector can be found associated with the double eigenvalue, the third eigenvector can only be some multiple of X_2. This means that the modal matrix will necessarily have two dependent columns and hence be a singular matrix.

***

In conclusion, it should be noted that if a matrix is not symmetric, it is not necessarily true that there will be k linearly independent eigenvectors associated with a repeated root of multiplicity k. In the next chapter, we will introduce a method of generating a third linearly independent vector that is not an actual eigenvector; however, it may be used in a similarity transformation to almost reduce the original matrix to diagonal form. This particular situation is referred to as the Jordan Canonical form.

EXERCISES

Find the eigenvalues and eigenvectors for the following matrices.

1. $\begin{bmatrix} -1 & -1 & -1 \\ -1 & -1 & 1 \\ -1 & 1 & -1 \end{bmatrix}$

2. $\begin{bmatrix} 1 & 0 & 1 \\ 0 & 2 & 0 \\ 1 & 0 & 1 \end{bmatrix}$

3. $\begin{bmatrix} 1 & -1 & 1 \\ -1 & 1 & -1 \\ 1 & -1 & 1 \end{bmatrix}$

ADDITIONAL EXERCISES

1. Determine the eigenvalues and eigenvectors of the rotation matrix

$$R = \begin{bmatrix} \cos\theta & \sin\theta \\ -\sin\theta & \cos\theta \end{bmatrix}.$$

2. For the following matrix, show that the eigenvalues are the roots of the equation $z^3 + a_1 z^2 + a_2 z + a_3 = 0$.

$$A = \begin{bmatrix} 0 & 1 & 0 \\ 0 & 0 & 1 \\ -a_3 & -a_2 & -a_1 \end{bmatrix}$$

Use the result of exercise 2 to evaluate the roots of the following polynomials.

3*. $z^3 + 2z^2 - z - 2 = 0$

4*. $z^4 - 10z^3 + 35z^2 - 50z + 24 = 0$

5. Given a second-order square matrix A that satisfies the equation $A^2 = I_2$, where I_2 is the second-order identity matrix, show that A is either $-I_2$, $+I_2$ or

$$A = \begin{bmatrix} a & re^{j\theta} \\ re^{-j\theta} & -a \end{bmatrix},$$

where $a^2 + r^2 = 1$ and a, r, and θ are real.

6.   Show that any orthogonal matrix of order two is necessarily one of the following four types:

$$R_1 = \begin{bmatrix} \cos\theta & \sin\theta \\ -\sin\theta & \cos\theta \end{bmatrix}, \quad R_2 = \begin{bmatrix} \cos\theta & \sin\theta \\ \sin\theta & -\cos\theta \end{bmatrix}$$

$$R_3 = \begin{bmatrix} \sin\theta & \cos\theta \\ -\cos\theta & \sin\theta \end{bmatrix}, \quad R_4 = \begin{bmatrix} \sin\theta & \cos\theta \\ \cos\theta & -\sin\theta \end{bmatrix}$$

7.   Given the two rotations described in exercise 1, section 8, consider a third rotation to generate another coordinate system W. Let this system be related to the Z system by a rotation through an angle of ω as illustrated in figure 5.9 below.

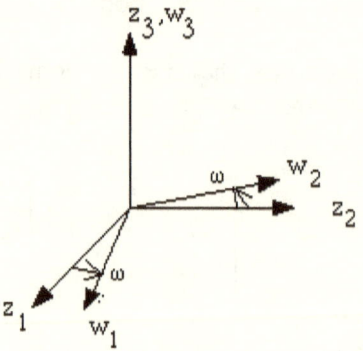

Fig. 5.9

Using matrix multiplication, derive the overall transformation matrix T for the relation $W = TX$. On a single sketch, show all the coordinate systems and their respective rotations. It should be noted that this problem deals with the well-known Euler's angles of classical mechanics.

8.*   Given the quadratic surface

$$x_1^2 + 3x_2^2 + 3x_3^2 - 2x_2x_3 = 1,$$

determine its principal axes by introducing a rotation of the coordinate axes by an orthogonal transformation.

9.   Prove in general that the product of orthogonal transformations is also orthogonal and demonstrate the result by using the rotation matrices $R(\theta)$ and $R(\phi)$ of exercise 1, section 8 above.

10.   Given $K(\mu)$ as the characteristic matrix of a matrix A, prove that when $\mu = \mu_k$, the columns of the adjoint matrix Adj $K(\mu_k)$ are the eigenvectors of M.

11.   A matrix is said to be positive definite if $X^T A X > 0$ for all vectors X. Show that a sufficient condition for the matrix A to be positive definite is for all its eigenvalues to be positive.

12.   Show that a necessary condition for a matrix A to be positive definite [additional exercise 10 above] is for all the diagonal elements of A to be positive.

13.   Prove that if an eigenvalue has the value of zero, then the matrix is singular.

Find the eigenvalues and eigenvectors for the following matrices.

14*. $\begin{bmatrix} -5 & 2 & 1 \\ 1 & -4 & 1 \\ -1 & 2 & -3 \end{bmatrix}$ 15*. $\begin{bmatrix} -4 & 1 & 1 \\ 2 & -5 & 1 \\ 2 & -1 & -3 \end{bmatrix}$ 16*. $\begin{bmatrix} -2 & 1 & 1 \\ 2 & -3 & 1 \\ 2 & 1 & -1 \end{bmatrix}$

17*. $\begin{bmatrix} 6 & -3 & -1 \\ -2 & 5 & -1 \\ 2 & -3 & 3 \end{bmatrix}$

18.*   Consider the linear algebraic equation $AX = B$. Using some of the third-order matrices in additional exercise 13 as the coefficient matrix, find solutions when the constant vector B is an eigenvector of A. Try several cases and discuss your results.

19.   Prove that the eigenvalues of any triangular matrix are the values on the main diagonal.

VI

THE CALCULUS OF MATRICES AND
MATRIX FUNCTIONS

---

1. INTRODUCTION

Since a fairly large number of applied problems are dynamic in their description, their solutions are conveniently obtained in terms of matrix functions. Specific examples are problems of population growth, time-dependent economic problems, the vibrations of mechanical and electrical systems, to name a few. In all such cases, the basic mathematical model of the system being studied is either a system of finite difference or differential equations. As previously stated, it is frequently useful to express the solution in terms of matrix functions. When numerical techniques are employed to obtain a solution over time, the algorithms used are almost always developed in terms of matrices and their functions.

In this chapter, the notion of a matrix function is discussed. Also to be demonstrated is how a matrix is considered as a variable in direct analogy with scalar variables. The definition of the derivative and integral of a matrix function is given. Methods for the evaluation of functions of matrices are discussed and illustrated. Chapter VII is devoted to the particular applications of matrix functions to finite difference and differential equations. Chapter VIII presents a number of typical applied problems in which these types of solutions are commonly encountered.

2. MATRIX POLYNOMIALS

Before introducing the idea of a function of a matrix, it is necessary to define a matrix polynomial and then to define and discuss infinite series of matrices. Let us first consider a polynomial of degree n of the algebraic variable z, where z may be real or complex. Such a polynomial has the general form

$$p(z) = a_0 + a_1 z + a_2 z^2 + \ldots + a_n z^n.$$

The argument of the polynomial $p(z)$ is the real or complex quantity z, and the operations of addition and multiplication of such quantities are known. It is therefore possible to compute the value of the polynomial $p(z)$ for any given value of its argument z since the coefficients a_i are also known quantities.

Now if the scalar argument z of the polynomial $p(z)$ is replaced by a square matrix A of the nth order, and if the coefficient a_0 is multiplied by the nth-order identity matrix, we obtain

$$p(A) = a_0 I_n + a_1 A + a_2 A^2 + \ldots + a_n A^n.$$

We define $p(A)$ to be a polynomial of the square matrix A of the nth degree. If the coefficients a_i and the elements of A are known, the polynomial $p(A)$ are computed numerically.

EXAMPLE 1

Evaluate the following matrix polynomial.

$$p(A) = I_2 + 2A + 5A^2,$$

where

$$A = \begin{bmatrix} 3 & 2 \\ 1 & 0 \end{bmatrix}.$$

SOLUTION

If the given matrix is substituted into the polynomial, the result is

$$p(A) = \begin{bmatrix} 1 & 0 \\ 0 & 1 \end{bmatrix} + 2 \begin{bmatrix} 3 & 2 \\ 1 & 0 \end{bmatrix} + 5 \begin{bmatrix} 3 & 2 \\ 1 & 0 \end{bmatrix}^2 = \begin{bmatrix} 62 & 34 \\ 17 & 11 \end{bmatrix}.$$

EXERCISES

1.* By direct computation, evaluate the matrix polynomial

$$p(A) = 2I_2 + 3A - A^2,$$

where

$$A = \begin{bmatrix} 3 & -1 \\ -1 & 3 \end{bmatrix}.$$

2.* Evaluate by direct computation the matrix polynomial

$$p(A) = I_2 - 3A + 3A^2 - A^3,$$

where A is the same matrix as in the first exercise above.

3.*   Consider the same polynomial as in the second exercise above. Show that the same result is obtained when the polynomial is evaluated using its factored form

$$p(A) = (I_2 - A)^3.$$

4.*   Evaluate both of the following forms of the matrix polynomial p(A) and verify they give the same result.

$$p(A) = 3I_3 + 5A - 2A^2$$
$$= (I_3 + 2A)(3I_3 - A),$$

where A is

$$A = \begin{bmatrix} 1 & -2 & -1 \\ -2 & 0 & 3 \\ -1 & 3 & 2 \end{bmatrix}.$$

***

3. INFINITE SERIES OF MATRICES

Let

(3.1) $$f(z) = a_0 + a_1 z + a_2 z^2 + \cdots = \sum_{k=0}^{\infty} a_k z^k$$

be an infinite series of the real or complex variable z.

Let the argument z of the infinite series (3.1) be replaced by the square matrix of the nth order A, and let the coefficient a_0 be multiplied by the identity matrix I_n of the nth order. We then obtain

(3.2) $$f(A) = a_0 I_n + a_1 A + a_2 A^2 + \cdots = \sum_{k=0}^{\infty} a_k A^k.$$

This is an infinite series of the matrix A. This series is convergent if the matrix $f(A)$ tends to a definite limit matrix L whose elements l_{ij} are finite as the number of terms in equation (3.2) becomes infinite. That is, if

$$f_p(A) = \sum_{k=0}^{p} a_k A^k$$

and

$$\lim_{p \to \infty} f_p(A) = L, \quad I_{ij} = \text{finite for all i and j}$$
,

then $f(A)$ is a convergent series. If the series (3.2) is a convergent one, it defines a function of the matrix A.

Examples of Matrix Series

The Geometric Series

(3.3) $$g(A) = I_n + aA + a^2 A^2 + \cdots = \sum_{k=0}^{\infty} a^k A^k$$

The Exponential Function

(3.4) $$\exp(A) = e^A = I_n + A + \frac{1}{2!}A^2 + \frac{1}{3!}A^3 + \cdots = \sum_{k=0}^{\infty} \frac{1}{k!}A^k$$

4.   THE MATRIX EXPONENTIAL FUNCTION

One of the most important functions that arises in the calculus of matrices is the matrix exponential function. This function is of great importance in the solution of linear differential equations. Equation (3.4) above defines the power series representation of the exponential function of a matrix A.

This infinite series that defines the matrix exponential function $\exp(A)$, may be shown to be convergent for all square matrices A. In order to establish the convergence of the series (3.4), let m_{max} be the greatest of the magnitudes of the n^2 elements m_{ij} of A so that

$$| m_{ij} | \le m_{max} \text{ for all } i \text{ and } j.$$

Then, since A is a square matrix of the nth order, each element in the matrix A^k will not exceed $n^{k-1}(m_{max})^k$ in numerical value.

Now if we let

$$S = \exp(A) = \sum_{k=0}^{\infty} \frac{1}{k!} A^k \,,$$

we see that all the elements s_{ij} of the matrix exponential function of A must satisfy the following inequality:

$$s_{ij} \leq 1 + m_{max} + \frac{n}{2!} m_{max}^2 + \frac{n^2}{3!} m_{max}^3 + \cdots = \frac{1}{n}(e^{n m_{max}} - 1) + 1$$

Since all the elements of $\exp(A)$ are bounded by this inequality, it follows that the matrix exponential function is convergent for all square matrices A.

By means of the series definition for $\exp(A)$, it can be shown that if A and B are commutative matrices so that $AB = BA$, we have

$$\exp(A)\exp(B) = \exp(B)\exp(A) = \exp(A + B).$$

It can also be shown that

$$\exp(A)\exp(-A) = \exp(-A)\exp(A) = I_n,$$

where I_n is the nth-order identity matrix. These relations express the fact that $\exp(-A)$ is the inverse of $\exp(A)$.

5.   THE MATRIX TRIGONOMETRIC AND HYPERBOLIC FUNCTIONS

Every scalar power series has its matrix analogue. However, the corresponding matrix power series have more complicated properties. By replacing the scalar argument in the series that define the trigonometric and hyperbolic functions by a square matrix A and realizing that $A^0 = I$, the identity matrix, it is possible to extend the definition of these functions to matrix functions. For example, the matrix sine function, $\sin(A)$, or the matrix cosine function, $\cos(A)$, are defined by

(5.1) $$\sin(A) = A - \frac{1}{3!}A^3 + \frac{1}{5!}A^5 - \cdots$$

and

(5.2) $$\cos(A) = I_n - \frac{1}{2!}A^2 + \frac{1}{4!}A^4 - \cdots$$

The usual trigonometric identities are not always satisfied by $\sin(A)$ and $\cos(A)$ for arbitrary matrices. If the matrix A in the defining equation (3.4) for the exponential matrix function is placed equal to jA, where j is the unit of imaginaries $\sqrt{-1}$, the following result is obtained:

$$
\begin{aligned}
\exp(jA) &= I - \frac{1}{2!}A^2 + \frac{1}{4!}A^4 - \cdots + j\left(A - \frac{1}{3!}A^3 + \frac{1}{5!}A^5 - \cdots\right) \\
&= \cos(A) + j\sin(A)
\end{aligned}
$$
(5.3)

Replacing A by $-A$ in (5.3) yields the relation

(5.4) $$\exp(-jA) = \cos(A) - j\sin(A)$$

If the equations (5.3) and (5.4) are added and subtracted, the following relations connecting the trigonometric functions with the exponential function are obtained.

(5.5) $$\sin(A) = \frac{\exp(jA) - \exp(-jA)}{2j}$$

(5.6) $$\cos(A) = \frac{\exp(jA) + \exp(-jA)}{2}$$

The matrix hyperbolic functions $\sinh(A)$ and $\cosh(A)$ are defined by the power series

(5.7) $$\sinh(A) = A + \frac{1}{3!}A^3 + \frac{1}{5!}A^5 + \cdots = \frac{\exp(A) - \exp(-A)}{2}$$

and

(5.8) $$\cosh(A) = I_n + \frac{1}{2!}A^2 + \frac{1}{4!}A^4 + \cdots = \frac{\exp(A) + \exp(-A)}{2}$$

It is possible to establish many of the properties of the matrix trigonometric and hyperbolic functions that are similar to the ordinary scalar trigonometric and hyperbolic functions. For example, if (5.3) and (5.4) are multiplied together, the result is

$$\exp(jA)\exp(-jA) = I_n = \cos^2(A) + \sin^2(A).$$

By the use of (5.5) and (5.6), the following equations are established:

$$2\sin(A)\cos(A) = \sin(2A)$$

$$\cos^2(A) - \sin^2(A) = \cos(2A)$$

Similarly, by the use of (5.7) and (5.8), the following result may be established.

$$\cosh^2(A) - \sinh^2(A) = I_n$$

Other identities analogous to those satisfied by the scalar hyperbolic and trigonometric functions are established in a similar manner.

The matrix J defined by

$$J = \begin{bmatrix} 0 & -1 \\ 1 & 0 \end{bmatrix}$$

behaves in matrix algebra in a manner similar to the unit of imaginaries $j = \sqrt{-1}$ in ordinary algebra. By direct multiplication, we have

$$J^2 = -I, \ J^3 = -J, \ J^4 = I, \ \text{etc.}$$

If in equation (5.1) we substitute $A = aJ$, where a is a scalar quantity, and compare the result with equation (5.7), it is easy to see that

$$\sin(aJ) = J\sinh(a).$$

If $A = aJ$ is substituted in (5.2) and the results are compared with equation (5.8), it can be seen that

$$\cos(aJ) = \cosh(a)I^2.$$

EXERCISES

1. Derive the result $2\sin(A)\cos(A) = \sin(2A)$ using the definitions given in equations (5.5) and (5.6).

2. Derive the result that $\sin(aJ) = J\sinh(a)$ where

$$J = \begin{bmatrix} 0 & -1 \\ 1 & 0 \end{bmatrix}$$

and a is a constant.

3.   Let D be the following diagonal matrix.

$$D = \begin{bmatrix} d_1 & 0 & \cdots & 0 \\ 0 & d_2 & \cdots & 0 \\ \vdots & \vdots & \ddots & \vdots \\ 0 & 0 & \cdots & d_3 \end{bmatrix} = \text{Diag}\ (d_1, d_2, \cdots, d_n)$$

Show that

$$\exp(D) = \begin{bmatrix} \exp(d_1) & 0 & \cdots & 0 \\ 0 & \exp(d_2) & \cdots & 0 \\ \vdots & \vdots & \ddots & \vdots \\ 0 & 0 & \cdots & \exp(d_n) \end{bmatrix}.$$

***

6.   THE CAYLEY-HAMILTON THEOREM

We now turn to a discussion of what is perhaps the most famous and important theorem in the algebra of matrices.

DEFINITION

Cayley-Hamilton Theorem

If A is a square matrix of the nth order and if $p(\mu)$ is its characteristic polynomial,

(6.1) $p(\mu) = \text{Det}\ (\mu I_n - A)$

$$= \mu^n + a_{n-1}\mu^{n-1} + a_{n-2}\mu^{n-2} + \ldots + a_0 = 0,$$

then the matrix A satisfies the polynomial equation

(6.2) $p(A) = A^n + a_{n-1}A^{n-1} + a_{n-2}A^{n-2} + \ldots + a_0 I_n = 0_n,$

where I_n and 0_n are the identity and zero matrix, respectively having the same order as A.

The Cayley-Hamilton theorem is often concisely stated in the following form:

> A matrix satisfies its own characteristic equation.

Symbolically, if $p\,(\mu)$ is the characteristic polynomial of A, then $p(A) = 0_n$. This is a very useful mnemonic device if it is properly understood.

The Eigenvalues of a Modified Matrix

Before we give a proof of the Cayley-Hamilton theorem, let us compute the eigenvalues of the matrix

(6.3) $N = A - aI_n$

when the eigenvalues of A are known. Now the eigenvalues of A are the roots of its characteristic equation,

$$p_A(\mu) = \text{Det}\,(\mu I - A) = 0.$$

The eigenvalues of the modified matrix N are the roots of its characteristic equation,

$$p_N(z) = \text{Det}\,(zI_n - N) = \text{Det}\,[(z + a)I - A] = 0.$$

When we compare this to the previous result, it is seen that if $z + a = \mu$, we then have

$$p_N(z) = p_A(\mu),\; z = \mu - a.$$

Since the roots of $p_A\,(\mu)$ are the eigenvalues μ_i of A, the roots of $p_N\,(z)$ are $z_i = \mu_i - a$, and these are the eigenvalues of the modified matrix N.

This result is used to establish the Cayley-Hamilton theorem for a matrix A that has distinct eigenvalues. To do this, let it be supposed that μ_i, $i = 1, 2, \ldots, n$ are the eigenvalues of A. In this case, the characteristic equation of A, equation (6.1), may be factored into the form

$$p(\mu) = (\mu - \mu_1)(\mu - \mu_2)\ldots(\mu - \mu_n) = 0.$$

Now consider the factored matrix polynomial

(6.4) $p(A) = (A - \mu_1 I_n)(A - \mu_2 I_n)\ldots(A - \mu_n I_n),$

where I_n is the nth-order identity matrix. In chapter V, it was shown that if an nth-order square matrix A has n distinct eigenvalues, then its modal matrix M may be constructed from the eigenvectors of A. The modal matrix diagonalizes A in the form

$$M^{-1}AM = D,$$

where D is the diagonal matrix of the eigenvalues of A.

We now let

(6.5) $\qquad N_i = A - \mu_i I_n.$

With this notation, equation (6.4) may be written in the form

$$p(A) = N_1 N_2 \ldots N_n.$$

Let this result be premultiplied by M^{-1} and postmultiplied by M, and let it be written in the form

(6.6) $\qquad M^{-1}{}_p(A)M = M^{-1}N_1 MM^{-1}N_2 M \ldots M^{-1}N_n M.$

If we define

$$D_i = M^{-1}N_i M,$$

then (6.6) may be written in the form

(6.7) $\qquad A^{-1}p(A)A = D_1 D_2 \ldots D_n$

The matrices D_i are the diagonal matrices of the modified matrices N_i. By comparing (6.5) to (6.3), we see that the modified matrix N_i has all its eigenvalues reduced by μ_i. Hence, the ith row of the diagonal matrix D_i has all its elements equal to zero. As a consequence of this fact, it can be seen that the product of diagonal matrices

$$D_1 D_2 \ldots D_n = 0_n.$$

Hence, (6.7) reduces to

$$M^{-1}p(A)M = 0_n$$

or

(6.8) $\qquad p(A) = M 0_n M^{-1} = 0_n$

Equation (6.8) shows that the matrix polynomial $p(A) = 0_n$ and establishes the Cayley-Hamilton theorem for matrices that have distinct eigenvalues. However, it can be shown that the relation (6.8) is valid in the general case in which some of the roots of the characteristic equation are repeated roots.[36]

EXERCISES

Verify the Cayley-Hamilton theorem for the following matrices.

1. $\begin{bmatrix} -5 & 4 \\ -4 & 5 \end{bmatrix}$
   2. $\begin{bmatrix} -3 & 1 \\ 1 & -3 \end{bmatrix}$
   3*. $\begin{bmatrix} -1 & 2 & 1 \\ 1 & 0 & 1 \\ -1 & 2 & 1 \end{bmatrix}$

4*. $\begin{bmatrix} -24 & 3 & 2 \\ 24 & -30 & 4 \\ 36 & -9 & -18 \end{bmatrix}$

***

7. REDUCTION OF POLYNOMIALS BY THE CAYLEY-HAMILTON THEOREM

Since a matrix satisfies its own characteristic equation in a matrix sense, as given by (6.2), it is possible to solve it for A^n in the form

$$A^n = -(a_{n-1}A^{n-1} + a_{n-2}A^{n-2} + \ldots + a_0 I_n).$$

A very important result of this is that it is possible to express any polynomial of the matrix A as a polynomial in A that does not exceed the degree $n-1$, where n is the order of the matrix A. This is a result that is unique to matrix polynomials and series. The most powerful implication is that any matrix function that is represented by an infinite matrix series is represented exactly by a polynomial whose degree is one less than the order of the matrix. There is no analogous relationship to this in ordinary scalar algebra.[37]

---

36. Bocher [1931].
37. The explanation lies in the fact that scalar algebra is infinite dimensional while matrices are finite dimensional. However, this topic is beyond the scope of the discussion in this text.

EXAMPLE 2

Evaluate the matrix defined by the following polynomial.

$$p_1(A) = A^4 + 6A,$$

where the matrix A is given by

(7.1)
$$A = \begin{bmatrix} 1 & -3 \\ 3 & 1 \end{bmatrix}.$$

SOLUTION

The characteristic equation of the matrix A is

$$p(A) = \text{Det}\,(\mu I_2 - A) = \mu^2 - 2\mu + 10 = 0_2.$$

As a consequence of the Cayley-Hamilton theorem, we know that the matrix A satisfies the equation

(7.2)
$$A^2 - 2A + 10I_2 = 0_2.$$

If equation (7.2) is solved for A^2, we have

(7.3)
$$A^2 = 2A - 10I_2$$

Now

$$A^4 = (A^2)^2 = (2A - 10I_2)^2 = 4A^2 - 40A + 100I_2$$

$$= 4(2A - 10I_2) - 40A + 100I_2 = -32A + 60I_2$$

Hence, the desired polynomial $p_1(A)$ is given by

$$p_1(A) = (-32A + 60I_2) + 6A = -26A + 60I_2.$$

$$= \begin{bmatrix} 34 & 78 \\ -78 & 34 \end{bmatrix}.$$

***

In general, for any second-order matrix A, it can be seen that A^2, A^3, A^4, ..., A^k, where k is an integer greater than or equal to 2, may be expressed in terms of a scalar times A and another scalar times the identity matrix in the form

$$A^k = a_k A + b_k I2, \ k = 2, 3, \ldots$$

The Inverse and General Negative Powers

If we premultiply the characteristic polynomial $p(A)$ by A^{-1} as given by (7.2), we obtain the equation

$$A^{-1}{}_p(A) = A^{n-1} + a_{n-1}A^{n-2} + a_{n-2}A^{n-3} + \ldots + a_0 A^{-1} = 0_n.$$

The equation may now be solved for A^{-1}; the result is

(7.4) $A^{-1} = -(A^{n-1} + a_{n-1}A^{n-2} + a_{n-2}A^{n-3} + \ldots + a_1 I_n)/a_0.$

This result expresses the inverse of the matrix A in terms of the coefficients a_i of its characteristic equation and the powers A^{n-1}, A^{n-2}, etc., of A. The expression (7.4) has been used as a practical method for the computation of the inverses of large matrices.[38]

EXAMPLE 3

For the same matrix A used in example 2 above, find A^{-1} and A^{-2} by the method just presented.

SOLUTION

The characteristic equation for the matrix A of example 2 is given by equation (7.2). If we multiply the characteristic equation through by A^{-1} and then solve for A^{-1}, we arrive at the following result.

(7.5) $A^{-1} = (2I_2 - A)/10$

If (7.5) is multiplied by itself, the result is

$$A^{-2} = (4I_2 - 4A + A^2)/100 = -(A + 3I_2)/50$$

***

[38.]   Hohn [1958].

By repeated multiplication of (7.5) by A^{-1} and the use of (7.3), it can be seen that negative powers of A may be expressed in the form

$$A^{-k} = c_k A + d_k I_2, \ k = 2, 3, \ldots .$$

EXAMPLE 4

Apply the Cayley-Hamilton theorem to reduce the following polynomial.

$$p_2(A) = A^3 + 2A^2 - 3I_2,$$

where the matrix A is given by

$$A = \begin{bmatrix} 4 & 5 \\ -1 & -2 \end{bmatrix}.$$

SOLUTION

The characteristic equation of this matrix is

$$p(\mu) = \text{Det} \ (\mu I_2 - A) = \mu^2 - 2\mu - 3 = 0$$

Hence, by the Cayley-Hamilton theorem, the matrix A satisfies the equation

$$p(A) = A^2 - 2A - 3I_2 = 0_2.$$

We have therefore

$$A^2 = 2A + 3I_2$$

and

$$A^3 = AA^2 = A(2A + 3I_2)$$

$$= 2A^2 + 3A = 2(2A + 3I_2)$$

$$= 7A + 6I_2$$

Now the required reduced polynomial is

$$p_2(A) = A^3 + 2A^2 - 3I_2 = (7A + 6I_2) + 2(2A + 3I_2) - 3I_2$$

$$= 11A + 9I_2 = 11 \begin{bmatrix} 4 & 5 \\ -1 & -2 \end{bmatrix} + \begin{bmatrix} 9 & 0 \\ 0 & 9 \end{bmatrix} = \begin{bmatrix} 53 & 55 \\ -11 & -13 \end{bmatrix}.$$

EXERCISES

1.*   Evaluate the matrix polynomial given in Exercise 1, Section 2 by the method presented in this Section.

2.*   Evaluate the matrix polynomial given in Exercise 2, Section 2 by the method presented in this Section.

3.*   Using the method of this section, evaluate the following matrix polynomial.

$$p(A) = 4I_2 - A + A^4,$$

where

$$A = \begin{bmatrix} 2 & -3 & 1 \\ -3 & 1 & 5 \\ 0 & -2 & 4 \end{bmatrix}.$$

***

8. THE INVERSION OF A MATRIX BY THE USE OF THE CAYLEY-HAMILTON THEOREM AND BOCHER'S FORMULAS

As shown in the previous section, equation (7.4) may be used for the inversion of a matrix. In addition, the coefficients of the characteristic equation are found by the application of Bocher's formulas, which were presented in section 11, chapter V. This avoids the need for expanding the characteristic determinant to find the characteristic equation. The following example illustrates this approach.

EXAMPLE 5

Find the inverse of the following matrix by using Bocher's formulas to determine the coefficients of the characteristic equation.

$$A = \begin{bmatrix} 2 & 2 & 0 \\ 0 & 3 & 1 \\ 1 & 0 & 1 \end{bmatrix}$$

SOLUTION

Bocher's formulas may be used to obtain the coefficients a_i of the characteristic equation of A. We first require the powers A^2 and A^3.

By direct multiplication, we obtain

$$A^2 = \begin{bmatrix} 4 & 10 & 2 \\ 1 & 9 & 4 \\ 3 & 2 & 1 \end{bmatrix}, \quad A^3 = \begin{bmatrix} 10 & 38 & 12 \\ 6 & 29 & 13 \\ 7 & 12 & 3 \end{bmatrix}.$$

We now compute the following traces:

$$s_1 = \text{Tr } A = 6, \; s_2 = \text{Tr } A^2 = 14, \; s_3 = \text{Tr } A^3 = 42$$

The coefficients a_i of the characteristic equation are now computed by the equations

$$a_2 = -s_1 = -6,$$

$$a_1 = -(a_1 s_1 + s_2)/2 = 11,$$

$$a_0 = -(a_2 s_1 + a_1 s_2 + s_3)/3 = -8.$$

We now use equation (8.4) to compute A^{-1}, and we obtain

$$A^{-1} = -\frac{1}{a_0}(A^2 + a_2 A + a_1 I_3) = \frac{1}{8}\begin{bmatrix} 3 & -2 & 2 \\ 1 & 2 & -2 \\ -3 & 2 & 6 \end{bmatrix}.$$

It is also of interest to recall that the determinant of A is given by

$$\text{Det } A = -a_0 = 8.$$

***

This method for the computation of the inverse of a matrix is straightforward and easily checked. The method has the further advantage of giving additional information besides A^{-1}. The determinant of A and the characteristic equation of A are obtained as byproducts of the computations.

EXERCISES

Calculate the inverses of the following matrices by the method of this section.

1. $\begin{bmatrix} 5 & -3 \\ -3 & 5 \end{bmatrix}$ 2. $\begin{bmatrix} 2 & 1 \\ 2 & 3 \end{bmatrix}$ 3. $\begin{bmatrix} -4 & 1 & 1 \\ 2 & -3 & 1 \\ 2 & 1 & -3 \end{bmatrix}$

4. $\begin{bmatrix} 1 & -1 & 0 \\ 0 & 2 & 1 \\ 3 & -2 & 1 \end{bmatrix}$

***

9. FUNCTIONS OF MATRICES AND SYLVESTER'S THEOREM

In section 3, it was shown how a function $f(z)$ of a scalar variable z could be generalized to the analogous matrix function by replacing the scalar z by the square matrix A in the infinite series that defines $f(z)$, as indicated in equation (3.2). In section 8, it was shown that as a consequence of the Cayley-Hamilton theorem, a polynomial of the form

$$p(A) = \sum_{k=0}^{m} a_k A^k \, , \, m > n - 1$$

may be reduced to the polynomial

$$p(A) = \sum_{k=0}^{n-1} b_k A^k \, ,$$

where n is the order of the matrix A.

It may also be shown that this reduction has an analog in the case where $f(A)$ is a matrix function defined by an infinite power series. The result is known in the mathematical literature as Sylvester's theorem, which may be expressed as follows.

DEFINITION

Sylvester's Theorem

If $f(x)$ is any function representable by a convergent infinite series and if A is a matrix of order n, then $f(A)$ is a matrix function of order n that is given by the following finite series.

(9.1) $$f(A) \;=\; \sum_{k=1}^{n} \frac{f(\mu_k)}{p'(\mu_k)}\, \text{Adj } K(\mu_k) \quad,$$

where the notation $p'(\mu_k)$ signifies

$$\frac{dp}{d\mu}(\mu)$$

evaluated at $\mu = \mu_k$ and $p'(\mu_k)$ is the characteristic equation of A. Also, $f(\mu_k)$ is the given function evaluated for each eigenvalue, and $K(\mu_k)$ is the adjoint of the characteristic matrix evaluated for each eigenvalue.

A proof of Sylvester's theorem will be found in the book by Frazer, Duncan, and Collar (1957). When the characteristic matrix and the eigenvalues of the matrix are known, Sylvester's theorem enables one to obtain expressions for matrix polynomials and matrix functions.

EXAMPLE 6

Apply Sylvester's theorem to evaluate the exponential function $\exp(A)$ when

$$A \;=\; \begin{bmatrix} 1 & 0 \\ 0 & 2 \end{bmatrix}$$

SOLUTION

In this case, we have

(9.2) $\qquad f(A) = \exp(A) = e^A$

In this case, the characteristic matrix of A is

$$K = (\mu I - A) = \begin{bmatrix} (\mu - 1) & 0 \\ 0 & (\mu - 2) \end{bmatrix}.$$

The adjoint Adj K is

$$\text{Adj } K = \begin{bmatrix} (\mu - 2) & 0 \\ 0 & (\mu - 1) \end{bmatrix}.$$

The characteristic polynomial of A is

$$p(\mu) = \text{Det } K(\mu)$$

$$= (\mu - 1)(\mu - 2) = \mu^2 - 3\mu + 2,$$

and,

$$p'(\mu) = 2\mu - 3.$$

The eigenvalues are clearly 1 and 2. Now since $p'(1) = -1$ and $p'(2) = 1$, we have, as a consequence of (9.1),

(9.3) $\qquad \exp(A) = e^2 \begin{bmatrix} 0 & 0 \\ 0 & 1 \end{bmatrix} - e \begin{bmatrix} -1 & 0 \\ 0 & 0 \end{bmatrix} = \begin{bmatrix} e & 0 \\ 0 & e^2 \end{bmatrix}.$

EXAMPLE 7

Compute A^{256} with the matrix

$$A = \begin{bmatrix} 1 & 0 \\ 0 & 3 \end{bmatrix}.$$

SOLUTION

To evaluate A^{256} by direct multiplication would be a very laborious task. However, the required power is computed by Sylvester's theorem very simply. In this case, the characteristic matrix of A is

$$K = (\mu I - A) = \begin{bmatrix} (\mu - 1) & 0 \\ 0 & (\mu - 3) \end{bmatrix}.$$

The characteristic polynomial of A is

$$p(\mu) = (\mu - 1)(\mu - 3) = \mu^2 - 4\mu + 3.$$

The eigenvalues of A are therefore $\mu_1 = 1$ and $\mu_2 = 3$. The adjoint of the characteristic matrix A is given by

$$\text{Adj } K = \begin{bmatrix} (\mu - 3) & 0 \\ 0 & (\mu - 1) \end{bmatrix}.$$

The derivative of the characteristic polynomial is, in this case,

$$p'(\mu) = 2\mu - 4.$$

Therefore, by Sylvester's theorem (9.1), we have

$$(9.4) \quad \begin{bmatrix} 1 & 0 \\ 0 & 3 \end{bmatrix}^{256} = 1^{256}\begin{bmatrix} 1 & 0 \\ 0 & 0 \end{bmatrix} + 3^{256}\begin{bmatrix} 0 & 0 \\ 0 & 1 \end{bmatrix} = \begin{bmatrix} 1 & 0 \\ 0 & 3^{256} \end{bmatrix}.$$

EXERCISES

1.*   Evaluate exp A by direct application of Sylvester's theorem for the following matrix.

$$A = \begin{bmatrix} 0 & 1 \\ 1 & 0 \end{bmatrix}$$

2.*   Evaluate exp A by direct application of Sylvester's theorem for the following matrix.

$$A = \begin{bmatrix} 7 & -2 & 4 \\ 2 & 3 & 0 \\ -3 & 1 & -1 \end{bmatrix}$$

3.*   Evaluate $\sin(A)$ by use of Sylvester's theorem for the following matrix.

$$A = \begin{bmatrix} 1 & -1 \\ -1 & 1 \end{bmatrix}$$

4.*   Evaluate $\sin(A)$ by use of Sylvester's theorem for the following matrix.

$$A = \begin{bmatrix} 1 & -2 & -1 \\ -4 & -1 & 5 \\ 3 & 0 & 1 \end{bmatrix}$$

***

10. THE USE OF THE CHARACTERISTIC EQUATION IN EVALUATING FUNCTIONS OF MATRICES

The use of Sylvester's theorem to evaluate functions of matrices is a laborious procedure. In this section, we show that a function of a matrix can be evaluated by the use of its own characteristic equation. In the case of matrix polynomials, this procedure is a generalization of the "remainder theorem" of ordinary algebra.

Let $f(x)$ be a given polynomial. Let $f(x)$ be divided by a polynomial $p(x)$ of lesser degree than $f(x)$.

$$\frac{f(x)}{p(x)} = q(x) + \frac{r(x)}{p(x)},$$

where $q(x)$ is the quotient and $r(x)$ is the remainder of the division. In this form, the degree of $r(x)$ is less than the degree of $p(x)$. Now if this is multiplied through by $p(x)$, the result is

(10.1) $f(x) = p(x)q(x) + r(x).$

Now let $p(a) = 0$; then if $x = a$, we have

$$f(a) = r(a).$$

EXAMPLE 8

Evaluate the polynomial $A^4 + 6A$ where

$$A = \begin{bmatrix} 1 & -3 \\ 3 & 1 \end{bmatrix}.$$

SOLUTION

The characteristic equation of A is

$$p(\mu) = \mu^2 - 2\mu + 10.$$

Now consider the scalar polynomial $f(x) = x^4 + 6x$, and divide it by the polynomial $p(x) = x^2 - 2x + 10$, the characteristic polynomial of A. The result of this division is

$$\frac{f(x)}{p(x)} = x^2 + 2x - 6 + \frac{-26x + 60}{p(x)}.$$

If this is multiplied by $p(x)$, the result is

$$f(x) = (x^2 + 2x - 6)p(x) + (-26x + 60).$$

Now if the matrix A is substituted in place of x we have $p(A) = 0$ by the Cayley-Hamilton theorem, and therefore,

$$f(A) = -26A + 60I.$$

***

This example suggests the following method for evaluating matrix polynomials:

If $f(A)$ is a polynomial of an nth-order matrix A and if the characteristic equation of A is $p(A) = 0$, then $f(A)$ is equal to $r(A)$, where $r(\mu)$ is the remainder when $f(\mu)$ is divided by $p(\mu)$.

The proof of this method follows from a consideration of the identity (10.1). If the scalar x is replaced by the matrix A, the identity still holds, and since $p(A) = 0$ by the Cayley-Hamilton theorem, we have $f(A) = r(A)$.

The above method applies only to the evaluation of matrix polynomials.

Matrix Functions

Now let it be supposed that it is desired to evaluate $f(A)$ where $f(\mu)$ is a function of μ, which has an infinite power series expansion in μ. To evaluate this function, we shall follow a procedure similar to that of (10.1). That is, we look for a polynomial $r(\mu)$ of degree $\leq (n-1)$ such that

$$(10.2) \qquad f(\mu) = p(\mu)q(\mu) + r(\mu),$$

where $p(\mu)$ is the characteristic polynomial of A and $q(\mu)$ is an analytic function of μ.

The polynomial $r(\mu)$ will have the general form

$$(10.3) \qquad r(\mu) = a_0 + a_1\mu + a_2\mu^2 + \ldots + a_{n-1}\mu^{n-1},$$

where the coefficients a_i are to be determined by the following procedure. In (10.2), we substitute the eigenvalues of A, $\mu_1, \mu_2, \ldots, \mu_n$ for μ. Now since $p(\mu)$ is the characteristic polynomial of A, we have $p(\mu_i) = 0$ for $i = 1, 2, \ldots, n$; by this procedure, we obtain the following equations from (10.2):

$$\begin{bmatrix} f(\mu_1) & = & r(\mu_1) \\ f(\mu_2) & = & r(\mu_2) \\ & \vdots & \\ f(\mu_n) & = & r(\mu_n) \end{bmatrix}$$

These comprise n linear equations that enable the n coefficients a_i of the polynomial $r(\mu)$ of (10.3) to be determined. Having determined $r(\mu)$ by this procedure, we may now solve (10.2) for $q(\mu)$ in the following form:

$$q(\mu) = \frac{f(\mu) - r(\mu)}{p(\mu)}$$

Now since equation (10.2) holds for all values of μ in the region of convergence of $f(\mu)$, we may substitute the matrix A for μ in (10.2) and get the following identity:

$$f(A) = p(A)q(A) + r(A)$$

If we now use the fact that $q(A) = 0$ as a consequence of the Cayley-Hamilton theorem, we find that

$$f(A) = r(A).$$

The use of this technique for the evaluation of matrix functions will now be illustrated by some examples.

EXAMPLE 9

Apply the method of this section to evaluate the exponential function $\exp(A)$ when

$$A = \begin{bmatrix} 1 & 0 \\ 0 & 2 \end{bmatrix}.$$

This is example 6 of the previous section.

SOLUTION

To do this, we assume that

$$e^A = a_0 I_2 + a_1 A, \quad A = \begin{bmatrix} 1 & 0 \\ 0 & 2 \end{bmatrix}.$$

In this case, the eigenvalues are $\mu_1 = 2$, $\mu_2 = 1$. Hence, the two equations to determinant a_0 and a_1 are

$$e^2 = a_0 + 2a_1$$
$$e = a_0 + a_1.$$

Hence,

$$a_1 = e^2 - e, \quad a_0 = 2e - e^2.$$

Therefore,

$$e^A = (2e - e^2)I_2 + (e^2 - e)A = \begin{bmatrix} e & 0 \\ 0 & e^2 \end{bmatrix}.$$

This is the result given by (9.2). It should be clear that this approach provides a somewhat easier method.

EXAMPLE 10

Repeat example 7, which is the problem of obtaining the power A^{256} of the matrix

$$A = \begin{bmatrix} 1 & 0 \\ 0 & 3 \end{bmatrix}.$$

SOLUTION

To obtain this power, we write

(10.4) $A^{256} = a_0 I_2 + a_1 A = \begin{bmatrix} (a_0 + a_1) & 0 \\ 0 & (a_0 + 3a_1) \end{bmatrix}.$

Since the eigenvalues of A were found to be $\mu_1 = 1$ and $\mu_2 = 3$, we have the following two equations from which to determine the constants a_0 and a_1.

$$1 = a_0 + a_1$$

$$3^{256} = a_0 + 3a_1$$

If these are substituted into (10.4), the result is

$$A^{256} = \begin{bmatrix} 1 & 0 \\ 0 & 3^{256} \end{bmatrix}.$$

This is the same result given by (11.3). Again, the process has been somewhat easier than the direct application of Sylvester's theorem.

EXAMPLE 11

As a third example, let it be required to evaluate the matrix exponential function e^A when the matrix

$$A = \begin{bmatrix} 0 & 1 \\ 1 & 0 \end{bmatrix}.$$

SOLUTION

In this case, the characteristic equation of A is $p(\mu) = \mu^2 - 1 = 0$, and the eigenvalues are $\mu_1 = 1$ and $\mu_2 = -1$. To evaluate e^A, we write

(10.5) $e^A = a_0 I_2 + a_1 A.$

In this case, the two equations to determine a_0 and a_1 are

$$e^1 = a_0 + a_1$$

and

$$e^{-1} = a_0 - a_1.$$

If we solve for a_0 and a_1 in the above equations, we obtain

$$a_0 = \frac{e^1 + e^{-1}}{2} = \cosh(1), \quad a_1 = \frac{e^1 - e^{-1}}{2} = \sinh(1).$$

If these values of a_0 and a_1 are now substituted into (10.5), the result is

(10.6) $e^A = \begin{bmatrix} \cosh(1) & \sinh(1) \\ \sinh(1) & \cosh(1) \end{bmatrix}.$

EXAMPLE 12

As a final example, let it be required to obtain a general expression for the positive integral powers of the matrix

$$T = \begin{bmatrix} a & b \\ c & d \end{bmatrix},$$

whose determinant is unity, so that $\mathrm{Det}\ T = (ad - bc) = 1$. In other words, we desire to find a simple expression for evaluating T^n when n is an integer.

SOLUTION

To evaluate the positive integral powers of T, we assume

(10.7) $T^n = a_0 I_2 + a_1 T.$

Since the determinant of T is unity, the characteristic equation of T is

$$p(\mu) = \mu^2 - (a + d)\mu + 1 = 0.$$

The eigenvalues of T therefore satisfy the equations

$$\mu_1 + \mu_2 = (a + d) = \operatorname{Tr} T$$

and

$$\mu_1\mu_2 = 1 = \operatorname{Det} T.$$

It is convenient to let $\mu_1 = e^m$ and $\mu_2 = e^{-m}$, which fulfills the last expression. Here, the parameter m is to be found. In this notation, the second equation above is satisfied identically, and the first equation may be written in the following form. Since

$$\frac{e^m + e^{-m}}{2} = \cosh(m) \quad,$$

we have

$$\cosh(m) = \frac{(a + d)}{2} = \frac{1}{2} \operatorname{Tr} T \quad.$$

In order to obtain the constants a_0 and a_1 in the expansion (10.7), we write the equations

$$(e^m)^n = e^{mn} = a_0 + a_1 e^m$$
$$(e^{-m})^n = e^{-mn} = a_0 + a_1 e^{-m} \quad.$$

The solutions of these equations are

$$a_0 = -\frac{\sinh[(n - 1)m]}{\sinh(m)} \quad, \quad a_1 = \frac{\sinh(nm)}{\sinh(m)} \quad.$$

If these values for a_0 and a_1 are now substituted into (10.7), we obtain the desired expression for the positive integral powers of the matrix T.

$$T^n = \frac{\sinh(nm)}{\sinh(m)} T - \frac{\sinh[(n - 1)m]}{\sinh(m)} I_2$$

It may be shown that this expression is valid for $n = 0, \pm1, \pm2, \pm3$, etc. It must be noted that the following restriction applies to this result. This result is very useful in the study of certain types of electric circuits as will be illustrated in chapter VIII.

***

In solving systems of differential equations, as will be seen in the next chapter, one frequently encounters the matrix exponential function in the form e^{At}. The matrix A is a square constant matrix, and t represents the independent variable in the problem. In applying either Sylvester's theorem or the method of this section to evaluating e^{At}, the reader should be aware of how the scalar independent variable t is handled.

To apply Sylvester's theorem, we write

$$f(A) = e^{At}.$$

Thus, (9.1) becomes in this case

$$e^{At} = \sum_{k=1}^{n} \frac{e^{\mu_k t}}{p'(\mu_k)} \, \text{Adj } K(\mu_k) \quad ,$$

where Adj $K(\mu_k)$ and $p'(\mu_k)$ remain as defined for (9.1).

For the method of this section to be applied, we must interpret the coefficients of (10.3), $a_0, \ldots, a_{n-1}$ as functions of t as follows:

$$a_0 = a_0(t), \, a_1 = a_1(t), \, \ldots, \, a_{n-1} = a_{n-1}(t)$$

EXAMPLE 13

Evaluate e^{At}, where A is the same matrix used in the first example of this section. That is,

$$A = \begin{bmatrix} 3 & 2 \\ 1 & 0 \end{bmatrix}.$$

SOLUTION

Based on the previous discussion and also noting that A is a second-order matrix, we write

(10.8) $e^{At} = a_0(t)I + a_1(t)A$

In example 1, we noted that the eigenvalues are 1 and 2; hence,

$$e^{2t} = a_0 + 2a_1$$
$$e^t = a_0 + a_1 .$$

From these equations, we solve for a_0 and a_1.

$$a_0 = 2e^t - e^{2t}, \quad a_1 = e^{2t} - e^t$$

Substituting these into (10.8) and rearranging terms gives the desired result.

$$e^{At} = e^t \begin{bmatrix} 1 & 0 \\ 0 & 0 \end{bmatrix} + e^{2t} \begin{bmatrix} 0 & 0 \\ 0 & 1 \end{bmatrix} = \begin{bmatrix} e^t & 0 \\ 0 & e^{2t} \end{bmatrix}$$

EXERCISES

1.*   Evaluate exp A by the method of this section where

$$A = \begin{bmatrix} 0 & 1 \\ 1 & 0 \end{bmatrix}.$$

2.*   Evaluate exp At by the method of this section where

$$A = \begin{bmatrix} 7 & -2 & 4 \\ 2 & 3 & 0 \\ -3 & 1 & -1 \end{bmatrix}.$$

3.*   Evaluate $\sin(A)$ by the method of this section for the following A matrix.

$$A = \begin{bmatrix} 1 & -1 \\ -1 & 1 \end{bmatrix}$$

4.*   Evaluate A^n by the method presented in this section with n being an integer.

$$A = \begin{bmatrix} 2 & -2 \\ 4 & -1 \end{bmatrix}$$

***

11. THE EVALUATION OF MATRIX FUNCTIONS BY THE APPLICATION OF THE SIMILARITY TRANSFORMATION

In a very large percentage of applications, the matrices describing a given problem are diagonalizable or, more particularly, real and symmetric. In these cases, there is a fairly direct method by which the function of a matrix may be evaluated. First consider an nth-order matrix A that is diagonalizable. This means that A has n linearly independent eigenvectors that in turn guarantee a nonsingular modal matrix. Also suppose that the eigenvalues and associated eigenvectors are known. Let D be the diagonal matrix whose diagonal elements are the eigenvalues of A. Also let M be the matrix of associated eigenvectors and thus is the modal matrix of A. From the basic definition of the eigenvalue problem, we have the following relation.

$$AM = MD$$

From this relation, the following two useful relations are obtained.

(11.1) $$M^{-1}AM = D \text{ and } A = MDM^{-1}$$

These relations are frequently referred to as similarity transformations.

Now let it be desired to evaluate an arbitrary function of the matrix A, which will be referred to as $f(A)$. For this function to exist, its scalar equivalent must be expressible as a converging infinite series. As we have demonstrated above, as a result of the Cayley-Hamilton theorem, the matrix function may be represented by a finite polynomial of order $n-1$. This means that

(11.2) $$f(A) = c_0 I_n + c_1 A + \ldots + c_{n-1} A^{n-1}.$$

In the previous section, we have introduced means of evaluating the constants c_i. To show the alternative approach, the matrix A and its powers will be replaced by the second relation of (11.1). In performing this substitution, we must make note of the relation that

$$A^2 = (MDM^{-1})(MDM^{-1}) = MD^2M^{-1}$$

and

$$A^3 = (MDM^{-1})(MDM^{-1})(MDM^{-1}) = MD^3M^{-1},$$

and in general, $A^k = MD^kM^{-1}$. Because of the properties of diagonal matrices, a diagonal matrix raised to any integral power is a diagonal matrix whose diagonal elements are those of the original matrix raised to the same power.

Now continuing with the replacement of A and its powers in (11.2) by the second relation of (11.1), the following relation may be obtained.

$$f(A) = c_0 I_n + c_1 MDM^{-1} + \ldots + c_{n-1} MD^{n-1} M^{-1}$$

Since we may write $I_n = MM^{-1} = MI_n M^{-1}$, the above relation may be expressed as

(11.3) $$f(A) = M[c_0 I_n + c_1 D + \ldots + c_{n-1} D^{n-1}] M^{-1},$$

where the modal matrix and its inverse have been factored out on the left and right sides respectively. Comparing the matrix expression in the brackets with equation (11.2), we can recognize that it is

$$f(D) = c_0 I_n + c_1 D + \ldots + c_{n-1} D^{n-1}.$$

Because D is a diagonal matrix with the eigenvalues μ_i of A on the main diagonal, (11.7) may be seen to reduce to

$$f(D) = \text{Diag}[\, f(\mu_i) \,].$$

This is a diagonal matrix with the scalar function $f(\mu_i)$ along the main diagonal.

Applying this result to equation (11.3) gives the desired formula by which the similarity transformation is used to evaluate the function of a matrix.

(11.4) $$f(A) = M \,\text{Diag}[\, f(\mu_i) \,] \, M^{-1}$$

EXAMPLE 14

Evaluate the matrix exponential function $\exp(A)$ where

$$A = \begin{bmatrix} 0 & 1 \\ 1 & 0 \end{bmatrix}.$$

This is the same problem as example 3 of the previous section.

SOLUTION

The eigenvalues and eigenvectors of this matrix are

$$\mu_1 = 1, \; E_1 = \begin{bmatrix} 1 \\ 1 \end{bmatrix}, \; \mu_2 = -1, \; E_2 = \begin{bmatrix} 1 \\ -1 \end{bmatrix}.$$

The modal matrix is

$$M = \begin{bmatrix} 1 & 1 \\ 1 & -1 \end{bmatrix}.$$

From (11.4), we may write

$$e^A = M \begin{bmatrix} e^1 & 0 \\ 0 & e^{-1} \end{bmatrix} M^{-1}$$

$$= \begin{bmatrix} 1 & 1 \\ 1 & -1 \end{bmatrix} \begin{bmatrix} e^1 & 0 \\ 0 & e^{-1} \end{bmatrix} \left(\frac{1}{-2}\right) \begin{bmatrix} -1 & -1 \\ -1 & 1 \end{bmatrix}.$$

Carrying out the matrix multiplications yields the final answer.

$$e^A = \begin{bmatrix} \cosh(1) & \sinh(1) \\ \sinh(1) & \cosh(1) \end{bmatrix}$$

In this answer, we have made use of the simplifying expressions

$$\frac{1}{2}(e^1 + e^{-1}) = \cosh(1)$$

$$\frac{1}{2}(e^1 - e^{-1}) = \sinh(1) \quad .$$

This answer is the same as given in equation (10.6).

***

It should be evident to the reader that this method is the preferred way for evaluating functions of matrices when the matrix can be reduced to diagonal form by a similarity transformation.

At this point, the reader might be wondering how this method would apply to evaluating the matrix function exp(At) that occurs frequently in the solution of matrix differential equations as noted above. The approach would be identical to the previous except that the elements on the diagonal matrix for exp(Dt) become exp($\mu_i t$).

Although this method is usually the easiest approach for evaluating matrix functions, it may still appear that extensive multiplications are involved in expanding (11.4) to obtain a desired expression. If the operations involved are carefully examined and the properties of diagonal matrices are taken into account, equation (11.4) may be written in a fairly simple expanded form. To do this, let the rows of M^{-1} be designated by R_i. Also recall that the columns of M are the eigenvectors E_i. Using this notation, equation (11.4) is expanded into the following convenient form.

$$f(A) = E_1 R_1 f(\mu_1) + E_2 R_2 f(\mu_2) + \cdots + E_n R_n f(\mu_n)$$

$$= \sum_{i=1}^{n} E_i R_i f(\mu_i)$$

This expansion is quite useful to evaluate matrix functions with matrix software programs that do not have any or limited symbolic capabilities.

EXAMPLE 15

Evaluate the matrix function $\exp(At)$, where

$$A = \begin{bmatrix} -3 & 2 & 2 \\ 4 & 2 & -2 \\ -6 & 0 & 4 \end{bmatrix}.$$

SOLUTION

The eigenvalues and associated eigenvectors of A are the following:

$$\mu_1 = 1, \ E_1 = \begin{bmatrix} 1 \\ 0 \\ 2 \end{bmatrix}, \qquad \mu_2 = -2, \ E_2 = \begin{bmatrix} -2 \\ 1 \\ -2 \end{bmatrix}$$

$$\mu_3 = 4, \ E_3 = \begin{bmatrix} 0 \\ 1 \\ -1 \end{bmatrix}$$

The modal matrix of A is

$$M = \begin{bmatrix} 1 & -2 & 0 \\ 0 & 1 & 1 \\ 2 & -2 & -1 \end{bmatrix}.$$

The inverse of the modal matrix is now

$$M^{-1} = \begin{bmatrix} -1/3 & 2/3 & 2/3 \\ -2/3 & 1/3 & 1/3 \\ 2/3 & 2/3 & -1/3 \end{bmatrix}$$

From this result, the respective rows are

$$R_1 = [\, -1/3 \; 2/3 \; 2/3 \,], \; R_2 = [\, -2/3 \; 1/3 \; 1/3 \,]$$

$$R_3 = [\, 2/3 \; 2/3 \; -1/3 \,]$$

From the relationship given above, we may now write the following.

$$\exp(At) = E_1 R_1 \exp(\mu_1 t) + E_2 R_2 \exp(\mu_2 t) + E_3 R_3 \exp(\mu_3 t)$$

$$= \begin{bmatrix} 1 \\ 0 \\ 2 \end{bmatrix} [\, -1/3 \; 2/3 \; 2/3 \,] \, e^t \; +$$

$$\begin{bmatrix} -2 \\ 2 \\ -2 \end{bmatrix} [\, -2/3 \; 1/3 \; 1/3 \,] \, e^{-2t} \; +$$

$$\begin{bmatrix} 0 \\ 1 \\ -1 \end{bmatrix} [\, 2/3 \; 2/3 \; -1/3 \,] \, e^{4t}$$

Performing the three matrix multiplications gives the following:

$$e^{At} = \frac{1}{3} \begin{bmatrix} -1 & 2 & 2 \\ 0 & 0 & 0 \\ -2 & 4 & 4 \end{bmatrix} e^t + \frac{1}{3} \begin{bmatrix} 4 & -2 & -2 \\ -2 & 1 & 1 \\ 4 & -2 & -2 \end{bmatrix} e^{-2t}$$

$$+ \frac{1}{3} \begin{bmatrix} 0 & 0 & 0 \\ 2 & 2 & -1 \\ -2 & -2 & 1 \end{bmatrix} e^{4t}$$

EXERCISES

1.*   Evaluate A^{500} by the method of this section if

$$A = \begin{bmatrix} 2 & -2 & 3 \\ 1 & 1 & 1 \\ 1 & 3 & -1 \end{bmatrix}.$$

2.*   Evaluate exp$A$ by means of a similarity transformation for the following matrix.

$$A = \begin{bmatrix} 0 & 1 \\ 1 & 0 \end{bmatrix}$$

3.*   Evaluate exp$A$ by means of a similarity transformation for the following matrix.

$$A = \begin{bmatrix} 7 & -2 \\ 2 & 3 \end{bmatrix}$$

4.*   Evaluate $\sin(A)$ by means of a similarity transformation for the following matrix.

$$A = \begin{bmatrix} 1 & -1 \\ -1 & 1 \end{bmatrix}$$

5.*   Evaluate $\exp(At)$ by means of a similarity transformation for the following matrix.

$$A = \begin{bmatrix} 1 & -2 & 1 \\ -2 & 3 & 0 \\ 1 & 0 & 1 \end{bmatrix}$$

12. DIFFERENTIATION AND INTEGRATION OF MATRICES

Before we begin the discussion of multiple eigenvalues and related topics, we will need to be able to take the derivative of a matrix. Therefore, at this point, let us introduce the topic of differentiation and integration of matrices in general.

Let $A(t)$ be a matrix whose elements $a_{ij} = a_{ij}(t)$ are functions of the scalar variable t.

(12.1)
$$A(t) = \begin{vmatrix} A_{11}(t) & A_{12}(t) & \cdots & A_{1n}(t) \\ A_{21}(t) & A_{22}(t) & \cdots & A_{2n}(t) \\ \cdot & \cdot & \cdots & \cdot \\ A_{m1}(t) & A_{m1}(t) & \cdot & A_{mn}(t) \end{vmatrix}$$

Then we define the derivative of $A(t)$ with respect to the variable t as

$$\frac{d}{dt}A(t) ,$$

which, in terms of elements, is

(12.2)
$$\frac{d}{dt}A(t) = \left[\frac{da_{ij}(t)}{dt}\right] = \dot{A}(t) = [\dot{a}_{ij}(t)]$$

In this expression, we have introduced the shorthand notation of using a dot above a term to denote the derivative with respect to the independent variable t. Equation (12.2) expresses the result that the derivative of a matrix whose elements are functions of a scalar variable t is defined to be the matrix of the derivatives of the elements with respect to the variable t.

By the use of the definition of the derivative of a matrix, it is easy to show that

$$\frac{d}{dt}\{A(t) + B(t)\} = \dot{A}(t) + \dot{B}(t) = [\dot{a}_{ij} + \dot{b}_{ij}]$$

$$\frac{d}{dt}A(t)B(t) = \dot{A}(t)B(t) + A(t)\dot{B}(t)$$

$$\frac{d}{dt}A^2(t) = \dot{A}(t)A(t) + A(t)\dot{A}(t)$$

$$\frac{d}{dt}A^{-1}(t) = -A^{-1}(t)\dot{A}(t)A(t)$$

If t is a real variable and A is a matrix whose elements are constants, then one obtains

$$\frac{d}{dt}t^kA = kt^{k-1}A$$

Consider the following matrix exponential function:

$$e^{At} = I + At + \frac{1}{2!} A^2t^2 + \frac{1}{3!} A^3t^3 + \cdots$$

Then if this expression is differentiated with respect to t, one obtains the following:

$$\frac{d}{dt}e^{At} = A + A^2t + \frac{1}{2!}A^3t^2 + \frac{1}{3!}A^4t^3 + \cdots$$

$$= Ae^{At} = e^{At}A$$

This result is of importance in the solution of linear differential equations with constant coefficients by the use of matrices as will be demonstrated in the following chapter.

The integral of the matrix $A(t)$ of (12.1) is defined as the matrix of the integrals of the various elements of the matrix. That is, we define the integral of $A(t)$ to be

$$\int A(t)\, dt = \left[\int a_{ij}(t)\, dt \right].$$

EXERCISES

1.   Evaluate the derivative of $\sin(At)$.

2.   Evaluate the derivative of $\cos(At)$.

3.   For the matrix A, evaluate the derivative of its characteristic matrix $K(\mu)$ with respect to the eigenvalue μ.

$$A = \begin{bmatrix} 2 & -1 \\ 3 & 4 \end{bmatrix}$$

4.   Determine the adjoint of the characteristic matrix found in the third exercise above, and compute its derivative with respect to the eigenvalue μ.

***

13. THE CASE OF MULTIPLE EIGENVALUES

When multiple eigenvalues occur, there are two important cases to consider. The first is the determination of the associated eigenvectors, and the second is the evaluation of matrix polynomials and functions.

To discuss the case of associated eigenvectors for the occurrence of multiple eigenvalues, several important aspects and statements based on the theory[39] of linear equations and vector spaces are introduced. For the following sequence of properties, we will refer to the following terms. If A is an nth-order matrix under consideration, its eigenvalues and associated eigenvectors are denoted by μ_i and E_i respectively. The characteristic matrix of A will be designated as

$$K(\mu) = \mu I_n - A.$$

The characteristic equation of A is

(13. 1) $p(\mu) = | K(\mu) | = 0 = (\mu - \mu_1)(\mu - \mu_2) \ldots (\mu - \mu_n).$

As a result of the Cayley-Hamilton theorem, we know that the matrix A satisfies its own characteristic equation, and we can therefore write

$$p(A) = 0_n = [A - \mu_1 I_n][A - \mu_2 I_n] \ldots [A - \mu_n I_n]$$

$$= K(\mu_1) K(\mu_2) \cdots K(\mu_n) = \prod_{i=1}^{n} K(\mu_i)$$

As a means of introducing properties to be presented, the following two example problems are solved.

EXAMPLE 16

Consider the matrix A below whose characteristic polynomial is $p(\mu) = (\mu - 3)^2(\mu - 12)$. This matrix has a unique eigenvalue of 12 and a double (repeated) eigenvalue of 3. Show that the characteristic matrix of A is singular for

---

[39]   The situation of multiple eigenvalues is covered at length in Frazer, Duncan, and Collar, *Elementary Matrices* [1957]. The example problems presented in this section are also taken from this reference.

each eigenvalue, and evaluate its rank in each case. Also find the eigenvectors by the Gaussian elimination method.

$$A = \begin{bmatrix} 7 & 4 & -1 \\ 4 & 7 & -1 \\ -4 & -4 & 4 \end{bmatrix}$$

SOLUTION

The characteristic matrix is

$$K(\mu) = \mu I_3 - A = \begin{bmatrix} \mu - 7 & -4 & 1 \\ -4 & \mu - 7 & 1 \\ 4 & 4 & \mu - 4 \end{bmatrix}.$$

For the unique eigenvalue of 12, we have

$$K(12) = \begin{bmatrix} 5 & -4 & 1 \\ -4 & 5 & 1 \\ 4 & 4 & 8 \end{bmatrix}.$$

Checking the rank shows that rank $K(12) = 2$; that can be seen by adding the first column to the second. The rank being two tells us that there is only one linearly independent eigenvector associated with this eigenvalue. The number of independent nontrivial solutions is the difference between the order—three in this case—and the rank. Also, because the rank is less than the order, the characteristic matrix is singular.

To find the first eigenvector E_1, we must solve the homogeneous equation $K(12)E_1 = 0$ for its nontrivial solution.

$$K(12)E_1 = \begin{bmatrix} 5 & -4 & 1 \\ -4 & 5 & 1 \\ 4 & 4 & 8 \end{bmatrix} \begin{bmatrix} e_1^1 \\ e_2^1 \\ e_3^1 \end{bmatrix} = \begin{bmatrix} 0 \\ 0 \\ 0 \end{bmatrix}$$

We have used the superscript one on the elements of the eigenvector to denote them as being elements of the first eigenvector, E_1. To find the nontrivial solution, the first step is to reduce the coefficient matrix to upper triangular form. This is accomplished by means of premultiplying by a series of elementary matrices. The first required elementary matrix is

$$EL_1 = \begin{bmatrix} 1 & 0 & 0 \\ 4/5 & 1 & 0 \\ -4/5 & 0 & 1 \end{bmatrix}.$$

Premultiplying the characteristic matrix by EL_1 gives the first modified K matrix, K_1.

$$EL_1 K(12) = K_1 = \begin{bmatrix} 5 & -4 & 4 \\ 0 & 1.8 & 1.8 \\ 0 & 7.2 & 7.2 \end{bmatrix}$$

The next elementary matrix is

$$EL_2 = \begin{bmatrix} 1 & 0 & 0 \\ 0 & 1 & 0 \\ 0 & -7.2/1.8 & 1 \end{bmatrix}.$$

Premultiplying K_1 by this yields the final modified K matrix.

$$K_2 = \begin{bmatrix} 5 & -4 & 1 \\ 0 & 1.8 & 1.8 \\ 0 & 0 & 0 \end{bmatrix}$$

To solve $K_2E_1 = 0$, the last element of E_1 may be set arbitrarily. For convenience, it is set to a value of 1. By back substitution, we find the following result for the desired eigenvector.

$$E_1 = \begin{bmatrix} -1 \\ -1 \\ 1 \end{bmatrix}$$

For the repeated eigenvalue of 3, the characteristic matrix becomes

$$K(3) = \begin{bmatrix} -4 & -4 & 1 \\ -4 & -4 & 1 \\ 4 & 4 & -1 \end{bmatrix}.$$

In this case, there is only one independent row; hence, the rank is only one. This means that there will be two linearly independent eigenvectors associated with the double eigenvalue. Again using Gaussian elimination (chapter II, section 4), we find the remaining two eigenvectors.

$$
E_2 = \begin{bmatrix} 1 \\ 0 \\ 4 \end{bmatrix}, \qquad E_3 = \begin{bmatrix} -1 \\ 1 \\ 0 \end{bmatrix}
$$

In this situation, the modal matrix for the given A matrix will have an inverse, and A may be diagonalized. Also, any function of A can be evaluated by means of a similarity transformation. Finally, the characteristic matrix is also singular. In fact, it should be noted, since finding eigenvalues requires solving for nontrivial solutions of the characteristic matrix, it must always be singular.

EXAMPLE 17

Consider the matrix A below whose characteristic polynomial is $p(\mu) = (\mu - 1)^2(\mu - 2)$. This matrix has a unique eigenvalue of 2 and a double (repeated) eigenvalue of 1. Show that the characteristic matrix of A is singular for each eigenvalue, and evaluate its rank in each case. Also find the eigenvectors.

$$
A = \begin{bmatrix} 2 & -2 & 3 \\ 10 & -4 & 5 \\ 5 & -4 & 6 \end{bmatrix}
$$

SOLUTION

The characteristic matrix is

$$
K(\mu) = \mu I_3 - A = \begin{bmatrix} \mu - 2 & 2 & -3 \\ -10 & \mu + 4 & -5 \\ -5 & 4 & \mu - 6 \end{bmatrix}.
$$

For the unique eigenvalue of 2, we have

$$
K(2) = \begin{bmatrix} 0 & 2 & -3 \\ -10 & 6 & -5 \\ -5 & 4 & -4 \end{bmatrix}.
$$

Checking the rank shows that rank $K(2) = 2$. As in the previous example, the matrix is singular, and there will be only one linearly independent eigenvector associated with this eigenvalue.

Employing Gaussian elimination yields the first eigenvector.

$$E_1 = \begin{bmatrix} 4 \\ 15 \\ 10 \end{bmatrix}$$

The reader should note that we have taken advantage of the fact that eigenvectors can only be determined up to an arbitrary multiplicative constant. Thus, E_1 has been scaled so that all of its elements are integers.

For the repeated eigenvalue of 1, the characteristic matrix becomes

$$K(1) = \begin{bmatrix} -1 & 2 & -3 \\ -10 & 5 & -5 \\ -5 & 4 & -5 \end{bmatrix}$$

In this situation, the rank is two, which means that there will be only one eigenvector associated with the double eigenvalue. As should be expected, this matrix is singular. Again employing Gaussian elimination, the remaining eigenvector is found to be

$$E_2 = \begin{bmatrix} 1 \\ 5 \\ 3 \end{bmatrix}.$$

***

Because we could only find two linearly independent eigenvectors for the given matrix in the second example above, the modal matrix cannot be constructed. This means that a similarity transformation is not available for evaluating any function of the matrix. However, there are means for dealing with such cases as we will see later on in this chapter.

These two examples provide illustrations of the following three properties.

Property 1

The characteristic matrix is necessarily singular for $\mu = \mu_i$, an eigenvalue. For distinct eigenvalues, the rank of $K(\mu_i) = n - 1$.

As we noted above, the characteristic matrix $K(\mu_i)$ is always singular when μ is an eigenvalue. In both of the above examples, the ranks of the characteristic matrices were two for the unique eigenvalues. This is one less than the order of the given matrix as stated in this property. This also means that there is only one associated eigenvector to be found as a solution to $K(\mu_i)E_i = 0_3$.

Property 2

If the rank of $K(\mu_i)$ is $n - q$, then at least q of the n eigenvalues are multiple and equal to μ_i.

In the example above, the rank of the characteristic matrix for the repeated eigenvalue was 1, for which q would have the value 2. Also, the number of linearly independent eigenvectors found in this case is q or 2. In example 17, the rank of the characteristic matrix for the repeated eigenvalue is 2, which means the q has the value of 1 for this case. Properties 1 and 2 only tell us that when q has a value of 1, the eigenvector may be unique, but only one associated eigenvector may be found.

Property 3

If μ_i is a multiple eigenvalue, the rank of $K(\mu_i)$ is not necessarily $< (n - 1)$.

The example above is a case where the rank of $K(\mu_i)$ is < 1 for the repeated eigenvalue. However, in the succeeding example, the rank is $n - 1$ for the repeated eigenvalue.

Before stating the next two properties, let us consider the following examples.

EXAMPLE 18

Determine the adjoint of the characteristic matrix for the matrix of example 16 above, and determine its rank for each eigenvalue. If the adjoint of the characteristic matrix becomes the zero matrix for a repeated eigenvalue, evaluate the rank of the derivative of the adjoint matrix for that eigenvalue. The derivative is taken with respect to the eigenvalue μ. Test all linear independent columns found in the adjoints and their derivatives to see if they are eigenvectors.

SOLUTION

From example 16, the given matrix and its characteristic matrix are the following.

$$A = \begin{bmatrix} 7 & 4 & -1 \\ 4 & 7 & -1 \\ -4 & -4 & 4 \end{bmatrix}, \quad K(\mu) = \begin{bmatrix} \mu - 7 & -4 & 1 \\ -4 & \mu - 7 & 1 \\ 4 & 4 & \mu - 4 \end{bmatrix}$$

From the characteristic matrix, we can determine its adjoint, which is

$$\text{Adj } K(\mu) = \begin{bmatrix} (\mu - 3)(\mu - 8) & 4(\mu - 3) & -(\mu - 3) \\ 4(\mu - 3) & (\mu - 3)(\mu - 8) & -(\mu - 3) \\ -4(\mu - 3) & -4(\mu - 3) & (\mu - 3)(\mu - 11) \end{bmatrix}$$

For the unique eigenvalue of 12, we have

$$\text{Adj } K(12) = \begin{bmatrix} 36 & 36 & -9 \\ 36 & 36 & -9 \\ -36 & -36 & 9 \end{bmatrix}$$

The rank of this matrix is one, which is another indication of the existence of a single eigenvector associated with the unique eigenvalue. In fact, it is important to note that all of the columns of this adjoint matrix are proportional to the eigenvector found for the eigenvalue $\mu_1 = 12$ in the example above. That is, all the columns of the above matrix are proportion to the eigenvector

$$E_1 = \begin{bmatrix} -1 \\ -1 \\ 1 \end{bmatrix}.$$

For the repeated eigenvalue of 3, we find

$$\text{Adj } K(3) = \begin{bmatrix} 0 & 0 & 0 \\ 0 & 0 & 0 \\ 0 & 0 & 0 \end{bmatrix}.$$

Since this is a zero or null matrix, we take the derivative of the adjoint matrix with respect to the eigenvalue to obtain the following result.

$$\frac{d}{d\mu} \text{Adj } K(\mu) = \text{Adj' } K(\mu) = \begin{bmatrix} 2\mu - 11 & 4 & -1 \\ 4 & 2\mu - 11 & -1 \\ -4 & -4 & 2\mu - 14 \end{bmatrix}$$

Evaluating this for the repeated eigenvalue we find

$$\text{Adj' } K(3) = \begin{bmatrix} -5 & 4 & -1 \\ 4 & -5 & -1 \\ -4 & -4 & -8 \end{bmatrix}$$

The rank of this matrix is 2 indicating that it contains two linearly independent columns. Since the third column is the sum of the first two, we will select them arbitrarily as potentially the remaining two eigenvectors.

$$E_2 = \begin{bmatrix} -5 \\ 4 \\ -4 \end{bmatrix}, \qquad E_3 = \begin{bmatrix} 4 \\ -5 \\ -4 \end{bmatrix}$$

It is now easy to verify that these two vectors are eigenvectors of the original matrix since they may be shown to be homogeneous solutions to the equation $K(3)E_i = 0_3$.

***

If we look back to example 16, we see that the two eigenvectors vectors found there are not identical to the second and third eigenvectors found above. It is easy to show that those vectors as well as these last two are all solutions to the homogeneous equation $K(3)E_i = 0_3$. This shows they are all eigenvectors, and any set of two may be used for the remaining two eigenvectors. To demonstrate this, let us form a modal matrix using the associated eigenvector found for the eigenvalue of 12 and the first vector from each of the two sets found for the repeated eigenvalue of 3.

$$M = \begin{bmatrix} -1 & 1 & -5 \\ -1 & 0 & 4 \\ 1 & 4 & -4 \end{bmatrix}$$

Performing a similarity transformation of the original A matrix produces the expected result.

$$M^{-1}AM = \begin{bmatrix} 12 & 0 & 0 \\ 0 & 3 & 0 \\ 0 & 0 & 3 \end{bmatrix}$$

EXAMPLE 19

Determine the adjoint of the characteristic matrix for the matrix of previous example, and determine its rank for each eigenvalue. If the adjoint of the characteristic matrix becomes the zero matrix for a repeated eigenvalue, evaluate the rank of the derivative of the adjoint matrix for that eigenvalue. The derivative is to be taken with respect to the eigenvalue μ. Test all linear independent columns found in the adjoints and their derivatives to see if they are eigenvectors.

SOLUTION

The given matrix and its characteristic equation are

$$A = \begin{bmatrix} 2 & -2 & 3 \\ 10 & -4 & 5 \\ 5 & -4 & 6 \end{bmatrix}, \quad K(\mu) = \begin{bmatrix} \mu-2 & 2 & -3 \\ -10 & \mu+4 & -5 \\ -5 & 4 & \mu-6 \end{bmatrix}.$$

As given in the previous example, the eigenvalues are $\mu_1 = 2$ and the repeated value $\mu_2 = \mu_3 = 1$. The adjoint of the characteristic matrix is

$$\text{Adj } K(\mu) = \begin{bmatrix} \mu^2-2\mu-4 & -2\mu & 3\mu+2 \\ 10\mu-35 & \mu^2-8\mu-3 & 5\mu+20 \\ 5\mu-20 & -4\mu-2 & \mu^2+2\mu+12 \end{bmatrix}.$$

For the unique eigenvalue of 2, we have

$$\text{Adj } K(2) \;=\; \begin{bmatrix} -4 & -4 & 8 \\ -15 & -15 & 30 \\ -10 & -10 & 20 \end{bmatrix}.$$

Clearly, the rank of this matrix is one since all columns are proportional to the column [4, 15, 10]. This is also the associated eigenvector found in the previous example.

For the repeated eigenvalue $\mu_2 = \mu_3 = 1$, we have

$$\text{Adj } K(1) \;=\; \begin{bmatrix} -5 & -2 & 5 \\ -25 & -10 & 25 \\ -15 & -6 & 15 \end{bmatrix}.$$

Since all columns are proportional, the rank is one. The single independent column is

$$E_2 \;=\; \begin{bmatrix} 1 \\ 5 \\ 3 \end{bmatrix}.$$

This is the eigenvector associated with the repeated eigenvalue as obtained in the former example.

Property 4

   If the rank of $K(\mu_i) = n - 1$, then the rank of the adjoint of the characteristic matrix is identically one, that is

$$\text{Rank } [\text{Adj } K(\mu_i)] = 1.$$

Property 5

   If the rank of $K(\mu_i) = n - q$ and $q > 1$, then Adj $K(\mu_i)$ and all its derivatives with respect to μ_1 up to the order $k = q - 2$ are null matrices.

Prior to showing the implications of these properties, there is one further important observation to make. From the definition of the inverse of a matrix, the following relation can be written.

$$K(\mu) \text{ Adj } K(\mu) = p(\mu)I_n$$

Now if we set $\mu = \mu_i$, where μ_i is an eigenvalue of A, this becomes

$$K(\mu_i) \text{ Adj } K(\mu_i) = 0_n.$$

This relation demonstrates that the columns of the matrix Adj $K(\mu_i)$ are the eigenvectors of A. This also provides another way by which the eigenvectors may be found as will be demonstrated in the following examples.

On the basis of the above properties, the following statements may be made.

Statement 1

If the rank of $K(\mu_i)$ is $n - q_i$, then there will be q_i linearly independent eigenvectors associated with the eigenvalue μ_i.

Statement 2

If μ_i is an eigenvalue of multiplicity r_i and the rank of $K(\mu_i)$ is $n - q_i$ such that $1 \le q_i \le r_i$, then there will be q_i linearly independent eigenvectors associated with μ_i. In other words, the number of linearly independent eigenvectors is always $n - \text{rank } K(\mu_i)$.

Statement 3

In the special case $q_i = r_i$, for each multiple eigenvalue μ_i, the modal matrix of A, M will be nonsingular, and hence, A is reducible to diagonal form by the similarity transformation

$$M^1AM.$$

To conclude this section, let us return to the example above and investigate the derivative of the adjoint matrix for the repeated eigenvalue. Taking the derivative of the adjoint of the characteristic matrix with respect to the eigenvalue gives

$$\frac{d}{d\mu} \text{Adj } K(\mu) = \text{Adj' } K(\mu) = \begin{bmatrix} 2\mu - 2 & -2 & 3 \\ 10 & 2\mu - 8 & 5 \\ 5 & -4 & 2\mu + 2 \end{bmatrix}.$$

For the repeated eigenvalue, we find

$$\text{Adj' } K(1) = \begin{bmatrix} 0 & -2 & 3 \\ 10 & -6 & 5 \\ 5 & -4 & 4 \end{bmatrix}.$$

The rank of this matrix is two, indicating the presence of two linearly independent columns. From the first column, we may identify [0, 2, 1]. One obtains a convenient second one by taking the second column scaled by -2 to give [1, 3, 2]. Testing these columns in the homogeneous eigenvector problem shows that neither of these two vectors are eigenvectors.

Now between the adjoint of the characteristic matrix, Adj $K(1)$, and its derivative, Adj' $K(1)$, we have found three columns, namely

$$\begin{bmatrix} 1 \\ 5 \\ 3 \end{bmatrix}, \quad \begin{bmatrix} 0 \\ 2 \\ 1 \end{bmatrix}, \quad \begin{bmatrix} 1 \\ 3 \\ 2 \end{bmatrix}.$$

The first is an eigenvector associated with the repeated eigenvalue 1. The other two are associated with the repeated eigenvalue but are not eigenvectors. We must now determine if all three are linearly independent. After a little observation, we may note that

$$\begin{bmatrix} 1 \\ 5 \\ 3 \end{bmatrix} = \begin{bmatrix} 0 \\ 2 \\ 1 \end{bmatrix} + \begin{bmatrix} 1 \\ 3 \\ 2 \end{bmatrix}.$$

This shows that between Adj $K(1)$ and Adj' $K(1)$, there are only two linearly independent vectors. This corresponds with the multiplicity of the eigenvalue being 2. Because the rank of the characteristic matrix $K(1)$ is two, yielding $q = 1$, there is only one eigenvector associated with the repeated eigenvalue.

In summary, suppose μ_i is a repeated eigenvalue of A of multiplicity r_i and it does not have r_i linearly independent associated eigenvectors ($q_i < r_i$). Statement 4 below formalizes the means of finding r_i linear independent columns to associate with an eigenvalue of multiplicity r_i. As we have seen, all of these columns will be eigenvectors associated with the repeated eigenvalue. In a later section, we will see the use of this complete set of vectors.

Statement 4

Let μ_i be an eigenvalue of A with multiplicity r_i, and let Adj $K(\mu)$ represent the adjoint of the characteristic matrix of A. Then it is true that r_i and only r_i linearly independent vectors are included in the sequence of r_i matrices Adj $K(\mu_i)$, Adj' $K(\mu_i)$, . . . , $\text{Adj}^{(r-1)} K(\mu_i)$. The linearly independent vectors of this set that are not the eigenvectors of A are called generalized eigenvectors. The above series of matrices are the adjoint of the characteristic matrix and its derivatives with respect to the variable μ up through the r_i^{th} minus first order evaluated for the repeated eigenvalue μ_i.

It must be noted that because of the existence of a number of good quality software programs that are available to perform matrix operations, the method presented above for evaluating eigenvectors for repeated eigenvalues is not necessary. These programs will indeed generate the eigenvectors found above automatically once the given matrix has been provided to the program. However, most of these programs will fail to find a complete independent vector basis in cases like that of the first example above. As the remainder of the discussion of multiple eigenvalues is completed, further comments on this circumstance will be made to help the reader understand more of the details of the situation.

EXERCISES

Apply the method of this section to find as many of the eigenvectors of the following matrices as possible.

1. $\begin{bmatrix} 7 & -2 \\ 2 & 3 \end{bmatrix}$ 2*. $\begin{bmatrix} 2 & -1 & 2 \\ 0 & -4 & 6 \\ 0 & 0 & -4 \end{bmatrix}$ 3*. $\begin{bmatrix} 1 & 0 & -3 \\ -3 & 4 & -3 \\ -3 & 0 & 1 \end{bmatrix}$

4*. $\begin{bmatrix} 76 & -26 & -35 \\ 21 & 27 & -35 \\ 42 & -56 & -15 \end{bmatrix}$, $\mu_{1,2,3} = -22, 55, 55$

***

14. THE EVALUATION OF MATRIX POLYNOMIALS AND FUNCTIONS WITH REPEATED EIGENVALUES

When repeated eigenvalues occur, the methods presented in the previous sections may not work for the evaluation of matrix functions. The methods are applicable when a matrix is diagonalizable by a similarity transformation as presented at the end of the section 11. For a matrix of order n, this approach is possible whenever n linearly independent eigenvectors can be found. This applies whether repeated eigenvalues occur or not.

As for evaluating matrix polynomials, the reduction by way of the Cayley-Hamilton theorem previously discussed is always applicable. However, in the case of repeated roots, a further reduction is possible in some instances.

To begin, let us examine the characteristic equation in more detail. If all the eigenvalues are distinct, the characteristic equation has the following form.

$$p(\mu) = |K(\mu)| = 0 = (\mu - \mu_1)(\mu - \mu_2) \ldots (\mu - \mu_n)$$

However, this form changes if one or more of the eigenvalues are repeated. For convenience, assume the first eigenvalue, μ_1 is repeated r times. In this situation, the above result becomes

$$p(\mu) = (\mu - \mu_1)^r(\mu - \mu_{r+1})(\mu - \mu_{r+2}) \ldots (\mu - \mu_n),$$

where we have implied the notation $\mu_1 = \mu_2 = \ldots = \mu_r$ or that the first r eigenvalues are designated by μ_1. Clearly, all the eigenvalues also satisfy the following reduced or minimum polynomial.

$$p_m(\mu) = (\mu - \mu_1)(\mu - \mu_{r+1})(\mu - \mu_{r+2}) \ldots (\mu - \mu 1_n)$$

By the Cayley-Hamilton theorem, we know that a given matrix satisfies its own characteristic polynomial. However, we now must ask the question if a matrix satisfies its own minimum polynomial. In other words, if a matrix A has a repeated eigenvalue μ_1 of multiplicity r so that its minimum polynomial is the one above, does it satisfy the corresponding minimum matrix polynomial below?

$$p_m(A) = [\, A - \mu_1 I_n \,][\, A - \mu_{r+1} I_n \,][\, A - \mu_{r+2} I_n \,] \cdots [\, A - \mu_n I_n \,] \overset{?}{=} O_n$$

$$= K(\mu_1)\, K(\mu_{r+1})\, K(\mu_{r+2}) \cdots K(\mu_n) = K(\mu_1) \prod_{i=r+1}^{n} K(\mu_i)$$

To investigate this possibility, consider the following two example problems.

EXAMPLE 20

For the matrix A of example 16 above, which has a unique eigenvalue of 12 and a double eigenvalue of 3, evaluate the product of characteristic matrices $K(12)K(3)$. Note that this product is the minimum matrix polynomial for A.

SOLUTION

From example 16, we have

$$K(12) = \begin{bmatrix} 5 & -4 & 1 \\ -4 & 5 & 1 \\ 4 & 4 & 8 \end{bmatrix}, \qquad K(3) = \begin{bmatrix} -4 & -4 & 1 \\ -4 & -4 & 1 \\ 4 & 4 & -1 \end{bmatrix}.$$

The product of these two matrices gives the following result.

$$K(12)K(3) = \begin{bmatrix} 5 & -4 & 1 \\ -4 & 5 & 1 \\ 4 & 4 & 8 \end{bmatrix} \begin{bmatrix} -4 & -4 & 1 \\ -4 & -4 & 1 \\ 4 & 4 & -1 \end{bmatrix} = \begin{bmatrix} 0 & 0 & 0 \\ 0 & 0 & 0 \\ 0 & 0 & 0 \end{bmatrix}$$

This matrix does satisfy its minimum polynomial.

EXAMPLE 21

For the matrix A of example 17 above, which has a unique eigenvalue of 2 and a double eigenvalue of 1, evaluate the product of characteristic matrices $K(2)K(1)$. Note that this product is the minimum matrix polynomial for A.

SOLUTION

From example 17 above, we have

$$K(2) = \begin{bmatrix} 0 & 2 & -3 \\ -10 & 6 & -5 \\ -5 & 4 & -4 \end{bmatrix}, \quad K(1) = \begin{bmatrix} -1 & 2 & -3 \\ -10 & 5 & -5 \\ -5 & 4 & -5 \end{bmatrix}.$$

The product of these two matrices is

$$K(2)K(1) = \begin{bmatrix} 0 & 2 & -3 \\ -10 & 6 & -5 \\ -5 & 4 & -4 \end{bmatrix}\begin{bmatrix} -1 & 2 & -3 \\ -10 & 5 & -5 \\ -5 & 4 & -5 \end{bmatrix} = \begin{bmatrix} -5 & -2 & 5 \\ -25 & -10 & 25 \\ -15 & -6 & 15 \end{bmatrix}$$

In this case, the matrix does not satisfy its minimum polynomial.

It is of interest in this case to evaluate the matrix product $K(2)K(1)^2$ since $K(1)$ is the characteristic matrix associated with the repeated eigenvalue whose multiplicity is 2. Evaluating this product gives

$$K(2)K(1)^2 = \begin{bmatrix} 0 & 2 & -3 \\ -10 & 6 & -5 \\ -5 & 4 & -4 \end{bmatrix}\begin{bmatrix} -1 & 2 & -3 \\ -10 & 5 & -5 \\ -5 & 4 & -5 \end{bmatrix}^2 = \begin{bmatrix} 0 & 0 & 0 \\ 0 & 0 & 0 \\ 0 & 0 & 0 \end{bmatrix}.$$

This result should be expected because this product is the characteristic matrix polynomial for the given matrix.

***

These two previous examples illustrate the following statement.

Statement 5

If for a given n^{th}-order matrix A, μ_1 is a repeated eigenvalue of multiplicity r and $q = n -$ rank $K(\mu_1)$, then A satisfies the following reduced matrix polynomial. Note that $1 \leq q \leq r$.

$$p_m(A) = [A - \mu_1 I_n]^{r-q+1}[A - \mu_{r+1}I_n][A - \mu_{r+2}I_n] \ldots [A - \mu_n I_n] = O_n$$

$$= K(\mu_1)^{r-q+1} K(\mu_{r+1}) K(\mu_{r+2}) \cdots K(\mu_n) = K(\mu_1) \prod_{i=r+1}^{n} K(\mu_i)$$

If more than one eigenvalue is repeated, the above expression becomes much more complex, but the form remains basically the same.

EXAMPLE 22

For the matrix A, below evaluate the matrix polynomial $f(A) = 2A^3 - 50I_3$.

$$A = \begin{bmatrix} 7 & 4 & -1 \\ 4 & 7 & -1 \\ -4 & -4 & 4 \end{bmatrix}$$

SOLUTION

This is the same matrix used in example 16, so we know that its characteristic equation is

$$p(\mu) = (\mu - 12)(\mu - 3)^2 = 0.$$

The minimum characteristic polynomial is

$$p_m(\mu) = (\mu - 12)(\mu - 3) = \mu^2 - 15\mu + 36 = 0.$$

By example 20 or statement 5 above, we know this matrix satisfies its minimum characteristic polynomial.

$$p_m(A) = A^2 - 15A + 36I_3 = 0_3$$

From this we find

$$A^2 = 15A - 36I_3.$$

The given polynomial may be evaluated as follows.

$$f(A) = 2A^3 - 50I_3 = 2A(A^2) - 50I_3$$

$$= 2A(15A - 36I_3) - 50I_3 = 30A^2 - 72A - 50I_3$$

$$= 30(15A - 36I_3) - 72A - 50I_3$$

$$= 378A - 1130I_3$$

Substituting the given a matrix yields the following result.

$$f(A) = 378 \begin{bmatrix} 7 & 4 & -1 \\ 4 & 7 & -1 \\ -4 & -4 & 4 \end{bmatrix} - 1130 \begin{bmatrix} 1 & 0 & 0 \\ 0 & 1 & 0 \\ 0 & 0 & 1 \end{bmatrix}$$

$$= \begin{bmatrix} 1516 & 1512 & -378 \\ 1512 & 1516 & -378 \\ -1512 & -1512 & 382 \end{bmatrix}$$

***

The above statement also provides us with the necessary basis for evaluating functions of matrices when repeated eigenvalues occur. If, for a given matrix of order n, a complete set of n linearly independent eigenvectors are obtained by any of the approaches presented, then any function of that matrix is most easily evaluated by means of a similarity transformation. In such situations, it is also true that the matrix satisfies its own minimum polynomial as the following example demonstrates.

EXAMPLE 23

For the matrix of example 16, evaluate the function $\exp(A)$ using the minimum polynomial.

SOLUTION

From example 16, the given matrix is

$$A = \begin{bmatrix} 7 & 4 & -1 \\ 4 & 7 & -1 \\ -4 & -4 & 4 \end{bmatrix},$$

and its characteristic polynomial is $(\mu - 12)(\mu - 3)^2$. As indicated in example 20 above, this matrix satisfies its own minimum polynomial. From example 16, we have $r = 2$ and $q = 2$. By statement 5 above, we can see that the reduced polynomial is the minimum polynomial. From the Cayley-Hamilton theorem, we know that $\exp(A)$ may be expressed as a second-order polynomial in the matrix A. In this case, the minimum polynomial is first order, which means that the desired function must also be expressible as a first-order polynomial in A as follows.

$$e^A = a_0 I_3 + a_1 A$$

We know that the eigenvalues are 12 and 3, so by the approach of section 10, the two equations needed to solve for the coefficients a_0 and a_1 are

$$e^{12} = a_0 + 12a_1$$
$$e^3 = a_0 + 3a_1 .$$

Solving for the coefficients gives

$$a_0 = \frac{1}{3}\left(4e^3 - e^{12}\right)$$

$$a_1 = \frac{1}{9}\left(e^{12} - e^3\right) .$$

Substituting these into the assumed form for the exponential function of A, and combining coefficients of the two exponentials leads us to the final answer.

$$e^A = \frac{1}{9}\begin{bmatrix} 5 & -4 & 1 \\ -4 & 5 & 1 \\ 4 & 4 & 8 \end{bmatrix} e^3 + \frac{1}{9}\begin{bmatrix} 4 & 4 & -1 \\ 4 & 4 & -1 \\ -4 & -4 & 1 \end{bmatrix} e^{12}$$

This answer is obtainable by means of a similarity transformation.

***

When repeated eigenvalues occur and the given matrix does not satisfy its own minimum polynomial, a complete set of eigenvectors does not exist, then a modified approach must be followed. This is illustrated in the next example.

EXAMPLE 24

Repeat example 23 except use the A matrix given in example 17, whose characteristic polynomial is $(\mu - 2)(\mu - 1)^2$.

SOLUTION

From example 17 we have

$$A = \begin{bmatrix} 2 & -2 & 3 \\ 10 & -4 & 5 \\ 5 & -4 & 6 \end{bmatrix}.$$

As we have already shown in example 21, the reduced polynomial for this matrix is its characteristic polynomial that is of second order. If we apply the same approach as in example 23, the function $\exp(A)$ is expanded as follows.

$$e^A = a_0 I_3 + a_1 A + a_2 A^2$$

For A, the eigenvalues are 2 and 1. With these we may write the two equations

$$e^2 = a_0 + 2a_1 A + 4a_2$$

$$e^1 = a_0 + a_1 + a_2 \qquad .$$

At this point, there are only two equations for the three unknown coefficients. The third equation is developed in the following manner. The above equations were obtained from the general equation

$$e^\mu = a_0 + a_1 \mu + a_2 \mu^2.$$

Each distinct eigenvalue was substituted into this equation to generate the specific equations above. The missing equation arises from the fact that the repeated eigenvalue is a double root of the characteristic polynomial. The above relation yields only one equation for the unknown coefficients. To obtain the next equation, we take the derivative of both sides of the general equation above with respect to μ. This gives the result below.

$$\frac{d}{d\mu}\left(e^\mu = a_0 + a_1\mu + a_2\mu^2\right)$$

$$e^\mu = a_1 + 2a_2\mu$$

The repeated eigenvalue is now substituted to yield the necessary third equation.

$$e^1 = a_1 + 2a_2$$

We now solve the three algebraic equations for the unknown coefficients and arrive at the following result.

$$a_0 = e^2 - 2e^1, \qquad a_1 = -2e^2 + 5e^1, \qquad a_2 = e^2 - 2e^1$$

If we substitute these results into the assumed expression for the desired matrix exponential function and collect coefficient of the terms e^1 and e^2, the following result is obtained.

$$e^A = -\left(2I_3 - 5A + 2A^2\right)e^1 + \left(I_3 - 2A + A^2\right)e^2$$

Substituting the given matrix A and evaluating all the terms leads to the final answer.

$$e^A = \begin{bmatrix} 10 & 6 & -13 \\ 40 & 26 & -55 \\ 25 & 16 & -34 \end{bmatrix} e^1 - \begin{bmatrix} 4 & 4 & -8 \\ 15 & 15 & -30 \\ 10 & 10 & -20 \end{bmatrix} e^2$$

EXERCISES

Evaluate $\exp(At)$ for the following matrices.

1. $\begin{bmatrix} 7 & -2 \\ 2 & 3 \end{bmatrix}$
2*. $\begin{bmatrix} 2 & -1 & 2 \\ 0 & -4 & 6 \\ 0 & 0 & -4 \end{bmatrix}$
3*. $\begin{bmatrix} 1 & 0 & -3 \\ -3 & 4 & -3 \\ -3 & 0 & 1 \end{bmatrix}$

4*. $\begin{bmatrix} 76 & -26 & -35 \\ 21 & 27 & -35 \\ 42 & -56 & -15 \end{bmatrix}$, $\mu_{1,2,3} = -22, 55, 55$

***

15. JORDAN CANONICAL FORM

As we have seen in the previous sections, when multiple eigenvalues occur, a matrix can still be reduced to diagonal form by means of a similarity transformation as long as the number of linearly independent eigenvectors associated with a multiple eigenvalue equals its multiplicity. The matrix considered in example 16 above was a matrix for which this was true. However, in example 19, in which the order of the matrix was 3, only two linearly independent eigenvectors could be found. In particular, for the double eigenvalue of this matrix, we were able to find only one associated eigenvector. We now wish to consider the case of such matrices in more detail.

Although such matrices cannot be reduced to pure diagonal form, it is true that any matrix A with repeated roots can be reduced to the following general pseudo or block diagonal form.

$$(16.1) \qquad J = \begin{bmatrix} D_1 & [0] & \cdots & [0] \\ [0] & D_2 & \cdots & [0] \\ \vdots & \vdots & \ddots & \vdots \\ [0] & [0] & \cdots & D_k \end{bmatrix}$$

This is the Jordan matrix, and the apparent diagonal elements D_i are square submatrices associated with each distinct value of an eigenvalue. If an eigenvalue μ_i of A is not repeated, then D_i is a one by one matrix whose value is just μ_i. If μ_i is an eigenvalue with a multiplicity r, then the matrix D_i is an rth-order matrix with the following form.

$$D_i = \begin{bmatrix} \mu_1 & 1 & 0 & \cdots & 0 \\ 0 & \mu_i & 1 & \cdots & 0 \\ 0 & 0 & \mu_i & \cdots & 0 \\ \vdots & \vdots & \vdots & \ddots & 1 \\ 0 & 0 & 0 & \cdots & \mu_i \end{bmatrix}$$

The ones shown above the main diagonal may actually be zeros or ones. The number of ones that occur is the multiplicity of the eigenvalue minus the number of linearly independent eigenvectors that are associated with that eigenvalue. Recall that the rank of the characteristic matrix evaluated for any eigenvalue is the order of the matrix minus the number of linearly independent eigenvectors associated with that eigenvalue. This means that if there are r linearly independent eigenvectors associated with an eigenvalue whose multiplicity is r, then the block D_i will be a pure diagonal matrix with the eigenvalue as the diagonal element.

The question now remains as to how one finds the Jordan matrix. It may be shown that any matrix can be reduced to Jordan canonical form by a modified similarity transformation such as

(15.2) $J = M^{-1} A M,$

where the matrix M is the modified modal matrix. The matrix M is constructed from all the linearly independent eigenvectors of A, which are found by the methods already presented. For repeated eigenvalues, all the eigenvectors and generalized eigenvectors found associated with that eigenvalue are used in the formation of the modified modal matrix. Recall that for an eigenvalue whose multiplicity is r, there must be a total of r linearly independent eigenvectors and generalized eigenvectors associated with it. If all repeated eigenvalues of A have complete sets of associated eigenvectors, no generalized eigenvectors, then M is the standard modal matrix and the Jordan matrix reduces to a pure diagonal matrix.

In the previous examples, we saw how the columns of the adjoint matrix and its derivatives yielded either the eigenvectors or generalized eigenvectors for a specific eigenvalue. However, the evaluation of the general adjoint of the characteristic matrix and its derivatives was an involved process even in the examples that were for third-order matrices. However, these operations may be reduced to an easier form by means of the following relations[40] that permit direct calculation of the desired matrices.

Statement 6

For a distinct eigenvalue μ_i, it may be shown that the following relation holds.

(16.3) $\text{Adj } K(\mu_i) = \prod_{j=1, \neq i}^{n} K(\mu_j)$

Now if μ_i is a repeated eigenvalue of multiplicity r, then the kth derivative of the adjoint of the characteristic matrix $K(M)$, with respect to μ evaluated for this eigenvalue, is given by the following expression.

(16.4)
$$\frac{d^k}{d\mu^k} \text{Adj } K(\mu) \Big|_{\mu = \mu_i} = k! \left[K(\mu_i)\right]^{r-k-1} \prod_{j=1, \neq i}^{n} K(\mu_j)$$
$$= \text{Adj}^{(k)} K(\mu_i)$$

---

40.   See Frazer, Duncan, and Collar [1957], p. 75.

EXAMPLE 25

As an illustration of the application of statement 6, find all the eigenvectors and generalized eigenvectors for the matrix employed in example 17. Also evaluate the Jordan matrix. The given matrix and its characteristic matrix are

$$(16.5) \quad A = \begin{bmatrix} 2 & -2 & 3 \\ 10 & -4 & 5 \\ 5 & -4 & 6 \end{bmatrix}, \quad K(\mu) = \begin{bmatrix} \mu-2 & 2 & -3 \\ -10 & \mu+4 & -5 \\ -5 & 4 & \mu-6 \end{bmatrix}.$$

SOLUTION

As we found before, this matrix has a double eigenvalue of 1 and a unique eigenvalue of 2. From equation (15.3), we may write for the unique eigenvalue

$$\text{Adj } K(2) = K(1)K(1) = K(1)^2.$$

Substituting the eigenvalue of 1 into the characteristic matrix given in (15.5), and squaring the result gives

$$\text{Adj } K(2) = \begin{bmatrix} -1 & 2 & -3 \\ -10 & 5 & -5 \\ -5 & 4 & -5 \end{bmatrix}^2 = \begin{bmatrix} -4 & -4 & 8 \\ -15 & -15 & 30 \\ -10 & -10 & 20 \end{bmatrix}.$$

The rank of this matrix is one, and the single independent column is the eigenvector associated with the unique eigenvalue $\mu_1 = 2$.

$$E_1 = \begin{bmatrix} 4 \\ 15 \\ 10 \end{bmatrix}$$

For the double eigenvalue $\mu_2 = \mu_3 = 1$, we find the adjoint of the characteristic matrix from (15.4) by setting $k = 0$ and $r = 2$.

$$\text{Adj } K(1) = K(1)^{2-0-1}K(2) = K(1)\,K(2) =$$

$$= \begin{bmatrix} -1 & 2 & -3 \\ -10 & 5 & -5 \\ -5 & 4 & -5 \end{bmatrix} \begin{bmatrix} 0 & 2 & -3 \\ -10 & 6 & -5 \\ -5 & 4 & -4 \end{bmatrix}$$

$$= \begin{bmatrix} -5 & -2 & 5 \\ -25 & -10 & 25 \\ -15 & -6 & 15 \end{bmatrix}$$

Again, the rank of this matrix is one indicating a single independent column. It may be easily seen that this column is the second eigenvector that was identified previously.

$$E_2 = \begin{bmatrix} 1 \\ 5 \\ 3 \end{bmatrix}$$

Now for the first derivative of the adjoint of the characteristic matrix, we set $k = 2$ in (15.4) to obtain the following result.

$$\text{Adj}^{(1)} K(1) = K(1)^{2-1-1}K(2) = K(2) = \begin{bmatrix} 0 & 2 & -3 \\ -10 & 6 & -5 \\ -5 & 4 & -4 \end{bmatrix}$$

Inspecting this matrix indicates that there are two linearly independent columns present since the rank is two. It may be seen that these two vectors are

$$G_1 = \begin{bmatrix} 0 \\ -2 \\ -1 \end{bmatrix}, \quad G_3 = \begin{bmatrix} 1 \\ 3 \\ 2 \end{bmatrix}.$$

As discussed in section 13, neither of these two vectors are eigenvectors, but they are called generalized eigenvectors. As we saw, the set of three vectors associated with this double eigenvalue, $E2$, $G1$, and $G2$, are linearly dependent. Any two of them are linearly independent. To form the modified modal matrix, we must take both eigenvectors as found and one of the generalized eigenvectors to form a complete set of three linearly independent vectors. We will choose the modified modal matrix to be

$$M = [E_1 \ E_2 \ G_1] = \begin{bmatrix} 4 & 1 & 0 \\ 15 & 5 & -2 \\ 10 & 3 & -1 \end{bmatrix}.$$

Using this matrix to evaluate the similarity transformation on A, we find the following Jordan canonical form.

$$J = M^{-1}AM = \begin{bmatrix} 2 & 0 & 0 \\ 0 & 1 & 1 \\ 0 & 0 & 1 \end{bmatrix}$$

***

As mentioned earlier, the Jordan canonical form does not occur very often in applied problems. In a large number of cases, the matrices of interest are real and symmetric, and in these situations, the matrix is always diagonalizable.

ADDITIONAL EXERCISES

1.*   Using the method of section 7, evaluate the following matrix polynomial.

$$p(A) = I - 2A + 5A^4,$$

where

$$A = \begin{bmatrix} 2 & 1 \\ 2 & 3 \end{bmatrix}.$$

2.*   Using the method of section 7, evaluate the following matrix polynomial.

$$p(A) = I + A - 3A^5,$$

where

$$A = \begin{bmatrix} -1 & 2 & 1 \\ 1 & 0 & 1 \\ -1 & 2 & 1 \end{bmatrix}.$$

3.*   Calculate the inverse of the following matrix by the method of section 7.

$$\begin{bmatrix} 1 & -1 & 1 \\ 5 & -4 & 1 \\ 4 & -3 & 0 \end{bmatrix}$$

4.*   Calculate the inverse of the following matrix by the method of section 7.

$$\begin{bmatrix} 1 & 0 & -2 & 1 \\ -1 & 2 & 1 & 3 \\ 2 & 1 & -1 & 0 \\ 1 & -2 & 3 & 2 \end{bmatrix}$$

5.*   Evaluate exp A by direct application of Sylvester's theorem for the following matrix.

$$\begin{bmatrix} -5 & 2 & 1 \\ 1 & -4 & 1 \\ -1 & 2 & -3 \end{bmatrix}$$

6.*   Evaluate sin(A) by use of Sylvester's theorem for the following matrix.

$$\begin{bmatrix} -4 & 1 & 1 \\ 2 & -5 & 1 \\ 2 & -1 & -3 \end{bmatrix}$$

7.   Evaluate A^n by the method of section 11 if

$$A = \begin{bmatrix} \cosh\theta & \sinh\theta \\ \sinh\theta & \cosh\theta \end{bmatrix} \text{ and } n = 0, \pm 1, \pm 2, \text{ etc.}$$

8.*   Repeat exercise 5 by using the method of section 11.

9.*   Repeat exercise 6 by using the method of section 11.

10.*   Use the method of section 11 to evaluate exp(At) where

$$A = \begin{bmatrix} 1 & 0 \\ 0 & 2 \end{bmatrix}.$$

11.   Show that the matrix below cannot be reduced to diagonal form if $a = 1$.

$$\begin{bmatrix} 1 & 1 \\ 0 & a \end{bmatrix}$$

12.*   Find all the eigenvectors and generalized eigenvectors, if any, for the following matrix, and transform it to Jordan canonical form.

$$\begin{bmatrix} 7 & 4 & -1 \\ 4 & 7 & -1 \\ -4 & -4 & 4 \end{bmatrix}$$

Apply appropriate methods to determine the eigenvalues and eigenvectors for the following matrices.

13*. $\begin{bmatrix} 7 & -1 & 1 \\ 1 & 7 & -1 \\ 2 & 0 & 6 \end{bmatrix}$ 14*. $\begin{bmatrix} 2 & 2 & -2 \\ 2 & -1 & -1 \\ -6 & -3 & 1 \end{bmatrix}$

15*. $\begin{bmatrix} 11 & 6 & -15 & -3 \\ 0 & -1 & 0 & 0 \\ 12 & 6 & -16 & -3 \\ 0 & 0 & 0 & 1 \end{bmatrix}$ 16*. $\begin{bmatrix} 26 & 33 & 24 & 18 \\ -6 & 95 & 28 & 36 \\ -3 & 33 & 53 & 18 \\ 0 & 0 & 0 & 29 \end{bmatrix}$

17.*   Find the eigenvalues and eigenvectors for the matrix below, and note anything of interest about the given matrix.

$$\begin{bmatrix} 1 & 2 & 4 \\ 2 & 4 & 8 \\ 4 & 8 & 16 \end{bmatrix}$$

18.*   Find the eigenvalues and eigenvectors for the matrix[41] below, and note anything of interest about the given matrix.

$$\begin{bmatrix} 0 & 1 & 0 \\ 0 & 0 & 1 \\ 0 & -1 & 0 \end{bmatrix}$$

---

[41]   R. Bronson [1991], p. 180

19.*   Find the eigenvalues and eigenvectors for

$$\begin{bmatrix} 1 & 2 & 3 \\ 0 & 1 & 2 \\ 0 & 0 & 1 \end{bmatrix}.$$

20.*   Determine the eigenvalues and eigenvectors for the matrices[42] A, B, and
         C below. Discuss the results for the C matrix in terms of the eigenvalues
         and eigenvectors obtained for matrices A and B. The matrix C is
         commonly known as a block diagonal matrix.

$$A = \begin{bmatrix} 1 & -1 \\ 3 & 5 \end{bmatrix}, \quad B = \begin{bmatrix} 1 & 4 \\ 1 & 1 \end{bmatrix}, \quad C = \left[\begin{array}{c:c} A & 0_2 \\ \hdashline 0_2 & B \end{array}\right]$$

---

[42]   R. Bronson [1991], p. 180

VII

LINEAR DIFFERENCE AND
LINEAR DIFFERENTIAL EQUATIONS

1. INTRODUCTION

As we have seen in earlier chapters', difference equations may be employed to represent a variety of problems in matrix terms. Specific examples presented included economic analysis of an economy and biological growth problems. In this chapter, we will examine difference equations from a more general perspective as well as discuss a few more examples.

Another topic that is somewhat related to difference equations and whose solutions strongly relate to the topic of eigenvalues and eigenvectors is that of linear differential equations. This subject is of great importance and interest especially in the physical sciences and engineering. A very high percentage of the problems studied in these fields deal with systems that are dynamic, that is, they are time dependent. The application of matrices to systems of differential equations and their solutions will be presented in some detail in this chapter. A variety of typical applications will be included as illustrations.

2. THE SOLUTION OF LINEAR DIFFERENCE EQUATIONS BY THE USE OF MATRICES

In this section, it will be shown that the solution of difference equations with constant coefficients may be simply effected by the use of matrix multiplication. To illustrate the general procedure, consider the following system of difference equations with constant coefficients:

(2.1)
$$p_{n+1} = 3p_n + 4q_n$$
$$q_{n+1} = p_n + 3q_n$$

It is required to determine p_n and q_n in terms of given initial values p_0 and q_0. In order to do this, let p_n and q_n be regarded as the component of a vector X_n defined by

(2.2)
$$X_n = \begin{bmatrix} p_n \\ q_n \end{bmatrix}.$$

With this notation, the difference equations (2.1) may be written in the form

$$(2.3) \qquad X_{n+1} = \begin{bmatrix} 3 & 4 \\ 1 & 3 \end{bmatrix} X_n .$$

If we now define the vector X_0 to be the vector of the initial values p_0, q_0, we have, as a consequence of (2.3),

$$(2.4) \qquad X_1 = M X_0, \quad M = \begin{bmatrix} 3 & 4 \\ 1 & 3 \end{bmatrix} .$$

Since by (2.3), $X_{n+1} = MX_n$, repeated multiplication will show that

$$(2.5) \qquad X_n = M^n X_0 .$$

This equation gives the solution for any n of the given problem expressed in terms of the nth power of the matrix M. To evaluate M^n, we use the method of chapter VI, section 13. The eigenvalues and modal matrix of M are

$$(2.6) \qquad \mu_1 = 1, \ \mu_2 = 5, \ E = \begin{bmatrix} 2 & 2 \\ -1 & 1 \end{bmatrix} .$$

Using a similarity transformation, we may write

$$(2.7) \qquad M^n = E \begin{bmatrix} 1^n & 0 \\ 0 & 5^n \end{bmatrix} E^{-1} .$$

Inverting the modal matrix gives

$$E^{-1} = \frac{1}{4} \begin{bmatrix} 1 & -2 \\ 1 & 2 \end{bmatrix} .$$

By carrying out the indicated multiplications in (2.7), the matrix M^n may be expressed in the following form:

$$M^n = \begin{bmatrix} \frac{1}{2}(5^n + 1) & (5^n - 1) \\ \frac{1}{4}(5^n - 1) & \frac{1}{2}(5^n + 1) \end{bmatrix}$$

Hence, if the matrix equation (2.5) is expanded, the following solution for the system of difference equations (2.1) is obtained:

$$p_n = \frac{1}{2}(5^n + 1)p_0 + (5^n - 1)q_0$$

$$p_n = \frac{1}{4}(5^n - 1)p_0 + \frac{1}{2}(5^n + 1)q_0$$

As a more general example of the use of matrices to solve a system of linear difference equations with constant coefficients, consider the following first-order system.

$$p_{n+1} = ap_n + bq_n$$

$$q_{n+1} = cp_n + dq_n,$$

where the coefficients a, b, c, and d are known constants.

DEFINITION

Order of a Finite Difference Equation

The order of a finite difference equation is the largest number of intervals between the highest and lowest state of a dependent variable appearing in the system equation.

EXAMPLE 1

Evaluate the order of the following three systems of difference equations.

a. $4x_{n+1} - 3x_n = 0$

b. $x_n - 2y_{n-1} = 3$
        $5x_n + x_{n-1} + y_n = -2$

c. $3p_{n+1} = 2q_n + p_{n-1}$
        $q_{n+1} = 2p_n - q_n + 5p_{n-1}$

SOLUTION

a.   The order is one since the highest state of the dependent variable appearing is $n+1$ and the lowest is n. The number of intervals covered is $(n+1) - n = 1$.

b.   The number of state intervals covered by the dependent variable x is $n - (n-1) = 1$. The number for the dependent variable y is $n - (n-1) = 1$. The order of this system is one.

c.   The number of state intervals covered by the dependent variable p is $(n+1) - (n-1) = 2$. The number for the dependent variable q is $(n+1) - n = 1$. The order of this system is two since the largest interval between states included is two.

***

Returning to the example of the general first-order system, let us assume that it is required to solve the system of difference equations for p_n and q_n in terms of some given initial values p_0 and q_0. This problem is a generalization of the problem expressed by equation (2.1). The coefficient matrix for the system is

$$M = \begin{bmatrix} a & b \\ c & d \end{bmatrix}.$$

The solution is given by equation (2.5) above and requires the evaluation of the matrix function M^n. The procedure for obtaining M^n is similar to that used in example 12 of chapter VI. Let the determinant of M be denoted by d_0 so that

$$d_0 = \det M = (ad - bc).$$

The matrix M may be written in the following form:

$$M = \sqrt{d_0} \begin{bmatrix} \dfrac{a}{\sqrt{d_0}} & \dfrac{b}{\sqrt{d_0}} \\ \dfrac{c}{\sqrt{d_0}} & \dfrac{d}{\sqrt{d_0}} \end{bmatrix} = \sqrt{d_0} \begin{bmatrix} p & q \\ r & s \end{bmatrix} = \sqrt{d_0}\, N$$

The elements of the matrix N are obtained from the elements of M by dividing each element of M by the square root of the determinant of M. The determinant of the matrix N is given by

$$\det N = (ps - rq) = \frac{(ad - bc)}{d_0} = 1.$$

In order to obtain the nth power of the matrix N, we may now use the result from chapter VI, section 10 since the determinant of N is unity. Hence,

$$(2.8) \qquad N^n = \frac{\sinh(mn)}{\sinh(m)} N - \frac{\sinh[m(n-1)]}{\sinh(m)} I_2 \quad ,$$

where

$$(2.9) \qquad \cosh(m) = \frac{(p+s)}{2} = \frac{(a+d)}{2\sqrt{d_0}} = \frac{1}{2\sqrt{d_0}} \operatorname{Tr} M \quad .$$

From the construction of the matrix N out of M, we have

$$N = \frac{1}{\sqrt{d_0}} M \quad .$$

If this result is now substituted into (2.8), one finds that

$$(2.10) \qquad M^n = d_0^{(n-1)/2} \frac{\sinh(mn)}{\sinh(m)} M - d_0^{n/2} \frac{\sinh[m(n-1)]}{\sinh(m)} I_2 \quad ,$$

where I_2 is the second-order identity matrix and the quantity m is defined by (2.9). The expression (2.10) is very useful in electric circuit theory. The solution of the given system of difference equations is expressed by equation (2.5).

EXAMPLE 2

As an illustration of the application of this method, consider the same matrix as given in (2.4) above, that is, let

$$M = \begin{bmatrix} 3 & 4 \\ 1 & 3 \end{bmatrix}$$

for which we wish to apply (2.10) to evaluate M^n.

SOLUTION

The determinant of M is $5 = d_0$. From equation (2.9), we have $\cosh(m) = 3/\sqrt{5} = 1.3416$. This gives $m = \cosh^{-1}(1.3416) = 0.8047$. Placing this value in (2.10) yields the following expression for M^n.

$$\begin{bmatrix} 3 & 4 \\ 1 & 3 \end{bmatrix}^n = 5^{(n-1)/2}\, \frac{\sinh(0.8047n)}{\sinh(0.8047)} \begin{bmatrix} 3 & 4 \\ 1 & 3 \end{bmatrix}$$

$$- 5^{n/2}\, \frac{\sinh[0.8047(n-1)]}{\sinh(0.8047)} \begin{bmatrix} 1 & 0 \\ 0 & 1 \end{bmatrix}$$

***

Difference Equation of Higher Order

In many applications of difference equations to practical problems, it is necessary to solve difference equations of a higher order than the first. To illustrate this approach, consider the following example.

EXAMPLE 3

Solve the second-order difference equation

(2.11) $\qquad p_{n+1} = 5p_n - 6p_{n-1}.$

SOLUTION

To solve this equation, two initial values must be specified because the equation is second order. Suppose that the initial values are specified as p_0 and p_1. The solution for p_n and $n > 1$ must be obtained from equation (2.11). This is most easily effected if we reduce the single second-order difference equation to two simultaneous first-order equations, which we have already seen how to solve. To accomplish this, a new variable q_{n+1} is introduced by the relation $q_{n+1} = p_n$. By this device, the problem is now reduced to solving the two simultaneous equations of the first order.

$$p_{n+1} = 5p_n - 6q_n$$
$$q_{n+1} = p_n$$

The initial value p_0 corresponds to q_1, and therefore, if we introduce the square matrix M given by

$$M = \begin{bmatrix} 5 & -6 \\ 1 & 0 \end{bmatrix}$$

and the column matrix as given in equation (2.2), that is,

$$X_n = \begin{bmatrix} p_n \\ q_n \end{bmatrix}$$

it can be seen that

(2.12) $X_{n+1} = M^n X_1$

since X_1 is now the column of the given initial values. The problem is therefore reduced to that of raising the matrix M to the integral power n. In this case, the eigenvalues of M are $\mu_1 = 3$ and $\mu_2 = 2$. As we did for the first example in this section, the matrix M^n may be evaluated by the application of a similarity transformation. Then if the matrix solution given by equation (2.12) is expanded, the resulting solution may be obtained in the following form:

$$p_n = (3^n - 2^n)p_1 - 6(3^{n-1} - 2^{n-1})p_0, \; n > 1$$

This is the solution of the difference equation (2.11) subject to the initial values p_1 and p_0.

***

The method illustrated by the above example may be generalized to solve the difference equation of the kth order with constant coefficients

$$p_{k+n} + a_1 p_{k+n-1} + a_2 p_{k+n-2} + \ldots + a_n p_n = 0,$$

when the initial values $p_0, p_1, p_2, \ldots, p_{k-1}$ are given. To solve this equation, construct the kth-order column matrix X_{n+1} whose elements are $p_{n+k}, p_{n+k-1}, \ldots, p_{n+1}$. The difference equation (2.27) may be shown to be equivalent to the matrix equation

(2.13) $X_{n+1} = MX_n,$

where the matrix M has the form

$$M = \begin{bmatrix} -a_1 & -a_2 & \cdots & -a_{k-1} & -a_k \\ 1 & 0 & \cdots & 0 & 0 \\ 0 & 1 & \cdots & 0 & 0 \\ \cdot & \cdot & \cdots & \cdot & \cdot \\ 0 & 0 & \cdots & 1 & 0 \end{bmatrix}.$$

The solution of the difference equation (2.13) may then be obtained by expanding the matrix equation (2.12) along the same lines as already presented.

EXERCISES

1.   Solve the difference equation

$$p_{k+1} - cp_k + p_{k-1} = 0$$

subject to given initial conditions values p_0 and p_1.

2.*   In exercise 1 above, let $c = 2$ and $p_0 = p_1 = 1$, and find the general solution. Also evaluate p_6.

3.*   Solve the following difference equation[43] if $p_0 = 1$ and $p_1 = 2$.

$$p_n = 2p_{n-1} - p_{n-2}$$

***

3.   STABILITY OF SOLUTIONS

If the solutions to the two example problems in the previous section are examined, it may be seen that in both cases, the elements of M^n grow without bound as n increases. The implication is that each state vector Xn is larger than the previous one. This situation also occurred for the example biological growth problem of chapter III, section 7, equation (7.2) which involved a Leslie population matrix.

In contrast, the heat transfer problems described in chapter III, section 5, the random walk and gambling problems described by Markov matrices in section 7

---

[43]   H. G. Campbell [1980], p. 202.

of the same chapter indicate that difference equations can converge to a steady state solution. In a similar fashion, it could happen that M^n could tend to the zero or null matrix as n increases.

It is of some importance to examine the reasons for this behavior and determine if there is some measure or means of predicting the behavior of the solution based on the examination of the transfer matrix M.

To begin our investigation, let us restate the basic finite difference problem. Specifically, we have some finite difference equation

(3.1) $$X_{n+1} = MX_n,$$

where M is the transfer or coefficient matrix and X_n is a state vector. As we have already shown, the solution may be written as

(3.2) $$X_{n+1} = M^n X_1,$$

where X_1 is a known initial state vector. The coefficient matrix M is almost always a real matrix, and in some cases, it is symmetric. For convenience, let us assume that the eigenvalues of M, μ_i are real but not necessarily distinct. Let us also assume that there are as many linear independent eigenvectors E_i as the order of M.

Now let the known initial state vector be developed as a linear expansion in terms of the eigenvectors of M. In specific terms, we are using the eigenvectors as a basis or coordinate system in which the initial state vector is being represented. We may therefore write

(3.3) $$X_1 = \sum_{k=1}^{m} c_k E_k ,$$

where m is the order of M. Equation (3.3) is frequently termed an eigenvector expansion of the vector X_1. If this expression is substituted in equation (3.2), we have

(3.4) $$X_{n+1} = \sum_{k=1}^{m} c_k M^n E_k .$$

Now since the vectors E_k are the eigenvectors of M, we may apply the following to equation (3.4)

(3.5) $$ME_k = \mu_k E_k \text{ and } M^n E_k = \mu_k^n E_k$$

The result is

$$(3.6) \qquad X_{n+1} = \sum_{k=1}^{m} c_k \mu_k^n E_k .$$

We may now draw some interesting and useful conclusions from this result.

Case 1

Unstable Process

If any one or more of the eigenvalues of the matrix M has a magnitude greater than one, then the succeeding vectors (increasing n) given by equation (3.2) become larger. The process is said to be divergent or unstable.

If the situation occurs that only one eigenvalue, say μ_k, has a magnitude greater than one, then it may be shown that each succeeding vector X_{n+1} is just μ_k times the previous.

This latter case occurs in biological growth problems described by a Leslie matrix, which were introduced in section 7 of chapter III. In these problems, our matrix M is the growth matrix G. For the example fourth-order problem presented in that section, the growth matrix has eigenvalues[44] of 1.39, −1.01, and −0.19 ± 0.33j. This means that each succeeding population state vector is 1.39 times larger than the previous or that the net growth rate is 39%.

A convergent or stable process occurs under the following condition.

Case 2

Stable Process

If all the eigenvalues of the matrix M have magnitudes less than one, then equation (3.6) indicates that the vector X_{n+1} goes to the null vector as n becomes large.

The only exception occurs when the matrix M has one or more zero eigenvalues.

---

[44]   These eigenvalues were calculated by the use of the Matrix Algebra Calculator program called MAX by Charles H. Jepsen and Eugene A. Herman, Brooks Cole Publishing Co., 1988.

Case 3

Marginal Stability

A marginally stable or metastable case occurs when one or more eigenvalues has a unity magnitude and all others have magnitudes less than one.

Consider a case in which one eigenvalue of the matrix M, say μ_i, has unity magnitude and all the others have magnitudes less than one. As n becomes large, it is easy to see that equation (3.6) gives the result

$$X_{n+1} = c_i E_i,$$

where E_i is the eigenvector associated with the eigenvalue $\mu_i = 1$. In this case, the iteration process has converged to the ith eigenvector.

Should the unity eigenvalue be a multiple value, equation (3.6) results in a vector, which is a linear combination of the eigenvectors associated with the unity value. This situation arose in the second example gambling problem given in section 7 of chapter III. For that problem, the matrix M is the fifth-order odds coefficient matrix G. For that matrix, the eigenvalues[45] are 1,1, 0.77, 0.3, and -0.17. For large n (3.6) yields

$$X_{n+1} = c_1 E_1 + c_2 E_2, \quad E_1 = \begin{bmatrix} 1 \\ 0 \\ 0 \\ 0 \\ 0 \end{bmatrix}, \quad E_2 = \begin{bmatrix} 0 \\ 0 \\ 0 \\ 0 \\ 1 \end{bmatrix},$$

where E_1 and E_2 are the eigenvectors associated with the double eigenvalue. From what we learned in that example problem, we may see that the coefficients c_1 and c_2 must represent the probabilities of either losing everything or winning respectively beginning from whatever the initial state vector is.

---

[45]   The eigenvalues and eigenvectors for this matrix were found by the use of MAX cited in the previous footnote.

EXERCISES

1.*   Given the system of difference equations

$$p_{n+1} = 0.6p_n + 0.5q_n$$

and

$$q_{n+1} = 0.4p_n + 0.2q_n,$$

a.   write the equations in matrix form and determine if the system is stable, and

b.   develop a general expression for the transition matrix to the nth power.

2.*   Bernadelli[46] studied a beetle "which lives three years only, and propagates in its third year." If the first age group survives with a probability of 1/2, and the second with a probability of 1/3, and the third produces six offspring, the growth matrix is

$$G = \begin{bmatrix} 0 & 0 & 6 \\ 1/2 & 0 & 0 \\ 0 & 1/3 & 0 \end{bmatrix}.$$

Show that $G^3 = I$, and follow the distribution of beetles for six years if the initial state has 1,000 beetles in each age group.

3.*   Suppose there is an epidemic in which every month, half of those who are well become sick, and a quarter of those who are sick die (the third and final state). Develop the transition matrix[47] and find the steady state for this Markov process.

***

4. DETERMINING EIGENVALUES AND EIGENVECTORS BY MATRIX ITERATION

An interesting and relatively easy method of determining the eigenvalue that has the largest magnitude and its associated eigenvector is by an iteration process. This procedure is based on starting with some initial guess of the eigenvector and establishing an iteration process for generating hopefully improved estimates. As structured, the procedure has the form of a difference equation.

---

[46]   Strang [1988], p 272.
[47]   Strang [1988], p 272.

Let us begin by restating the standard eigenvalue problem for a given matrix M of order m.

(4.1) $$ME_i = \mu_i E_i,$$

where we are using the notation that E_i is the eigenvector associated with the eigenvalue μ_i. Assume that the matrix M is real and that the eigenvalues are ordered such that μ_1 has the largest magnitude. We also must assume that E_1 is real, which must be true if M is symmetric but not necessarily true for a general M.

Now consider that we guess some initial trial vector for E_i, and let it be called X_0. If this trial vector is substituted into the left-hand side of equation (4.1), a new vector is generated.

$$X_1 = MX_0$$

Let this process be repeated N times. The result may be written as

(4.2) $$X_n = MX_{n-1} = M^2 X_{n-2} = \ldots = M^n X_0,$$

which is a standard difference equation.

As we did in the previous section, assume that the initial vector is replaced by its eigenvector expansion.

$$X_0 = \sum_{k=1}^{m} c_k E_k$$

Substituting this expression into (4.2) and utilizing equation (3.5) gives the result

$$X_n = \sum_{k=1}^{m} c_k M^n E_k = \sum_{k=1}^{m} c_k \mu_k^n E_k .$$

Now let μ_1^n be factored out of the expansion as follows.

$$X_n = \mu_1^n \left\{ c_1 E_1 + \sum_{k=2}^{m} c_k \left(\frac{\mu_k}{\mu_1} \right)^n E_k \right\}$$

Since μ_1 is the eigenvalue of greatest magnitude, all the terms represented by the summation will become negligibly small as n becomes large. For large n, we then have the following:

$$\lim_{n \to \infty} X_n = \mu_1^n c_1 E_1$$

Now consider the next iteration, which gives the following:

$$X_{n+1} = \mu_1^{n+1} c_1 E_1 = \mu_1 X_n$$

This expression shows that with each iteration, the elements of the next vector are all μ_1 times the corresponding elements of the previous. This suggests the following procedure for determining the largest eigenvalue. After guessing an initial trial vector, the easiest to use is a vector of all ones; it is multiplied by the matrix M raised to a large power, say ten or twenty. The resulting vector is then multiplied by M to generate the next iterate. The ratio of any two corresponding vector elements provides an estimate of the largest eigenvalue. This process is repeated with a higher power until the result for the eigenvalue does not change at a desired number of significant figures.

A slight variation of this method which allows observation of the convergence is to normalize each trial vector prior to computing the next iterate. Normalizing is usually accomplished by scaling a given vector by its norm. The choice of norm is arbitrary, but the L_∞ norm is most frequently employed due to its ease of evaluation. A scaled vector is then given by the following:

$$\tilde{X}_n = \frac{1}{L_\infty} X_n$$

EXAMPLE 4

Consider the biological growth problem of section 7 of chapter III whose Leslie matrix is

$$G = \begin{bmatrix} 0 & 2 & 1 & 1 \\ 0.70 & 0 & 0 & 0 \\ 0 & 0.85 & 0 & 0 \\ 0 & 0 & 0.35 & 0 \end{bmatrix},$$

Use the above process to determine the largest eigenvalue and its associated eigenvector.

SOLUTION

As indicated in the above discussion, let the first trial vector be

$$X_0 = \begin{bmatrix} 1 \\ 1 \\ 1 \\ 1 \end{bmatrix}.$$

The unnormalized and normalized (using the L_∞ norm) first iterate is the following:

$$X_1 = \begin{bmatrix} 4 \\ 0.70 \\ 0.85 \\ 0.35 \end{bmatrix}, \quad \tilde{X}_1 = \begin{bmatrix} 1 \\ 0.18 \\ 0.21 \\ 0.09 \end{bmatrix}$$

Repeating the process 23 times[48] yields the result $\mu_1 = 1.39$ to two-decimal place accuracy. The eigenvector associated with this eigenvalue is

$$E_1 = \begin{bmatrix} 1 \\ 0.50 \\ 0.31 \\ 0.08 \end{bmatrix}.$$

EXERCISES

1.*   Use matrix iteration to determine the largest eigenvalue and its associated eigenvector for the following matrices.

a. $\begin{bmatrix} 1 & 1 & 1 \\ 1 & 2 & 2 \\ 1 & 2 & 3 \end{bmatrix}$ b. $\begin{bmatrix} 2 & -2 & 3 \\ 10 & -4 & 5 \\ 5 & -4 & 6 \end{bmatrix}$ c. $\begin{bmatrix} 1 & 2 & 5 \\ 4 & 1 & 4 \\ 0 & 3 & 7 \end{bmatrix}$

---

[48]   This calculation was performed in the spreadsheet program EXCEL by means of the built-in matrix function.

2.*   Consider the following Leslie matrix. Apply iteration to find the largest eigenvalue and the steady state population (eigenvector) to which the system will converge after a long period of time.

$$\begin{bmatrix} 0 & 7 & 6 \\ \dfrac{1}{2} & 0 & 0 \\ 0 & \dfrac{1}{3} & 0 \end{bmatrix}$$

***

5.   SYSTEMS OF DIFFERENTIAL EQUATIONS

In applied problems, one frequently finds that the mathematical representation of dynamical problems is expressed in terms of one or more differential equations such as

$$a_n y^{(n)} + a_{n-1} y^{(n-1)} + \ldots + a_1 y^{(1)} + a_0 y = f(x)$$

or

$$a_1 x_1'' + a_2 x_1 + a_3 x_2 = 0$$

$$b_1 x_2'' + b_2 x_2 + b_3 x_1 = 0.$$

In these equations, we have made use of the common notation

$$y^{(i)} = i^{th} \text{ derivative of } y \text{ with respect to } x$$

and

$$x'' = 2^{nd} \text{ derivative of } x \text{ with respect to its}$$
$$\text{independent variable.}$$

Clearly, these are only examples, and a large variety of other possible combinations and forms occur.

The systems we will be concerned with in this text will be restricted to linear differential equations. By this we mean that in the above two equations, the coefficients of the dependent variables are constants or functions of the independent variable. If functions occur on the right-hand side of the equations, they must also be restricted to be functions of the independent variable. In linear equations, the dependent variable and any of its derivatives must always be of first order, i. e., squares, cubes, etc., of these terms must be excluded.

The functions appearing on the right-hand side of differential equation are commonly called the forcing, driving, or input functions. If the forcing functions do not exist and the equations are all equal to 0, the system of equations is said to be homogeneous in the same manner as linear algebraic equations. If nonzero forcing functions appear, the system is termed nonhomogeneous.

6. ASSOCIATION OF MATRICES WITH LINEAR DIFFERENTIAL EQUATIONS

Matrix algebra provides a most concise notation for the formulation of linear differential equations that involve several dependent variables. The advantage of concise expression is very important in intricate problems that involve several dependent variables. Such problems are now receiving extensive treatment in the theory of vibrations of structures, electrical circuits, etc. These problems involve linear differential equations; the application of matrices to the solution of these equations will now be discussed.

Any single or set of differential equations of higher than the first order may be easily reduced to a system of first-order differential equations by the appropriate introduction of new independent variables. As an illustration, consider the following example.

EXAMPLE 5

Reduce the following differential equation to two simultaneous first-order differential equations.

(6.1)
$$\frac{d^2x}{dt^2} + x = 0$$

SOLUTION

In order to express this equation as a set of first-order equations, let $x = x_1$ and a second variable x_2 be defined by

$$x_2 = \frac{dx_1}{dt}.$$

With this, notation equation (6.1) becomes

$$\frac{dx_2}{dt} = -x_1$$

Therefore, the original second-order differential equation (6.1) may be written in the following matrix form:

$$\frac{d}{dt}\begin{bmatrix} x_1 \\ x_2 \end{bmatrix} = \begin{bmatrix} 0 & 1 \\ -1 & 0 \end{bmatrix}\begin{bmatrix} x_1 \\ x_2 \end{bmatrix} \text{ or } X' = MX$$

where the prime has been used to denote differentiation with respect to the independent variable t.

***

By following the procedure suggested above, it can be shown that any set of homogeneous linear differential equations with constant coefficients can be reduced to the following form.

$$x_1' = a_{11}x_1 + a_{12}x_2 + \ldots a_{1n}x_n$$

$$x_2' = a_{21}x_1 + a_{22}x_2 + \ldots a_{2n}x_n$$

$$\ldots$$

$$x_n' = a_{n1}x_1 + a_{n2}x_2 + \ldots a_{nn}x_n,$$

where the prime denotes differentiation with respect to the independent variable. If we let X represent the nth-order column matrix whose elements are the dependent variables $x_i(t)$, and A the nth-order square matrix of the constant coefficients a_{ij}, the set of differential equations above may be expressed in the following compact form.

$$X' = AX$$

Under the restrictions of the case we are considering, the elements a_{ij} of the square matrix A are constants. In the general case of linear differential equations, the elements of A are functions of the independent variable t so that $a_{ij} = a_{ij}(t)$. In the general case in which the elements of A are functions of t, we may write $A(t)$ so the set of linear equations with variable coefficients may be written in the form

$$X' = A(t)X.$$

One additional type of system frequently encountered is a nonhomogeneous system due to the occurrence of forcing functions. The general of such a system is the following:

$$x_1' = a_{11}x_1 + a_{12}x_2 + \ldots a_{1n}x_n + f_1(t)$$

(6.2) $$x_2' = a_{21}x_1 + a_{22}x_2 + \ldots a_{2n}x_n + f_2(t)$$

. . .

$$x_n' = a_{n1}x_1 + a_{n2}x_2 + \ldots a_{nn}x_n + f_n(t)$$

In this system, the $f_i(t)$ are the forcing functions.

Let us define a forcing function matrix as follows.

$$F = \begin{bmatrix} f_1(t) \\ f_2(t) \\ \ldots \\ f_n(t) \end{bmatrix}$$

Utilizing this definition along with the previous definitions, the system (6.2) may be expressed in matrix terms as

(6.3) $X' = AX + F.$

From the theory of ordinary differential equations, it may be recalled that the solution of a single first-order differential equation included one undetermined coefficient while an nth-order equation required the inclusion of n undetermined coefficients in the solution. From this we state that the general solution of (6.3) must include n undetermined coefficients.

The general solution is comprised of two separate parts, the homogeneous solution and the particular integral. The homogeneous solution is the solution to the homogeneous equation, that is, the solution for the case $F = [0]$. This solution includes the required n unknown or undetermined coefficients. The particular solution is the solution that includes or represents the response to the forcing function. A particular solution is obtained when the general solution is required to satisfy a prescribed set of initial or boundary conditions.[49] In the case of the

---

[49.] The term *initial condition* is typically used in problems when time is the independent variable. Boundary conditions are referred to when a spatial dimension is the independent variable.

equation above, there should be n conditions available to evaluate the n coefficients or constants occurring in the general solution. In most problems represented by (6.3), initial conditions are specified at a particular point in time, say $t = t_0$. The initial conditions prescribe the values of the dependent variables in (6.3) at the time t_0. In other words, the following initial condition vector must be given or known.

$$X_0 = \begin{vmatrix} x_1(t_0) \\ x_2(t_0) \\ \ldots \\ x_n(t_0) \end{vmatrix} = X(t = t_0)$$

In most cases, the initial time from which the solution time is measured is arbitrarily set to zero. Although we have referred to the independent variable as being time in the above, it can be any other appropriate quantity. As already noted, the independent variable in applied problems is either time or a space (distance or displacement) variable.

EXERCISES

Reduce the following systems of differential equations to a first-order matrix differential equation. Include the appropriate initial condition vector. Express the answer to the matrix problem in terms of the matrix exponential function. Do not evaluate this function. The independent variable is assumed to be x.

1. $y'' + y' - 12y = 0, y(0) = 3, y'(0) = 15$

2. $y'' + 8y' + 16y = 0, y(0) = 3, y'(0) = -14$

3. $y''' - 3y'' - y' + 3y = 0, y(0) = 0, y'(0) = 1, y''(0) = -1$

***

7. SOLUTION OF HOMOGENEOUS LINEAR DIFFERENTIAL EQUATIONS WITH CONSTANT COEFFICIENTS

To begin our discussion of the solution of linear first-order systems of differential equations such as in (6.3), in this section, we shall restrict our attention to homogeneous systems with constant coefficients. Included in these considerations will be the existence of a prescribed set of initial conditions. In this case, we are concerned with the system

(7.1) $X' = AX,$

subject to a set of known initial conditions

(7.2) $X_0 = X(t = t_0) =$ constant known vector.

In (7.1), A is a matrix of constants.

By analogy with the solution of a single first-order differential equation, we shall assume a solution of the form

(7.3) $X = \exp(At)C = e^{At}C,$

where C is a vector of unknown constants. From chapter VI, section 13, we have for the derivative of (7.3)

(7.4) $X' = (e^{At}C)' = Ae^{At}C.$

Substituting (7.3) and (7.4) into (7.1) gives

$$Ae^{At}C = Ae^{At}C,$$

which is an identity indicating that equation (7.3) is the solution. As the solution of linear differential equations are unique, the solution we assumed in (7.3) is indeed the only solution. This solution will be further justified in a later section.

To evaluate the constant vector C, we must apply the initial conditions as stated in (7.2). Substituting (7.3) into (7.2) gives

$$exp(At_0)C = X_0$$

or

$$C = \exp(-At_0)X_0.$$

By substituting this result into (7.3), we have the particular solution

$$X = \exp\{A(t - t_0)\}X_0.$$

As already stated, the initial time t_0 is frequently set to zero for convenience. In this case, the above solution becomes

(7.5) $X = \exp(At)X_0 = e^{At}X_0.$

The matrix function e^{At} may be evaluated by any of the methods described in chapter V.

EXAMPLE 6

Solve the differential equation (7.1) where

$$A = \begin{bmatrix} 1 & 0 \\ 0 & 2 \end{bmatrix} \quad \text{and} \quad X_0 = \begin{bmatrix} 1 \\ -1 \end{bmatrix}.$$

Also we will set $t_0 = 0$.

SOLUTION

From (7.5), the solution is written as

$$X = \exp\left(\begin{bmatrix} 1 & 0 \\ 0 & 2 \end{bmatrix} t\right) \begin{bmatrix} 1 \\ -1 \end{bmatrix}.$$

From chapter VI, section 12 we may find

$$\exp\left(\begin{bmatrix} 1 & 0 \\ 0 & 2 \end{bmatrix} t\right) = \begin{bmatrix} e^t & 0 \\ 0 & e^{2t} \end{bmatrix}.$$

Hence, the solution is

$$X = \begin{bmatrix} e^t & 0 \\ 0 & e^{2t} \end{bmatrix} \begin{bmatrix} 1 \\ -1 \end{bmatrix} = \begin{bmatrix} e^t \\ -e^{2t} \end{bmatrix} = \begin{bmatrix} 1 \\ 0 \end{bmatrix} e^t - \begin{bmatrix} 0 \\ 1 \end{bmatrix} e^{2t}.$$

***

As another illustration, consider a second-order constant coefficient differential equation given in the next example.

EXAMPLE 7

Solve the following differential equation subject to the stated initial conditions.

(7.6) $y'' + 3y' + 2y = 0$

$$y(0) = 0, \, y'(0) = 4$$

SOLUTION

In order to put this equation in the form of (7.1), we must transform it to two simultaneous first-order equations. To accomplish this, the following new dependent variables will be defined. Let $x_1 = y$ and $x_2 = y' = x_1'$. Substituting these into (7.6) gives

$$x_2'' + 3x_2 + 2x_1 = 0.$$

The desired system of first-order equations is now

(7.7)
$$x_1' = x_2$$

$$x_2' = -2x_1 - 3x_2.$$

The initial conditions expressed in (7.15) may be written in terms of the new variables as follows:

(7.8) $$x_1(0) = y(0) = 0, \; x_2(0) = y'(0) = 4$$

Introducing matrix notation into (7.7) and (7.8), we have the following system

$$X' = \begin{bmatrix} 0 & 1 \\ -2 & -3 \end{bmatrix} X, \quad X_0 = \begin{bmatrix} 0 \\ 4 \end{bmatrix}.$$

From (7.5), the solution to this system is

$$X = \exp \left(\begin{bmatrix} 0 & 1 \\ -2 & -3 \end{bmatrix} t \right) \begin{bmatrix} 0 \\ 4 \end{bmatrix}.$$

The final answer is determined by evaluating the matrix function. The result is

(7.9) $$X = \begin{bmatrix} 4e^{-t} & -4e^{-2t} \\ -4e^{-t} & -8e^{-2t} \end{bmatrix} = 4 \begin{bmatrix} 1 \\ -1 \end{bmatrix} e^{-t} - 4 \begin{bmatrix} 1 \\ -2 \end{bmatrix} e^{-2t}.$$

EXERCISES

Solve the following matrix differential equations by use of the matrix exponential function, and reduce the answer to direct matrix form by means of any convenient method.

1. $$X' = \begin{bmatrix} 0 & 1 \\ 1 & 0 \end{bmatrix} X, \quad X_0 = \begin{bmatrix} 2 \\ 0 \end{bmatrix}$$

2. $$X' = \begin{bmatrix} 1 & 1 \\ 1 & -1 \end{bmatrix} X, \quad X_0 = \begin{bmatrix} 0 \\ 3 \end{bmatrix}$$

3. $$X' = \begin{bmatrix} 1 & -2 \\ -4 & -1 \end{bmatrix} X, \quad X_0 = \begin{bmatrix} 1 \\ 1 \end{bmatrix}$$

4. $$X' = \begin{bmatrix} 2 & 1 \\ 2 & 3 \end{bmatrix} X, \quad X_0 = \begin{bmatrix} 1 \\ -1 \end{bmatrix}$$

5. $$X' = \begin{bmatrix} 5 & 2 & 1 \\ 1 & -4 & 1 \\ -1 & 2 & -1 \end{bmatrix} X, \quad X_0 = \begin{bmatrix} 0 \\ 1 \\ 0 \end{bmatrix}$$

***

8.   EIGENVECTOR EXPANSION

In the last section, we saw examples of how a system of linear constant coefficient differential equations could be solved utilizing the matrix functions. In the last example, it was indicated how the solution could be related to the eigenvalues and eigenvectors of the coefficient matrix. If the solutions for the first two examples are examined, it may be seen that these solutions are also linear combinations of the eigenvalues and eigenvectors of the coefficient matrix.

What we have been dealing with is simply an application of linear vector theory. The coefficient matrix of a differential equation defines the space in which the solution must exist. Given a basis or coordinate system for that space, then any homogeneous solution must be expressed as a linear combination of the basis vectors. Clearly, the eigenvectors are a convenient basis in which all solution may be expressed. As we have also seen in the final example in the previous section, the eigenvalues and eigenvectors do have physical meaning in applied problems.

The method referred to in the last example is called the eigenvector expansion method, which we shall now present in a somewhat more formal manner. It must be remembered that even if this approach is not taken in solving a system of differential equations, any other method such as utilizing and evaluating the matrix exponential function will always lead to the same solution in terms of the eigenvalues and eigenvectors of the coefficient matrix.

To develop this method, consider the system of n linear first-order equations

(8.1) $$X' = AX,$$

where the coefficient matrix is a constant matrix. Let us also assume that the coefficient matrix is real and symmetric. This may appear to be a fairly limiting restriction, but it turns out that a large number of applied problems indeed are described by real symmetric matrices.

Now introduce a new variable Y through the following transformation.

(8.2) $$X = MY$$

Substituting this relation into (8.1) and premultiplying by the inverse of M gives

(8.3) $$Y' = M^{-1}AMY.$$

If we now stipulate that M must be the modal matrix of A, that is, the columns of M are the eigenvectors of A, the following relationship holds.[50]

$$M^{-1}AM = D,$$

where D is a diagonal matrix with the eigenvalues of A on the main diagonal. Equation (8.3) may now be written as

(8.4) $$Y' = DY.$$

Because D is a diagonal matrix, the system is said to be uncoupled. In other words, a typical scalar equation from (8.4) is

$$y_i' = \mu_i y_i.$$

In this equation, none of the other dependent variables appear.

---

[50]   For details, please refer to chapter V, section 5.

Applying the method presented in the previous section, the solution of (8.4) is

$$(8.5) \qquad Y = e^{D't}C,$$

where C is an undetermined constant vector of order n, and we have assumed the independent variable is represented by t. From the theory of matrix functions as discussed in chapter VI, the matrix exponential function becomes

$$e^{Dt} = \begin{bmatrix} \ddots & & \\ & e^{\mu_i t} & \\ & & \ddots \end{bmatrix},$$

which is a diagonal matrix whose diagonal elements are $\exp(\mu_i t)$. Based on this form, equation (8.5) may be written as

$$Y = \begin{bmatrix} \ddots & & \\ & e^{\mu_i t} & \\ & & \ddots \end{bmatrix} C = \begin{bmatrix} c_1 e^{\mu_1 t} \\ c_2 e^{\mu_2 t} \\ \vdots \\ c_n e^{\mu_n t} \end{bmatrix}.$$

This equations gives the solution in the transformed coordinates. The solution for X may be obtained by utilizing (8.2) to transform this result back to the original variables. Thus,

$$(8.6) \qquad X = MY = M \begin{bmatrix} \ddots & & \\ & e^{\mu_i t} & \\ & & \ddots \end{bmatrix} C.$$

From the definition of the modal matrix, we recall

$$M = [\, E_1 E_2 \ldots E_n \,].$$

Substituting this relation into (8.6) and expanding the matrix products, it may be shown that

$$(8.7) \qquad X = MY = M \begin{bmatrix} \ddots & & \\ & e^{\mu_i t} & \\ & & \ddots \end{bmatrix} C = \sum_{i=1}^{n} c_i E_i e^{\mu_i t}.$$

This form of the solution to (8.1) shows how it may be expressed as a linear combination of the eigenvectors of the coefficient matrix. In this form, the solution is called an eigenvector expansion. The constants c_i appearing in this solution are the arbitrary constants of integration and are evaluated by imposing initial conditions if they are prescribed. Since the original equation is a homogeneous equation, this solution is called the general homogeneous solution.

If at $t = 0$ we have the initial conditions

$$X(t{=}0) = X_0,$$

the constants c_i may be determined from (8.6) as follows. At $t = 0$,

$$X_0 = MC,$$

from which

(8.8) $$C = M^{-1}X_0.$$

As an illustration of this method, let us reconsider the differential equation of the second example in the previous section.

EXAMPLE 8

Solve the following matrix differential equation subject to the stated initial conditions.

$$X' = \begin{bmatrix} 0 & 1 \\ -2 & -3 \end{bmatrix} X$$

$$X_0 = \begin{bmatrix} 0 \\ 4 \end{bmatrix}$$

SOLUTION

The eigenvalues and eigenvectors of the coefficient matrix are

$$\mu_1 = -1, \; E_1 = \begin{bmatrix} 1 \\ -1 \end{bmatrix}, \; \mu_2 = -2, \; E_2 = \begin{bmatrix} 1 \\ -2 \end{bmatrix}.$$

From (8.7), the general homogeneous solution is expressed as

$$(8.9) \qquad X = c_1 \begin{bmatrix} 1 \\ -1 \end{bmatrix} e^{-t} + c_2 \begin{bmatrix} 1 \\ -2 \end{bmatrix} e^{-2t} .$$

To evaluate the constants, we apply (8.8), which yields

$$C = \begin{bmatrix} 1 & 1 \\ -1 & -2 \end{bmatrix}^{-1} \begin{bmatrix} 0 \\ 4 \end{bmatrix} = \begin{bmatrix} 4 \\ -4 \end{bmatrix} .$$

Substituting this result into (8.9) gives the particular solution

$$X = 4 \begin{bmatrix} 1 \\ -1 \end{bmatrix} e^{-t} - 4 \begin{bmatrix} 1 \\ -2 \end{bmatrix} e^{-2t} ,$$

which agrees with the previously obtained solution given in equation (7.9).

***

At this point, it is of interest to review a few characteristics that were previously noted of the general form of the eigenvector solution as given by (8.7). This solution is a linear combination of the eigenvectors of the coefficient matrix of the differential equation. In terms of our earlier discussion about linear spaces, we may state that the system of linearly independent eigenvectors form a basis for the solution space. The coefficients c_i represent the projections of a particular solution on the coordinate or basis vectors. The particular solution is a uniquely defined vector in the n dimensional space of eigenvectors which represent the complete general homogeneous solution.

As a concluding example that includes complex eigenvalues and eigenvectors, consider the following problem.

EXAMPLE 9

Solve the following matrix differential equation subject to the stated initial conditions.

$$(8.10) \qquad X' = AX,$$

where

$$A = \begin{bmatrix} 0 & 1 \\ -5 & -2 \end{bmatrix}, \text{ and } X_0 = \begin{bmatrix} 1 \\ -1 \end{bmatrix}.$$

SOLUTION

For this coefficient matrix, the eigenvalues and associated eigenvectors are

$$\mu_1 = -1 + 2j, \; E_1 = \begin{bmatrix} 1 \\ -1 + 2j \end{bmatrix}, \; \mu_2 = -1 + 2j, \; E_2 = \begin{bmatrix} 1 \\ -1 - 2j \end{bmatrix}.$$

In this case, the eigenvalues and eigenvectors form complex conjugate pairs. It should be recalled that $j = \sqrt{-1}$. In terms of an eigenvector expansion, the solution to (8.10) may be written as

$$(8.11) \qquad X = c_1 \begin{bmatrix} 1 \\ -1 + 2j \end{bmatrix} e^{(-1 + 2j)t} + c_2 \begin{bmatrix} 1 \\ -1 - 2j \end{bmatrix} e^{(-1 - 2j)t} .$$

Imposing the initial conditions, we find

$$C = \begin{bmatrix} c_1 \\ c_2 \end{bmatrix} = \begin{bmatrix} 1 & 1 \\ -1 + 2j & -1 - 2j \end{bmatrix}^{-1} \begin{bmatrix} 1 \\ -1 \end{bmatrix} .$$

From this expression, we find that the values of the two arbitrary coefficients are $c_1 = c_2 = 1/2$. Substituting these values into (8.11) and expanding the exponential functions leads us to the following result.

$$X = \frac{1}{2} \begin{bmatrix} 1 \\ -1 + 2j \end{bmatrix} e^{-t} e^{2jt} + \frac{1}{2} \begin{bmatrix} 1 \\ -1 - 2j \end{bmatrix} e^{-t} e^{-2jt}$$

This solution is adequate; however, since the coefficient matrix A is a real matrix, we might suspect that the answer should be expressible in real terms. It must be realized that this is not always true as A in not real and symmetric, but it is a valid exercise to attempt to put the last result into real terms. To do this, the first step to try is replace the complex exponential functions through the use of Euler's equation.

$$e^{\pm jx} = \cos(x) \pm j \sin(x)$$

Making this substitution and collecting coefficients of all like functions of time yields the following real answer.

$$X = \begin{bmatrix} 1 \\ -1 \end{bmatrix} e^{-t} \cos(2t) + \begin{bmatrix} 0 \\ -2 \end{bmatrix} e^{-t} \sin(2t)$$

EXERCISES

Solve the following matrix differential equations by use of the matrix exponential function, and reduce the answer to direct matrix form by means of any convenient method.

1.
$$X' = \begin{bmatrix} 0 & 1 \\ 1 & 0 \end{bmatrix} X, \quad X_0 = \begin{bmatrix} 2 \\ 0 \end{bmatrix}$$

2.
$$X' = \begin{bmatrix} 1 & 1 \\ 1 & -1 \end{bmatrix} X, \quad X_0 = \begin{bmatrix} 0 \\ 3 \end{bmatrix}$$

3.
$$X' = \begin{vmatrix} 1 & -2 \\ -4 & -1 \end{vmatrix} X, \quad X_0 = \begin{vmatrix} 1 \\ 1 \end{vmatrix}$$

4.
$$X' = \begin{vmatrix} 2 & 1 \\ 2 & 3 \end{vmatrix} X, \quad X_0 = \begin{vmatrix} 1 \\ -1 \end{vmatrix}$$

5.
$$X' = \begin{bmatrix} 5 & 2 & 1 \\ 1 & -4 & 1 \\ -1 & 2 & -1 \end{bmatrix} X, \quad X_0 = \begin{bmatrix} 0 \\ 1 \\ 0 \end{bmatrix}$$

***

9.   EIGENVECTOR EXPANSION FOR MULTIPLE EIGENVALUES

At times in applied problems, one encounters situations where the coefficient matrix of a system of differential equations possesses multiple or repeated eigenvalues. To illustrate the solution, consider a general first-order homogeneous system

(9.1) $X' = AX$

with the initial condition

$$X(t = 0) = X_0.$$

For convenience, let us assume that the coefficient matrix A is of order n and has one eigenvalue, μ, which is repeated r times. The remainder of the eigenvalues are assumed to be distinct. That is,

$\mu_1 = \mu_2 = ... = \mu_r = \mu$, and μ_i for $i = r + 1, \ldots, n$ are distinct.

It may be shown[51] that the solution to (9.1) in this case is given by the following:

$$X = \left[c_1 E_1 + c_2 E_2 t + c_3 E_3 \, t^2 + \cdots + c_r E_r t^{r-1} \right] e^{\mu t} +$$

(9.2) $$c_{r+1} E_{r+1} e^{\mu_{r+1} t} + \cdots + c_n E_n e^{\mu_n t}$$

$$= \left[\sum_{k=1}^{r} c_k t^{k-1} E_k \right] e^{\mu t} + \sum_{k=r+1}^{n} c_k E_k e^{\mu_k t}$$

EXAMPLE 10

Solve the following example differential equation subject to the stated initial conditions.

(9.3) $$X' = AX = \begin{bmatrix} 2 & 2 & -6 \\ 2 & -1 & -3 \\ -2 & -1 & 1 \end{bmatrix} X, \quad X_0 = \begin{bmatrix} 1 \\ -1 \\ 1 \end{bmatrix}$$

SOLUTION

The eigenvalues[52] and eigenvectors of the coefficient matrix are the following:

$$\mu_1 = \mu_2 = -2, \quad E_1 = \begin{bmatrix} 2 \\ -1 \\ 1 \end{bmatrix}, \quad E_2 = \begin{bmatrix} 2 \\ 5 \\ 3 \end{bmatrix},$$

$$\mu_3 = 6, \quad E_3 = \begin{bmatrix} 2 \\ 1 \\ -1 \end{bmatrix}$$

Following the general solution given in (9.2), an assumed solution will be written as follows.

(9.4) $$X = [c_1 E_1 + c_2 E_2 t] e^{-2t} + c_3 E_3 e^{6t}$$

---

[51] See Hochstadt {1964], p. 54.
[52] The eigenvalues and eigenvectors have been found using the package MATLAB previously cited. It should be noted that only two eigenvectors were obtained by means of this software. The third is found as shown in the example.

These constants may be evaluated by the application of the given initial condition at $t = 0$ from equation (9.3). Because of the form of the solution for the repeated eigenvalue, the constant c_2 will disappear at $t = 0$ as it appears multiplied by t. This problem may be circumvented by calculating the initial value of the derivative of the solution at $t = 0$. From (9.3) we have

$$X'_0 = AX_0 = \begin{bmatrix} 2 & 2 & -6 \\ 2 & -1 & -3 \\ -2 & -1 & 1 \end{bmatrix} \begin{bmatrix} 1 \\ -1 \\ 1 \end{bmatrix} = \begin{bmatrix} -6 \\ 0 \\ 0 \end{bmatrix}.$$

Taking the time derivative of (9.4) gives

$$X' = -2c_1 E_1 e^{-2t} - 2c_2 E_2 t e^{-2t} + c_2 E_2 e^{-2t} + 6c_3 E_3 e^{6t}.$$

At $t = 0$, this becomes

$$X'_0 = -2c_1 E_1 + c_2 E_2 + 6E_3 = \begin{bmatrix} -4 & 2 & 12 \\ 2 & 5 & 6 \\ -2 & 3 & -6 \end{bmatrix} \begin{bmatrix} c_1 \\ c_2 \\ c_3 \end{bmatrix}.$$

Equating the two results for the value of the first derivative at $t = 0$ yields

$$\begin{bmatrix} -6 \\ 0 \\ 0 \end{bmatrix} = \begin{bmatrix} -4 & 2 & 12 \\ 2 & 5 & 6 \\ -2 & 3 & -6 \end{bmatrix} \begin{bmatrix} c_1 \\ c_2 \\ c_3 \end{bmatrix}.$$

From this we find that $c_1 = 3/4$, $c_2 = 0$, $c_3 = -1/4$. Substituting these into the general solution (9.4) gives the final answer to be

$$X = \frac{3}{4} \begin{bmatrix} 2 \\ -1 \\ 1 \end{bmatrix} e^{-2t} - \frac{1}{4} \begin{bmatrix} 2 \\ 1 \\ -1 \end{bmatrix} e^{6t}.$$

***

It should be noted that this same solution may be obtained directly by application of a similarity transformation. The solution to the original equation $X' = AX$, and the given initial conditions may be expressed as follows.

(9.5) $X = \exp(At)X_0$

By means of a similarity transformation, the exponential matrix function becomes

$$\exp(At) = EDE^{-1},$$

where E is the modal matrix of A or the matrix of eigenvectors

$$E = \begin{bmatrix} 2 & 2 & 2 \\ -1 & 5 & 1 \\ 1 & 3 & -1 \end{bmatrix}.$$

The matrix D is a diagonal matrix of associated eigenfunctions, that is, $\exp(\mu_i t)$, which must be modified to account for the occurrence of the repeated eigenvalue. The D matrix is

$$D = \begin{bmatrix} e^{-2t} & 0 & 0 \\ 0 & te^{-2t} & 0 \\ 0 & 0 & e^{6t} \end{bmatrix}.$$

If we evaluate the similarity transformation indicated above and collect coefficient matrices for like time functions, the following result is obtained:

$$\exp(At) = \frac{1}{4}\begin{bmatrix} 2 \\ -1 \\ 1 \end{bmatrix}\begin{bmatrix} 1 & -1 & 1 \end{bmatrix}e^{-2t}$$

$$+ \frac{1}{8}\begin{bmatrix} 2 \\ 5 \\ 3 \end{bmatrix}\begin{bmatrix} 0 & 1 & 1 \end{bmatrix}te^{-2t} + \frac{1}{8}\begin{bmatrix} 2 \\ 1 \\ -1 \end{bmatrix}\begin{bmatrix} 2 & 1 & -3 \end{bmatrix}e^{6t}$$

Substituting this into the solution (9.5) above can be shown to give the identical answer as previously derived.

EXERCISES

Solve the following matrix differential equations by use of the matrix exponential function, and reduce the answer to direct matrix form by means of an eigenvector expansion.

1.
$$X' = \begin{bmatrix} 1 & 1 \\ 1 & -1 \end{bmatrix}X, \quad X_0 = \begin{bmatrix} 0 \\ 3 \end{bmatrix}$$

2. $$X' = \begin{bmatrix} 1 & -2 \\ -4 & -1 \end{bmatrix} X, \quad X_0 = \begin{bmatrix} 1 \\ 1 \end{bmatrix}$$

3*. $$X' = \begin{bmatrix} 5 & 2 & 1 \\ 1 & -4 & 1 \\ -1 & 2 & -1 \end{bmatrix} X, \quad X_0 = \begin{bmatrix} 0 \\ 1 \\ 0 \end{bmatrix}$$

4.* Solve the differential equation[53] subject to the given initial condition.

$$X' = \begin{bmatrix} 78 & -60 & 15 \\ 150 & -117 & 30 \\ 200 & -160 & 43 \end{bmatrix} X, \quad X_0 = \begin{bmatrix} 1 \\ -1 \\ 2 \end{bmatrix}$$

***

10. THE ADJOINT METHOD

Up to this section, we have focused our attention on the solution of homogeneous differential equations. In many applications, one actually has to deal with nonhomogeneous equations, which were introduced in equation (6.3). In this section, a method for obtaining explicit closed form solutions will be presented. Numerical methods for integrating matrix differential equations of all types will be presented in the next chapter. A general nonhomogeneous matrix differential equation is written in the following form.

(10.1) $\qquad X' = AX + F,$

where A is the coefficient matrix that will be assumed to be a constant matrix, X the unknown vector, and F the input or forcing function vector. It is assumed that this system is of the nth order.

In addition to (10.1), initial conditions may be prescribed as

(10.2) $\qquad\qquad\qquad X(t = 0) = X_0.$

For convenience, the independent variable will be considered to be t. It does not effect the following development whether this variable represents time or space.

---

[53]   Gere, J. M. and W. Weaver [1983], p. 150

The method of solving (10.1) subject to (10.2) that follows is called the adjoint method and may be applied to problems in which the coefficient matrix is a function of the independent variable as well. To solve equation (10.1), the method begins by seeking a solution to the following homogeneous system.

(10.3) $$Y' = -A^T Y$$

In this equation, the new unknown matrix is a square matrix of the nth order. Equation (10.3) is called the adjoint equation to (10.1), and the initial condition is taken to be

(10.4) $Y(t = 0) = I$, the nth-order identity matrix.

To develop the desired solution to (10.1), we proceed as follows. Premultiply (10.1) by Y^T, and postmultiply the transpose of (10.3) by X. These operations give the result

$$Y^T X' = Y^T A X + Y^T F$$

and

$$(Y^T)' X = - Y^T A X.$$

These two equations are now added to give

$$Y^T X' + (Y^T)' X = Y^T F.$$

A brief inspection of the left-hand side of this equation enables us to recognize the two terms as being the derivative of a product. Thus, the last may be written as

$$(Y^T X)' = Y^T F.$$

If this equation is now integrated over the interval 0 to t, we find that after application of the initial conditions (10.2) and (10.4),

$$\int_0^t (Y^T X)' \, dt = (Y^T X)_0^t + \int_0^t Y^T F \, dt$$

or

$$(Y^T X) = X_0 + \int_0^t Y^T F \, dt .$$

Premultiplying through by $(Y^T)^{-1}$ yields the desired solution to (10.1).

(10.5)
$$X = (Y^T)^{-1}X_0 + (Y^T)^{-1}\int_0^t Y^T F \, dt$$

It must be noted that in order to apply this solution, one must first solve the homogeneous system given by (10.3) and (10.4).

Since the coefficient matrix A has been assumed to be a constant matrix, the solution to (10.3) for Y subject to the initial condition (10.4) may be written in terms of a matrix exponential function as follows.

$$Y = \exp(-A^T t) = e^{-A^T t}$$

From this result, we have

$$Y^T = e^{-At}$$

and

$$(Y^T)^{-1} = e^{At}.$$

Substituting these results into (10.5) yields the desired solution for the case of constant coefficients.

(10.6)
$$X = e^{At}X_0 + e^{At}\int_0^t e^{-A\eta}F \, d\eta \quad,$$

where the symbol η is a dummy variable of integration.

The solution given by equation (10.6) illustrates the classic form for solutions of differential equations. The solution is a linear combination of the homogeneous solution (first term) and the so-called particular integral (second term), which brings in the response to the forcing function. In applied terminology, these terms are frequently referred to as the free and forced responses respectively.

EXAMPLE 11

Find the general solution to the following differential equation.

$$X' = \begin{bmatrix} 1 & 2 \\ -3 & -4 \end{bmatrix} X + \begin{bmatrix} e^{-t} \\ t \end{bmatrix}$$

SOLUTION

For this equation,

$$A = \begin{bmatrix} 1 & 2 \\ -3 & -4 \end{bmatrix}, \quad F = \begin{bmatrix} e^{-t} \\ t \end{bmatrix}.$$

To apply (10.6), we must evaluate the matrix function e^{-At}. Employing any of the methods presented in chapter VI, we find

$$e^{At} = \begin{bmatrix} 3 & 2 \\ -3 & -2 \end{bmatrix} e^{-t} - \begin{bmatrix} 2 & 2 \\ -3 & -3 \end{bmatrix} e^{-2t},$$

and hence,

$$e^{-At} = \begin{bmatrix} 3 & 2 \\ -3 & -2 \end{bmatrix} e^{t} - \begin{bmatrix} 2 & 2 \\ -3 & -3 \end{bmatrix} e^{2t}.$$

Substituting these two relations into (10.6) and performing the indicated integration gives the final solution.

$$X = \left(\begin{bmatrix} 3 & 2 \\ -3 & -2 \end{bmatrix} e^{-t} - \begin{bmatrix} 2 & 2 \\ -3 & -3 \end{bmatrix} e^{-2t} \right) X_0 +$$

$$\frac{1}{4} \begin{bmatrix} -6 \\ 5 \end{bmatrix} + \frac{1}{2} \begin{bmatrix} 2 \\ -1 \end{bmatrix} t + \begin{bmatrix} 0 \\ 1 \end{bmatrix} e^{-t} +$$

(10.7)

$$3 \begin{bmatrix} 1 \\ -1 \end{bmatrix} te^{-t} + \frac{1}{4} \begin{bmatrix} 6 \\ -9 \end{bmatrix} e^{-2t}$$

EXERCISES

1. In the solution (10.7) given above, write the two second-order matrices as a product of a second-order column vector times a second-order row vector. Identify the column vectors.

Reduce the following systems of differential equations to a first-order matrix differential equation, and determine the solution. The independent variable is assumed to be x.

2. $y'' - 4y' + 3y = e^{2x}$, $y(0) = 0$, $y'(0) = 0$

3. $y''' + 2y'' - 4y' - 8y = xe^{-x}$,

   $y(0) = 1$, $y'(0) = 0$, $y''(0) = 0$.

Apply the adjoint method to obtain the solution for the following differential equations.

4*. $$X' = \begin{bmatrix} 2 & 1 \\ 2 & 3 \end{bmatrix} X + \begin{bmatrix} e^{-t} \\ 0 \end{bmatrix}, \quad X_0 = \begin{bmatrix} -1 \\ 1 \end{bmatrix}$$

5*.[54] $$X' = \begin{bmatrix} -2 & 2 \\ -5 & 1 \end{bmatrix} X + \begin{bmatrix} e^t \\ e^{2t} \end{bmatrix}$$

***

11. SOLUTION BY TRANSFORMATION TO NORMAL COORDINATES

As seen in all of the previous examples, the solutions to either homogeneous or nonhomogeneous matrix differential equations always have the eigenvalues and eigenvectors of the coefficient matrix appearing in their solutions. This occurs whether the solution was obtained by means of expressing the answer in terms of matrix functions, which were then evaluated, or directly by the means of an eigenvalue expansion. The adjoint method just presented for solving nonhomogeneous equations does not utilize any properties of eigenvalues and eigenvectors in establishing the general solution. However, when applied to the example problems, the eigenvalues and eigenvectors appear as an integral part of the solution. The reason for this may hopefully be illustrated as the following

---

[54] Derick and Grossman [1976], p328.

alternative method for solving either homogeneous or nonhomogeneous matrix differential equations is presented.

To present this method, consider the following standard nonhomogeneous matrix differential equation with given initial condition.

(11.1) $X' = AX + F, X(t = 0) = X_0$

Let the eigenvalues and associated eigenvectors of the coefficient matrix A be μ_1, E_1, μ_2, E_2, ... , μ_n, E_n where n is the order of the system. Now let M be the modal matrix of A, that is $M = [E_1\ E_2 \ldots E_n]$. A new unknown vector Y is introduced by the transformation $X = MY$. Substituting this change of variable into equation (11.1) and premultiplying the result through by M^{-1} yields

(11.2) $M^{-1}MY' = Y' = M^{-1}AMY + M^{-1}F.$

Since M is the modal matrix of A the product $M^{-1}AM$ produces a diagonal matrix D whose diagonal elements are the eigenvalues of A. For convenience, let us also define a new forcing function vector G as $M^{-1}F$. Equation (11.2) may be seen to have the following form.

(11.3) $Y' = DY + G = \begin{bmatrix} \ddots & & \\ & \mu_i & \\ & & \ddots \end{bmatrix} Y + G$

Because D is diagonal, this equation simply represents a series of n scalar differential equations of the form

(11.4) $y^{i\prime} = \mu_i y_i + g_i,$

where g_i is the ith element of G. The initial conditions for (11.3) are given by $Y_0 = M^{-1}X_0.$

Equation (11.3) is said to be the uncoupled form of the original system (11.1). From (11.1), an equation for the derivative of a single variable involves a linear combination of all the other variables. In contrast, each equation of (11.3)—that is, (11.4)—involves only a single variable. Physically, equation (11.4) represents the vibration or oscillation of the system in each of its individual modes.

From the theory of ordinary differential equations, the solution to (11.4) may be expressed in general terms as follows.

$$y_i(t) = y_i(0) e^{\mu_i t} + e^{\mu_i t} \int_0^t e^{-\mu_i \eta} g_i(\eta) d\eta$$

By the use of diagonal matrices, this result may be expressed in matrix form.

$$Y = \begin{bmatrix} \ddots & & \\ & e^{\mu_i t} & \\ & & \ddots \end{bmatrix} \{Y_0 +$$

$$\int_0^t \begin{bmatrix} \ddots & & \\ & e^{-\mu_i \eta} & \\ & & \ddots \end{bmatrix} M^{-1} F(\eta) \, d\eta\}$$

where the original forcing function vector has been substituted. For convenience of notation, let us define the two diagonal matrices as $\text{Diag}\{\exp(\mu_i t)\}$ and $\text{Diag}\{\exp(-\mu_i t)\}$. With these definitions, (11.6) becomes the following:

$$(11.5) \quad Y = \text{Diag}\{e^{\mu_i t}\} Y_0 + \text{Diag}\{e^{\mu_i t}\} \int_0^t \text{Diag}\{e^{-\mu_i \eta}\} M^{-1} F(\eta) \, d\eta$$

If we premultiply this expression by M and note that $X = MY$ as well as $Y_0 = M^{-1}X_0$, the following form of the solution may be obtained.

$$X = MY = M \, \text{Diag}\{e^{\mu_i t}\} M^{-1} X_0 +$$

$$M \, \text{Diag}\{e^{\mu_i t}\} \int_0^t \text{Diag}\{e^{-\mu_i \eta}\} M^{-1} F(\eta) \, d\eta$$

If we introduce an nth-order identity matrix inside the integral in front of the term $\text{Diag}\{\exp(-\mu_i t)\}$, the equation will not be affected. Since $M^{-1} M = I$, the last equation may be written as follows.

$$(11.6) \quad X = MY = M \, \text{Diag}\{e^{\mu_i t}\} M^{-1} X_0 +$$

$$M \, \text{Diag}\{e^{\mu_i t}\} M^{-1} \int_0^t M \, \text{Diag}\{e^{-\mu_i \eta}\} M^{-1} F(\eta) \, d\eta$$

From the theory of matrix functions previously discussed, we recall that

$$e^{At} = M \text{ Diag} \{e^{\mu_i t}\} M^{-1}.$$

This permits (11.6) to be written as the following:

$$X = e^{At} X_0 + \{e^{At}\} \int_0^t e^{-A\eta} F(\eta) \, d\eta$$

This solution is exactly the same as given in the previous section by equation (10.6).

EXAMPLE 12

Determine the solution the first example problem of the previous section using the method of this section. The given differential equation is

$$X' = \begin{bmatrix} 1 & 2 \\ -3 & -4 \end{bmatrix} X + \begin{bmatrix} e^{-t} \\ t \end{bmatrix}.$$

SOLUTION

For this equation,

$$A = \begin{bmatrix} 1 & 2 \\ -3 & -4 \end{bmatrix}, \quad F = \begin{bmatrix} e^{-t} \\ t \end{bmatrix}.$$

The eigenvalues and eigenvectors of A are

$$\mu_1 = -1, \quad E_1 = \begin{bmatrix} 1 \\ -1 \end{bmatrix}, \quad \mu_2 = -2, \quad E_1 = \begin{bmatrix} -2 \\ 3 \end{bmatrix}.$$

The resulting modal matrix and its inverse are

$$M = \begin{bmatrix} 1 & -2 \\ -1 & 3 \end{bmatrix}, \quad M^{-1} = \begin{bmatrix} 3 & 2 \\ 1 & 1 \end{bmatrix}.$$

The modified forcing function vector G is

$$G = M^{-1}F = \begin{bmatrix} 3 & 2 \\ 1 & 1 \end{bmatrix}\begin{bmatrix} e^{-t} \\ t \end{bmatrix} = \begin{bmatrix} 3e^{-t} + 2t \\ e^{-t} + t \end{bmatrix}.$$

Substituting into equation (11.5) and carrying out the matrix multiplication gives the following:

$$Y = \begin{bmatrix} e^{-t} & 0 \\ 0 & e^{-2t} \end{bmatrix}Y_0 + \begin{bmatrix} e^{-t} & 0 \\ 0 & e^{-2t} \end{bmatrix}\int_0^t \begin{bmatrix} 3 + 2\eta e^{\eta} \\ e^{\eta} + \eta e^{2\eta} \end{bmatrix}d\eta$$

Performing the indicated integration and then the matrix multiplication provides the general solution for Y.

$$Y = \begin{bmatrix} e^{-t} & 0 \\ 0 & e^{-2t} \end{bmatrix}Y_0 + \begin{bmatrix} 2t - 2 + 2e^{-t} + 3te^{-t} \\ \dfrac{t}{2} - \dfrac{1}{4} - \dfrac{3e^{-2t}}{4} + e^{-t} \end{bmatrix}$$

It has been left as an exercise for the reader to verify that this answer is identical with the one given by (10.7).

EXERCISES

1. Verify that the solution to the example problem above is identical to the one given by equation (10.7).

2. Reduce the following differential equation and its associated initial conditions to a first-order system, and determine its solution.

$$Y'' - 4y' + 3y = e^{2x}, \ y(0) = 0, \ y'(0) = 0$$

Solve the following differential equations.

3*. $\qquad X' = \begin{bmatrix} 2 & 1 \\ 2 & 3 \end{bmatrix}X + \begin{bmatrix} e^{-t} \\ 0 \end{bmatrix}, \ X_0 = \begin{bmatrix} -1 \\ 1 \end{bmatrix}$

$$4*.^{55} \qquad X' = \begin{bmatrix} -2 & 2 \\ -5 & 1 \end{bmatrix} X + \begin{bmatrix} e^t \\ e^{2t} \end{bmatrix}$$

***

12. SPECIAL CONSIDERATION OF SECOND-ORDER HOMOGENEOUS EQUATIONS

As has been indicated throughout this chapter, the recommended approach for solving differential equations in matrix form is to reduce them to a first-order system. There are, however, some situations in which this may not be the best way of proceeding. In many applied physical problems in which there is no energy dissipation, that is, conservative systems, an alternative form of solving the system equations is suggested. In general, such systems are characterized by the following equation.

(12.1) $AX'' + BX = [0],$

where the coefficient matrices are real and symmetric. For convenience, let us assume that the initial conditions are

(12.2) $X(t = 0) = X_0, X''(t = 0) = V_0.$

To establish the solution of (12.1), it is convenient to premultiply that equation through by A^{-1}. If we define a new matrix W as $A^{-1}B$, the resulting equation becomes

(12.3) $X'' + WX = [0].$

Even though the coefficient matrices A and B are real and symmetric, the new matrix W is not necessarily real symmetric. A number of situations do occur in which A is a diagonal matrix, which means that W would be a real symmetric matrix also. Some of these cases will be seen in chapter VIII.

The general solution to (12.3) is based on the knowledge of solutions to scalar second-order equations of the same form. By using the matrix sine and cosine functions, the general solution is expressed as follows.

$$X = \sin(W^{1/2}t) \, C_1 + \cos(W^{1/2}t)C_2$$

---

[55]   Derick and Grossman [1976], p328.

In this solution, C_1 and C_2 are two arbitrary constant column vectors that may be evaluated by matching the stated initial conditions. This solution may be verified by substitution into (12.3).

If we now impose the first initial condition from (12.2), we find that $C_2 = X_0$. To meet the second initial condition, the first derivative of X is required, which is

$$X' = W^{1/2}\cos(W^{1/2}t)C_1 - W^{1/2}\sin(W^{1/2}t)C_2.$$

If we now impose the second initial condition on this result and solve for C_1, we find that $C_1 = W^{-1/2}V_0$. If the values of the two constant vectors are substituted into the general solution, the particular solution will be found.

(12.4) $X = \sin(W^{1/2}t)W^{-1/2}V_0 + \cos(W^{1/2}t)X_0$

EXAMPLE 13

Solve the differential equation given by (12.3) for which the W matrix is the following.

$$W = \begin{bmatrix} 9 & -6 & 1 \\ -3 & 10 & -3 \\ 1 & -6 & 9 \end{bmatrix}$$

SOLUTION

For this matrix, the eigenvalues and associated eigenvectors are[56] the following:

$$m_1 = 16,\ E_1 = \begin{bmatrix} 1 \\ -1 \\ 1 \end{bmatrix},\ m_2 = 8,\ E_2 = \begin{bmatrix} 1 \\ 0 \\ -1 \end{bmatrix},$$

$$m_3 = 4,\ E_3 = \begin{bmatrix} 1 \\ 1 \\ 1 \end{bmatrix}$$

---

[56]   The MATLAB program has been used to evaluate the eigenvalues and eigenvectors for this example problem.

For the W matrix, the modal matrix is

$$E = \begin{bmatrix} 1 & 1 & 1 \\ -1 & 0 & 1 \\ 1 & -1 & 1 \end{bmatrix}.$$

The sine and cosine matrix function appearing in the solution may be evaluated by utilizing a similarity transformation as developed in section 13 of chapter VI. By this method, the following relations may be obtained.

(12.5)

$$\sin(W^{1/2}t) = \frac{1}{4}\begin{bmatrix} 1 & -2 & 1 \\ -1 & 2 & -1 \\ 1 & -2 & 1 \end{bmatrix}\sin(4t)$$

$$+\frac{1}{4}\begin{bmatrix} 2 & 0 & -2 \\ 0 & 0 & 0 \\ -2 & 0 & 2 \end{bmatrix}\sin(2\sqrt{2}\,t)$$

$$+\frac{1}{4}\begin{bmatrix} 1 & 2 & 1 \\ 1 & 2 & 1 \\ 1 & 2 & 1 \end{bmatrix}\sin(2t)$$

(12.6)

$$\cos(W^{1/2}t) = \frac{1}{4}\begin{bmatrix} 1 & -2 & 1 \\ -1 & 2 & -1 \\ 1 & -2 & 1 \end{bmatrix}\cos(4t)$$

$$+\frac{1}{4}\begin{bmatrix} 2 & 0 & -2 \\ 0 & 0 & 0 \\ -2 & 0 & 2 \end{bmatrix}\cos(2\sqrt{2}\,t)$$

$$+\frac{1}{4}\begin{bmatrix} 1 & 2 & 1 \\ 1 & 2 & 1 \\ 1 & 2 & 1 \end{bmatrix}\cos(2t)$$

To evaluate the particular solution above, we also need and expression for $W^{-1/2}$, which may be evaluated by the same approach. The result is the following:

(12.7)

$$W^{-1/2} = \frac{1}{16}\begin{bmatrix} 1 & -2 & 1 \\ -1 & 2 & -1 \\ 1 & -2 & 1 \end{bmatrix} + \frac{\sqrt{2}}{16}\begin{bmatrix} 2 & 0 & -2 \\ 0 & 0 & 0 \\ -2 & 0 & 2 \end{bmatrix} +$$

$$\frac{1}{8}\begin{bmatrix} 1 & 2 & 1 \\ 1 & 2 & 1 \\ 1 & 2 & 1 \end{bmatrix} = \frac{1}{16}\begin{bmatrix} 3+2\sqrt{2} & 2 & 3-2\sqrt{2} \\ 1 & 6 & 1 \\ 3-2\sqrt{2} & 2 & 3+2\sqrt{2} \end{bmatrix}$$

The next step is to substitute all of the above expressions into the particular solution given in (12.4). This appears to involve a great deal of matrix multiplication to carry out. However, the task is not as complicated if the three recurring coefficient matrices in (12.5), (12.11), and (12.7) are recognized to be the product of a column vector times a row vector. A few minutes of examination will lead to the following observation.

$$\begin{bmatrix} 1 & -2 & 1 \\ -1 & 2 & -1 \\ 1 & -2 & 1 \end{bmatrix} = \begin{bmatrix} 1 \\ -1 \\ 1 \end{bmatrix} \begin{bmatrix} 1 & -2 & 1 \end{bmatrix}$$

$$\begin{bmatrix} 2 & 0 & -2 \\ 0 & 0 & 0 \\ -2 & 0 & 2 \end{bmatrix} = \begin{bmatrix} 1 \\ 0 \\ -1 \end{bmatrix} \begin{bmatrix} 2 & 0 & -2 \end{bmatrix}$$

$$\begin{bmatrix} 1 & 2 & 1 \\ 1 & 2 & 1 \\ 1 & 2 & 1 \end{bmatrix} = \begin{bmatrix} 1 \\ 1 \\ 1 \end{bmatrix} \begin{bmatrix} 1 & 2 & 1 \end{bmatrix}$$

It is of importance to recognize that the columns are the respective eigenvalues of the W matrix and the rows are the respective rows of the inverse of the modal matrix. This means that the rows are the rows of the adjoint of the modal matrix.

If we now substitute all these relations into (12.4), complete all the indicated matrix multiplications, and collect coefficients of like terms leads to the following expression for the particular solution.

$$X = \frac{1}{16} \begin{bmatrix} 1 \\ -1 \\ 1 \end{bmatrix} \begin{bmatrix} 1 & -2 & 1 \end{bmatrix} \{\sin(4t)V_0 + \cos(4t)X_0\}$$

$$+ \frac{1}{16} \begin{bmatrix} 1 \\ 0 \\ -1 \end{bmatrix} \begin{bmatrix} 2 & 0 & -2 \end{bmatrix} \{\sin(2\sqrt{2}t)V_0 + \cos(2\sqrt{2}t)X_0\}$$

$$+ \frac{1}{8} \begin{bmatrix} 1 \\ 1 \\ 1 \end{bmatrix} \begin{bmatrix} 1 & 2 & 1 \end{bmatrix} \{\sin(2t)V_0 + \cos(2t)X_0\}$$

EXERCISES

1.*   Solve the following differential equation using matrix trigonometric functions.

$$X'' + \begin{bmatrix} \frac{13}{3} & -\frac{14}{3} & \frac{1}{3} \\ -3 & 6 & -3 \\ -\frac{4}{3} & -\frac{4}{3} & \frac{8}{3} \end{bmatrix} X = [0], \quad X_0 = \begin{bmatrix} 1 \\ 0 \\ 2 \end{bmatrix}, \quad X'_0 = \begin{bmatrix} 0 \\ 0 \\ -1 \end{bmatrix}$$

2.*   Solve the following differential equation using matrix functions.

$$X'' = \begin{bmatrix} 2 & 2 & -2 \\ -10 & 6 & 6 \\ 10 & -2 & -6 \end{bmatrix} X, \quad X_0 = \begin{bmatrix} 1 \\ -1 \\ 1 \end{bmatrix}, \quad X'_0 = \begin{bmatrix} -1 \\ 0 \\ 1 \end{bmatrix}$$

***

13. SOME FINAL REMARKS ABOUT SECOND ORDER EQUATIONS

As it will be seen in the next chapter, many applied problems result in nonhomogeneous second-order matrix differential equations with constant coefficients of the following form.

(13.1) $AX'' + BX' + CX = F$

As discussed earlier in this chapter, the general solution to this equation is written as the sum of two solutions, the homogeneous solution and the particular integral.

(13.2) $X = X_h + X_{pi}$

The homogeneous solution is the solution to (13.1) with the forcing function vector F set equal to the null vector. If the system is n^{th} order, then there will be $2n$ arbitrary constants appearing in this solution. The particular integral is the solution to (13.1) with the forcing function included and thus reflects its effect on the solution. If initial conditions are prescribed, they are used to evaluate the arbitrary constants that appear in the homogeneous solution.

Many problems studied are conservative systems, which means that they do not contain any elements that dissipate energy. In these systems, the coefficient matrix B disappears and the homogeneous solution takes the form of sine and cosine functions as described in the previous system. In any physical system that is nonconservative, the presence of the B matrix ensures that the homogeneous solution will always be exponentially damped. This means that in the presence of energy dissipation the homogeneous solution will always decay to zero with time.

These effects have led to the common terminology of transient and steady state responses of systems. A transient response refers to a solution of equations like (13.1) in the form of (13.2), which has been evaluated to satisfy any given initial conditions. The steady state response refers to the system behavior after all exponentially decaying terms have vanished. For nonconservative systems, this means that the homogeneous solution has disappeared and the steady state solution is simply the particular integral. In solving system equations, these considerations are sometimes helpful in developing the solution by avoiding the calculation of unnecessary terms.

ADDITIONAL EXERCISES

1.*   The well known Fibonacci sequence is the series 0, 1, 1, 2, 3, 5, 8, 13, 21, etc., such that each succeeding number is the sum of the two previous numbers. The initial values are 0 and 1.

   a.   Write a difference equation for generating the nth term.

   b.   Reduce your answer to part a to a first-order matrix equation.

   c.   Find a general expression for the nth power of the transition matrix of this problem.

   d.   Show that as n becomes large, the ratio of two succeeding numbers approaches the "golden mean" value of 1.618.

2.*   As an English professor who teaches two sections of English 1A every term, you have made the following general observations. Every week, 20% of the students in section A and 30% of those in section B drop the course. Also, each week, 8% transfer to the other section.

   a.   Determine the transition matrix for this system if the state vector represents the enrollment of each section and the number dropped each week.

   b.   For a fifteen-week term, what will your final enrollments be if you initially have 100 students in each section?

3.*   The U-Rent-Um car rental agency has outlets in Los Angeles, San Francisco, Las Vegas, and Phoenix. After several months of operation, you have made the following observations about the operations for monthly transfers:

In LA, 20% go to SF, 20% go to LV, and 10% go to Phoenix, with the balance remaining in L.A.

In SF, 20% go to LA, 15% go to LV, and 5% go to Phoenix, with the balance remaining.

In LV, 30% go to LA, 20% go to SF, and 10% go to Phoenix, with the balance remaining.

In Phoenix, 5% go to SF, 5% go to LA, and 25% go to LV, with the balance remaining.

a.   Develop the monthly transition matrix for this car rental service.

b.   If there are initially 100 cars in each city, what is the steady state distribution of vehicles?

4.*   Find the largest values of the constants a, b, and c for which these transition matrices[57] are stable or marginally stable. Discuss your findings.

$$\begin{bmatrix} a & -0.8 \\ 0.8 & 0.2 \end{bmatrix}, \quad \begin{bmatrix} b & 0.8 \\ 0 & 0.2 \end{bmatrix}, \quad \begin{bmatrix} c & 0.8 \\ 0.2 & c \end{bmatrix}$$

Solve the following matrix differential equations by use of the matrix exponential function, and reduce the answer to direct matrix form by means of any convenient method.

5*.
$$X' = \begin{bmatrix} -2 & 1 & 1 \\ 2 & -3 & 1 \\ 2 & -1 & -1 \end{bmatrix} X, \quad X_0 = \begin{bmatrix} 2 \\ 2 \\ -2 \end{bmatrix}$$

6*.
$$X' = \begin{bmatrix} 1 & 1 & -2 \\ -1 & 2 & 1 \\ 0 & 1 & -1 \end{bmatrix} X, \quad X_0 = \begin{bmatrix} 1 \\ 1 \\ -1 \end{bmatrix}$$

7*.
$$X' = \begin{bmatrix} 1 & -1 & 4 \\ 3 & 2 & -1 \\ 2 & 1 & -1 \end{bmatrix} X, \quad X_0 = \begin{bmatrix} 2 \\ -1 \\ 0 \end{bmatrix}$$

---

[57]   Strang, [1988], p. 274

8,* 9,* 10.* Solve the matrix differential equations given in exercises 5, 6, 7 above by the use of an eigenvector expansion.

Apply the adjoint method to obtain the solution for the following differential equations.

11*.
$$X' = \begin{bmatrix} 2 & 1 \\ 2 & 3 \end{bmatrix} X + \begin{bmatrix} e^{-t} \\ 0 \end{bmatrix}, \quad X_0 = \begin{bmatrix} -1 \\ 1 \end{bmatrix}$$

12*.
$$X' = \begin{bmatrix} -2 & -2 \\ -5 & 1 \end{bmatrix} X + \begin{bmatrix} e^t \\ e^{2t} \end{bmatrix}, \quad \text{Derick and Grossman, p328}$$

13*.
$$X' = \begin{bmatrix} -2 & 1 & 1 \\ 2 & -3 & 1 \\ 2 & -1 & -1 \end{bmatrix} X + \begin{bmatrix} 1 \\ e^{-2t} \\ 0 \end{bmatrix}, \quad X_0 = \begin{bmatrix} 1 \\ -1 \\ 0 \end{bmatrix}$$

14*.
$$X' = \begin{bmatrix} 1 & 1 & -2 \\ -1 & 2 & 1 \\ 0 & 1 & -1 \end{bmatrix} X + \begin{bmatrix} e^t \\ e^{2t} \\ e^{3t} \end{bmatrix}, \quad \text{Derrick and Grossman, p328}$$

15. Verify that the differential equation (8.10) is satisfied by the solution (8.27) and that the solution meets the stated initial condition.

16.* Solve the differential equation[58] subject to the given initial condition.

$$X' = \begin{bmatrix} 1 & 6 & 3 \\ 0 & 3 & 1 \\ -1 & 3 & 3 \end{bmatrix} X, \quad X_0 = \begin{bmatrix} 1 \\ 0 \\ -1 \end{bmatrix}$$

17. Verify that the solution of the example problem of section 9, equation (9.3), is correct and that it also satisfies the stated initial conditions.

18.* Solve the following second-order differential equation by reducing it to a first-order system.

$$X'' = \begin{bmatrix} 2 & 2 & -2 \\ -10 & 6 & 6 \\ 10 & -2 & -6 \end{bmatrix} X, \quad X_0 = \begin{bmatrix} 1 \\ -1 \\ 1 \end{bmatrix}, \quad X'_0 = \begin{bmatrix} -1 \\ 0 \\ 1 \end{bmatrix}$$

---

[58] Gere, J. M. and W. Weaver [1983], p. 153.

VIII

APPLICATION OF DIFFERENCE AND
DIFFERENTIAL EQUATIONS

1. INTRODUCTION

In the previous chapter, some of the basic approaches to solving linear difference equations and linear differential equations in matrix form were presented. This chapter is devoted to illustrating how these methods may be applied to a number of practical problems of interest. Clearly, the full range of possible applications cannot be covered, but hopefully, the ones presented will be of interest to most readers and indicate the breadth of applicability of these methods.

2. INCOMPRESSIBLE FLOW OR PROCESS PROBLEMS

In the area of hydraulics or flow of incompressible fluids and chemical processes, differential equations are useful in analyzing certain types of problems. In particular, they may be used to model the concentration of one or more chemical constituents in a chemically reacting or flowing system. Even the study of electrolyte concentrations in biological systems may be studied in exactly the same manner.

To illustrate this area of application, a single illustration will be presented. Figure 8.1 below shows a relatively simple steady state flow system for which it is desired to analyze the concentration of salt as it circulates through the system.

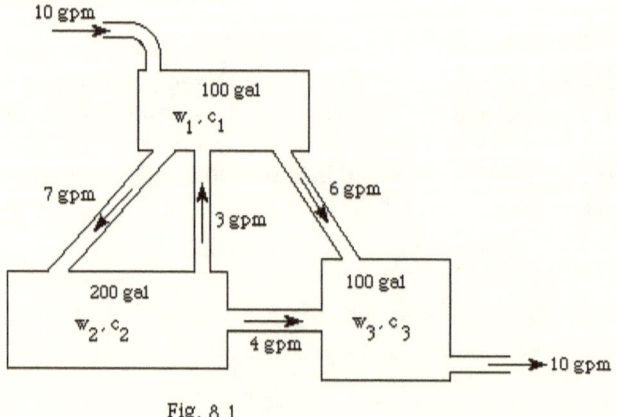

Fig. 8.1

434

In this system, all the water flow rates are steady, and the input of 10 gallons per minute (gpm) is always pure water. It is assumed the tanks remain full at their indicated capacities. The flow rates between tanks and at the outlet are also shown on the figure.

The term w indicates the weight of salt in a tank in pounds at any given time. The term c is the salt concentration in a given tank, and it is given at any time by the ratio of the weight of salt in a tank to the tanks capacity. The units for salt concentration is therefore pounds per gallon. It is also assumed that each tank has thorough mixing so that its concentration is uniform. The concentration of the fluid leaving any tank is assumed to be equal to the salt concentration in the tank.

The problem is to study how the concentrations of salt vary with time if the system is filled with pure water at $t = 0$ and 100 lb of salt are added to the first tank at that instant. At the initial time then, the initial salt concentrations are $c_1 = 1$ lb/gal, $c_2 = c_3 = 0$ lb/gal.

To develop the differential equations for the concentrations, conservation of mass must be applied to the salt in each tank. For example, applying conservation to the first tank requires that the time rate of increase of salt in that tank must equal the net mass flow rate of salt into the tank. The total amount of salt in a tank at a given instant of time is the concentration of salt times the tank volume. This means that the rate of increase of salt in the first tank is written as the following:

(2.1)
$$\frac{d}{dt}(V_1 c_1) = (100 \text{ gal}) \frac{d\{c_1 \text{ (lb/gal)}\}}{dt} = 100\, c_1'$$
$$= \text{ rate of increase of salt in lb/min}$$

As stated, this term represents the rate of increase of salt in the first tank. To this tank, there are attached four lines. In any line, the flow rate of salt is the salt concentration times the volume flow rate. The main inlet is supplying only pure water at a rate of 10 gpm so no salt is added from the source. The first tank is supplying liquid to the two other tanks at the rate of 7 gpm to the second tank and 6 gpm to the third tank. This is a net loss of salt at the rate of 13 gpm $\times c_1$ (lb/gal) $= 13c_1$ lb salt/min. The return line from the second tank brings in salt at the rate of $3c_2$ lb salt/min. The net inflow rate of salt in pounds per minute is then $3c_2 - 13c_1$. Balancing (2.1) with this flow yields the desired conservation equation for the first tank.

(2.2) $100\, c_1' = -13\, c_1 + 3\, c_2$

Dividing by the volume of the first tank gives

(2.3) $c_1' = -0.130c_1 + 0.030c_2.$

Repeating this procedure for the other two tanks yield the additional conservation equations.

(2.4)

$$c_2' = 0.035c_1 - 0.035c_2$$

$$c_3' = 0.060c_1 + 0.040c_2 - 0.100c_3$$

Introducing a concentration vector C, these equations may be expressed in the following matrix differential equation form.

(2.5)
$$C' = AC,$$

where the coefficient matrix is

(2.6)
$$A = \begin{bmatrix} -0.130 & 0.030 & 0 \\ 0.035 & -0.035 & 0 \\ 0.060 & 0.040 & -0.100 \end{bmatrix}.$$

The stated initial condition stated above is

(2.7)
$$C_0 = \begin{bmatrix} 1 \\ 0 \\ 0 \end{bmatrix}.$$

The solution to Equation (2.5) is given by the following expression.

(2.8)
$$C = \exp(At)C_0$$

The eigenvalues[59] and eigenvectors of the coefficient matrix are the following:

(2.9)
$$\mu_1 = -0.140, \quad E_1 = \begin{bmatrix} 6 \\ -2 \\ -7 \end{bmatrix}, \quad \mu_2 = -0.100, \quad E_1 = \begin{bmatrix} 0 \\ 0 \\ 1 \end{bmatrix}$$

$$\mu_3 = -0.025, \quad E_3 = \begin{bmatrix} 6 \\ 21 \\ 16 \end{bmatrix}$$

---

[59]. The previously cited MATLAB program was used to evaluate these eigenvalues and eigenvectors and perform the necessary matrix operations in solving this example.

Using the similarity transformation method to evaluate the matrix exponential function and applying the initial condition leads to the final answer for the concentration vector.

(2.10)
$$C = e^{-0.140t}\begin{bmatrix} 0.9132 \\ -0.3044 \\ -1.0654 \end{bmatrix} + e^{-0.100t}\begin{bmatrix} 0 \\ 0 \\ 0.8333 \end{bmatrix} + e^{-0.025t}\begin{bmatrix} 0.0870 \\ 0.3045 \\ 0.2320 \end{bmatrix}$$

Another type of incompressible fluid flow application occurs in the area of hydraulics or pipe flow. At low fluid velocities, the pressure drop along a fluid-filled pipe is linearly related to the volume flow rate of fluid passing through the pipe. This is depicted in figure 8.2 below.

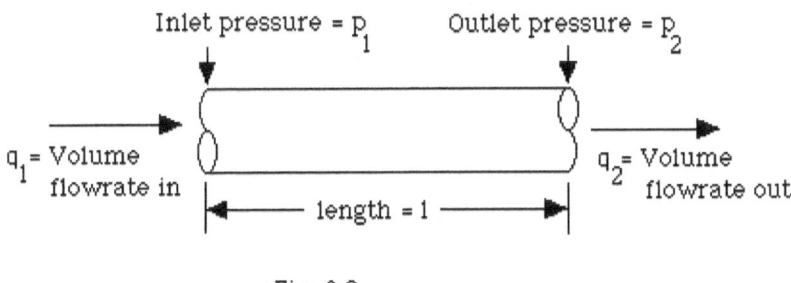

Fig. 8.2

Because the fluid is incompressible and the pipe cannot store fluid, the volume flow rate out must equal the volume flow rate in. That is, $q_1 = q_2 = q$. If the hydraulic resistance of the pipe is r_p, the outlet pressure is related to the inlet pressure and the flow rate of fluid passing through the pipe as follows.

(2.11) $$p_2 = p_1 - q r_p$$

The pressure drop along the pipe in the direction of the flow is $p_1 - p_2$. From equation (2.11), we obtain the basic linear pipe-flow relation.[60]

(2.12) $p_1 - p_2 = qr_p$

This simple model permits the analysis of flow through complex piping systems. The initial pressure is usually supplied by a pump or by gravity acting on a liquid in an elevated tank. The final outlet pressure is almost always atmospheric pressure. Between the inlet (p_1 is known) and the outlet (p_2 is known), there may be an extensive network of pipes of various sizes and lengths connected in patterns of series and parallel paths. Knowing p_1 and p_2, the flow rates in every segment and pressures at every intermediate junction can be found. The overall model for this type of problem is a system of linear algebraic equations. This situation is illustrated in the following example.

EXAMPLE 1

Establish the matrix equations for the pipe-flow system below. Assume that the inlet and outlet pressures p_1 and p_4, as well as the pipe resistances r_1, r_2, r_3, and r_4 are known.

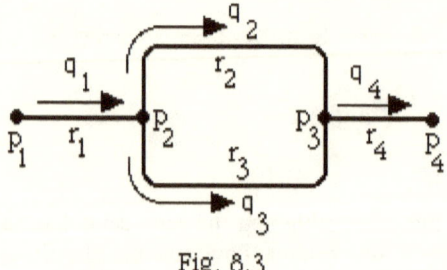

Fig. 8.3

SOLUTION

In this problem, there are two unknown junction or node pressures—p_2 and p_3. There are also three unknown flow rates—q_1, q_2, and q_3—because

---

60.   Unfortunately, the assumption of linear flow only covers a limited range of actual applications. These are called laminar flow problems. The majority of pipe-flow problems like the ones presented fall into the flow regime termed turbulent flow. In these situations, the model used to depict the flow is nonlinear in that the pressure drop varies as the square of the volume flow rate through the pipe.

conservation of flow requires that $q_1 = q_4$. Using equation (2.12) as the flow model, the flow in each pipe can be expressed by the following set of relations.

$$q_1 r_1 = p_1 - p_2$$
$$q_2 r_2 = p_2 - p_3$$
$$q_3 r_3 = p_2 - p_3$$
$$q_4 r_4 = p_3 - p_4$$

Applying conservation of flow at the junction (node) where p_2 is measured gives the following relation.

$$q_1 = q_2 + q_3$$

Replacing the flow rates by the above relations and performing some algebraic manipulations, the following equation may be obtained.

$$p_1 = \left(1 + \frac{r_1}{r_2} + \frac{r_1}{r_3}\right) p_2 - \left(\frac{r_1}{r_2} + \frac{r_1}{r_3}\right) p_3$$

Now at the junction where p_3 is measured, we have

$$q_1 = q_2 + q_3.$$

Following the same steps leads to the next relation

$$p_4 = -\left(\frac{r_4}{r_2} + \frac{r_4}{r_3}\right) p_2 - \left(1 + \frac{r_4}{r_2} + \frac{r_4}{r_3}\right) p_3.$$

These equations may be expressed in matrix form as follows.

$$\begin{bmatrix} \left(1 + \frac{r_1}{r_2} + \frac{r_1}{r_3}\right) & -\left(\frac{r_1}{r_2} + \frac{r_1}{r_3}\right) \\ -\left(\frac{r_4}{r_2} + \frac{r_4}{r_3}\right) & \left(1 + \frac{r_4}{r_2} + \frac{r_4}{r_3}\right) \end{bmatrix} \begin{bmatrix} p_2 \\ p_3 \end{bmatrix} = \begin{bmatrix} p_1 \\ p_4 \end{bmatrix}$$

or

$$R \begin{bmatrix} p_2 \\ p_3 \end{bmatrix} = \begin{bmatrix} p_1 \\ p_4 \end{bmatrix},$$

where R is a coefficient matrix of pipe resistances.

This is a system of two linear algebraic equations for the two unknown pressures p_2 and p_3. The inlet and outlet pressures p_1 and p_4 are known. Once the pressures have been determined, the flow rates can be found if they are also desired.

***

This example problem shows how linear algebraic equations are used to solve pipe-flow problems. When tanks or reservoirs are included in the flow network, the resulting mathematical model becomes a system of differential equations.

When a tank or reservoir is present in a piping system, the pressure at the location of the tank is controlled by the height of fluid in the tank. This is illustrated in figure 8.4 below.

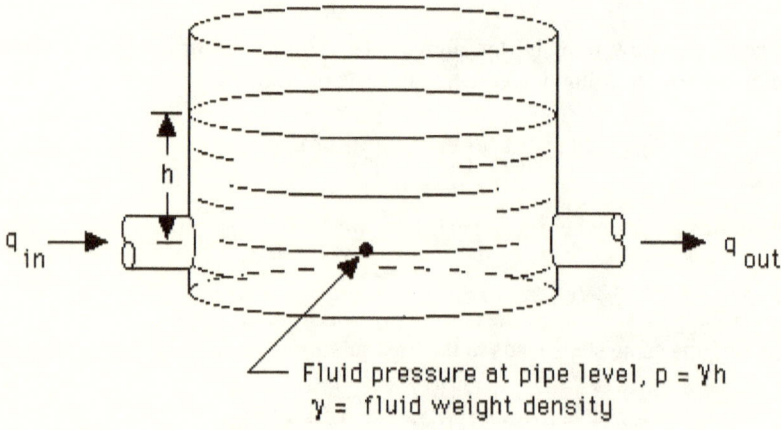

Fluid pressure at pipe level, p = γh
γ = fluid weight density

Fig. 8.4

The effective volume of fluid in the tank, V, is related to the tanks cross sectional area, a, and the height of fluid in the tank, h, by the relation $v = ah$. Assuming the fluid is incompressible, conservation of fluid requires that the rate at which the volume of fluid in the tank increases must equal the input flow rate minus the output flow rate. In equation form, this is

$$\frac{dv}{dt} = q_{in} - q_{out}.$$

Since the height of fluid above the pipe is the only variable, this equation may be rewritten as follows.

$$\frac{dh}{dt} = \frac{1}{h}(q_{in} - q_{out})$$

EXAMPLE 2

Establish the matrix differential equation for the unknown fluid heights in the two-tank pipe system shown below.

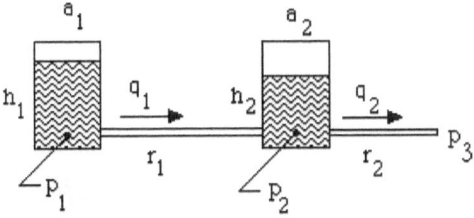

SOLUTION

Applying conservation of flow to the first tank gives the relation

$$\frac{dv_1}{dt} = -q_1.$$

For the pipe connecting the two tanks, the flow equation is

$$q_1 = \frac{p_1 - p_2}{r_1}.$$

If we now make use of the relations $v_1 = a_1 h_1$, $p_1 = \gamma h_1$, $p_2 = \gamma h_2$, the above relations may be combined to give the differential equation describing the rate of change of fluid level in the first tank.

$$\frac{dh_1}{dt} = -\frac{\gamma}{a_1 r_1} h_1 + \frac{\gamma}{a_1 r_1} h_2$$

Now the conservation of flow equation for the second tank is

$$\frac{dv_2}{dt} = q_1 - q_2.$$

Assume that the final outlet pressure, p_3, is atmospheric pressure, and assign a value of 0 to it. This is a very standard procedure for which all other pressures are measured relative to atmospheric pressure. If measured in this way, the pressures are called gage pressures. With this assumption and making the similar substitutions as for the first tank equation, the differential equation describing the fluid height in the second tank becomes

$$\frac{dh_2}{dt} = \frac{\gamma}{a_2 r_2} h_1 - \frac{\gamma}{a_2}\left(\frac{1}{r_1} + \frac{1}{r_2}\right) h_2.$$

In matrix form, the two differential equations become the following system equation.

$$\frac{d}{dt}\begin{bmatrix} h_1 \\ h_2 \end{bmatrix} = \begin{bmatrix} -\dfrac{\gamma}{a_1 r_1} & \dfrac{\gamma}{a_1 r_1} \\ \dfrac{\gamma}{a_2 r_2} & -\dfrac{\gamma}{a_2}\left(\dfrac{1}{r_1} + \dfrac{1}{r_2}\right) \end{bmatrix}\begin{bmatrix} h_1 \\ h_2 \end{bmatrix}$$

EXAMPLE 3

Determine the solution to the problem of example 2 above if the specific values of the physical quantities are given below. The solution will give the liquid level in both tanks as a function of time.

weight density = γ = 10,000 N/m^3
area of first tank = a_1 = 10 m^2
area of second tank = a_2 = 6 m^2
resistance of first pipe = r_1 = 3 × 10^5 Ns/m^5
resistance of second pipe = r_2 = 2 × 10^5 Ns/m^5
initial liquid level in first tank = 10 m
initial liquid level in second tank = 2 m

SOLUTION

Substituting the physical values into the differential equation describing the system gives the following specific system equation.

$$\frac{d}{dt}\begin{bmatrix} h_1 \\ h_2 \end{bmatrix} = \begin{bmatrix} -3.33 \times 10^{-3} & 3.33 \times 10^{-3} \\ 8.33 \times 10^{-3} & -1.38 \times 10^{-2} \end{bmatrix}\begin{bmatrix} h_1 \\ h_2 \end{bmatrix}$$

The eigenvalues and associated eigenvectors for the coefficient matrix are the following:

$$\mu_1 = -0.0011, \; E_1 = \begin{bmatrix} 0.8354 \\ 0.5496 \end{bmatrix}; \; \mu_2 = -0.0160, \; E_2 = \begin{bmatrix} -0.2544 \\ 0.9671 \end{bmatrix}$$

If we define the system coefficient matrix as A, the solution may be expressed as follows.

$$H = \exp(At)H_o,$$

where H is the matrix of liquid levels and H_o is the initial condition vector describing the liquid levels in the tanks at the time $t = 0$. Using the stated values, we have $H_o = [10 \; 2]^T$. Employing a similarity transform to evaluate the matrix exponential function, we find the following:

$$e^{At} = M^{-1} \text{Diag}\left[e^{\mu_i t}\right] M$$

$$= \begin{bmatrix} 0.8354 & -0.2544 \\ 0.5496 & 0.9671 \end{bmatrix} \begin{bmatrix} e^{-0.0011t} & 0 \\ 0 & e^{-0.0160t} \end{bmatrix} \begin{bmatrix} 0.8354 & -0.2544 \\ 0.5496 & 0.9671 \end{bmatrix}^{-1}$$

$$= \begin{bmatrix} 0.8524 & 0.2242 \\ 0.5608 & 0.1475 \end{bmatrix} e^{-0.0011t} + \begin{bmatrix} 0.1476 & -0.2242 \\ -0.5608 & 0.8525 \end{bmatrix} e^{-0.0160t}$$

If we postmultiply this by the given initial condition vector, the time solution for the liquid levels in the tanks is found to be

$$H(t) = \begin{bmatrix} 8.972 \\ 5.903 \end{bmatrix} e^{-0.0011t} + \begin{bmatrix} 1.028 \\ -3.903 \end{bmatrix} e^{-0.0160t}$$

This solution has been plotted below as an aid to visualizing the liquid levels as the tanks empty. It is interesting to note that the second tank begins filling initially until the liquid level of the first tank has dropped to the same level. At that point, both tanks simply continue to drain.

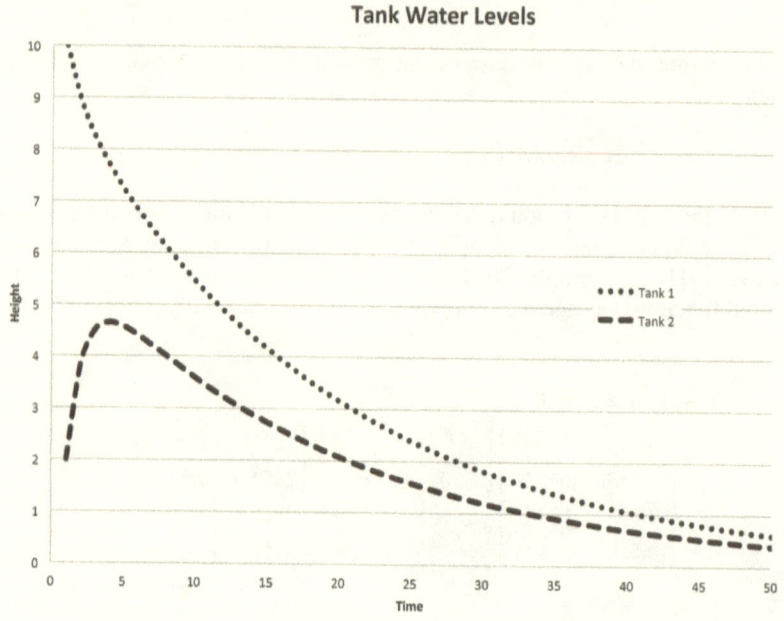

Tank Water Levels

EXERCISES

1.   For the salt concentration problem given in figure 8.1,

   a.   sketch $c_3(t)$ versus t,
   b.   find out how long it will take for the salt concentration in the final outlet to fall to a level of 0.01 lb salt/gal of water, and
   c.   sketch $c_2(t)$ versus t.

2.   Consider the problem shown in figure 8.1 with the change in flow rates indicated below.

   a.   Determine the dynamic equations for this system.

b. Find the solution given the same initial conditions as for the example problem in the chapter.

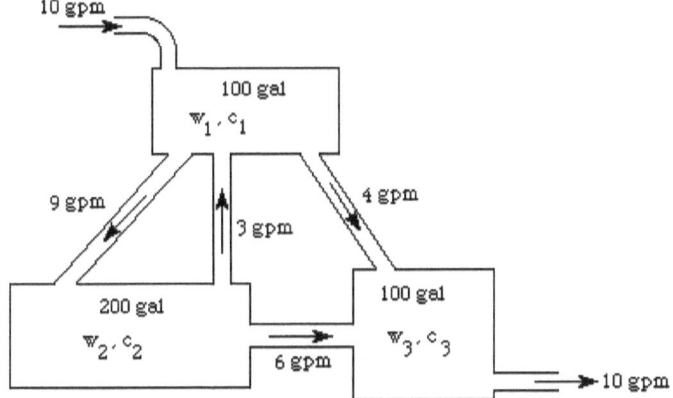

3.   A 400 gallon (gal) tank[61] is half-filled with water in which 50 pounds of salt is dissolved. Fresh water enters the tank at the rate of 1 gallon per minute (gpm). The well-stirred mixture drains into a second tank at the rate of 2 gpm. This second tank initially contains 150 gal of fresh water. The mixture in the second tank is kept well stirred and is allowed to drain at the rate of 2 gpm. Half of this outflow is returned to the first tank, and the rest is discarded. Find the amount of salt (pounds) in each tank after 1 hour has elapsed.

4.   For the two-tank problem of examples 2 and 3 above, determine and plot the solution if the initial liquid level of the first tank is 2 m and that of the second tank is 10 m. Discuss the meaning of your solution.

5.   Set up the matrix differential equation describing the liquid levels in the three-tank system of figure 8.5 below.

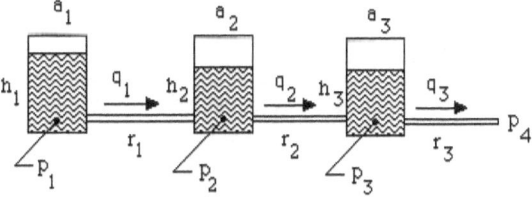

Figure 8.5

---

[61] Cullen (1979), problem 8, page 270.

***

3.   THE ANALYSIS OF ELECTRICAL CIRCUITS

The application of matrix notation for writing differential equations describing the behavior of electrical circuits is an extremely useful tool. To begin the analysis, consider the generalized circuit shown in figure 8.6 in which specific circuit elements have not been shown.

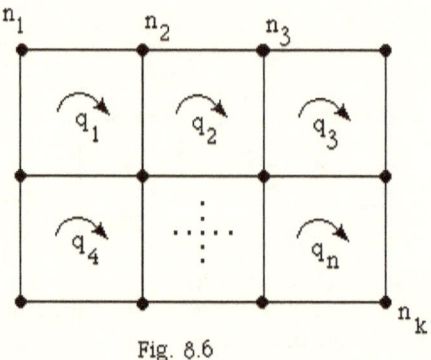

Fig. 8.6

As implied in this figure, the circuit is visualized as composed of a number of small conducting loops or mesh much like a fishnet. Each intersection is called a node and has also been designated in figure 8.6. The path connecting any two nodes is called a branch. Instead of using the circulating currents i_k in each mesh, which is the most common approach, the circulating charges q_k have been indicated simply to avoid dealing with integrals in the basic circuit equations. These two variables are connected by the relation that the time rate of change of the charge flow is the electrical current.

Application of Kirchoff's voltage law to a general circuit as depicted in figure 8.6 will yield the following matrix differential equation for the circulating charge vector q.

(3.1) $LQ'' + RQ' + SQ = V$

The coefficient matrices and the V vector for this equation may be constructed directly from examination of a given circuit. These matrices are defined as follows:

$$
l_{ij} = \begin{cases} \text{sum of all inductances in } i^{th} \text{ mesh when } i = j \\ \\ \text{-sum of all inductances in branch} \\ \text{between } i^{th} \text{ and } j^{th} \text{ mesh when } i \neq j \end{cases}
$$

$$
r_{ij} = \begin{cases} \text{sum of all resistances in } i^{th} \text{ mesh when } i = j \\ \\ \text{-sum of all resistances in branch} \\ \text{between } i^{th} \text{ and } j^{th} \text{ mesh when } i \neq j \end{cases}
$$

$$
s_{ij} = \begin{cases} \text{sum of all elastances in } i^{th} \text{ mesh when } i = j \\ \\ \text{-sum of all elastences in branch} \\ \text{between } i^{th} \text{ and } j^{th} \text{ mesh when } i \neq j \end{cases}
$$

$$
v_i = \text{sum of all voltage sources in } i^{th} \text{ mesh}
$$

It must be noted that an elastance is the reciprocal of a capacitance.

In many cases, it is desired to evaluate the steady state behavior of electrical circuits. This has already been discussed in section 3 of chapter III, and the reader is referred to that section for specific details.

To solve (3.1), it must be reduced to a first-order differential equation. This may be accomplished by introducing the current vector as a second variable vector. The symbol I will be used to designate the current vector. From the definition of charge, we have $I = Q'$. Substituting these variables into (3.1) and rearranging terms yields the following equation.

$$
I' = -L^{-1}RI - L^{-1}SQ + L^{-1}V
$$

Defining a state vector X as

(3.2)
$$
X = \begin{bmatrix} I \\ Q \end{bmatrix}
$$

allows the system equations to be written in a partitioned form as

(3.3)
$$
X' = \begin{bmatrix} -L^{-1}R & -L^{-1}S \\ I & 0 \end{bmatrix} X + L^{-1} V
$$
.

Defining the coefficient matrix as A and a forcing function vector by $F = L^{-1}V$, this equation takes the standard form of a nonhomogeneous, linear, and first-order differential equation that can be solved by the adjoint method discussed in chapter VII, section 10. The solution is

(3.4)
$$X = e^{At}X_0 + e^{At}\int_0^t e^{-A\eta}F\, d\eta$$
.

EXAMPLE 4

Solve for the circulating charges in the electrical circuit given in figure 8.7 below.

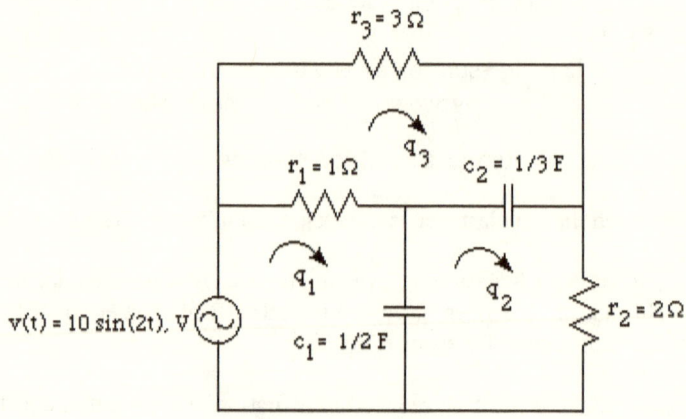

Fig. 8.7

SOLUTION

Since there are no inductances in this figure, the general equation of (3.1) is reduced to the following.

(3.5) $RQ' + SQ = V$

Applying the above construction rules, the matrices R, S, and V may be seen to be the following.

$$R = \begin{bmatrix} 1 & 0 & -1 \\ 0 & 2 & 0 \\ -1 & 0 & 4 \end{bmatrix}, \quad S = \begin{bmatrix} 2 & -2 & 0 \\ -2 & 5 & -3 \\ 0 & -3 & 3 \end{bmatrix},$$

$$V = 10 \sin(2t) \begin{bmatrix} 0 \\ 1 \\ 0 \end{bmatrix}$$

Premultiplying (3.5) by the inverse of the resistance matrix and rearranging terms allows the system equation for the given circuit to be written as

$$Q' = AQ + R^{-1}V,$$

where A is

$$A = \begin{bmatrix} -8/3 & 11/3 & -1 \\ 1 & -5/2 & 3/2 \\ -2/3 & 5/3 & -1 \end{bmatrix}.$$

The eigenvalues and eigenvectors[62] of A are

$$\mu_1 = 0, \; E_1 = \begin{bmatrix} 1 \\ 1 \\ 1 \end{bmatrix}, \; \mu_2 = -0.9604, \; E_2 = \begin{bmatrix} 1.0000 \\ 0.3938 \\ -0.2625 \end{bmatrix}$$

$$\mu_3 = -5.2603, \; E_3 = \begin{bmatrix} 1.0000 \\ -0.5861 \\ 0.3907 \end{bmatrix}.$$

In order to employ the general solution given by equation (3.4), the matrix function e^{At} must be evaluated. Following the method of section 13 of chapter VI, the following result is obtained. It should also be noted that since the circuit

---

[62.] The eigenvalues and other matrix calculations for this example and the others in this section were carried out using the program MATLAB previously cited.

equation for this example is already first order, the general solution may be applied directly in terms of the circulating charge vector.

$$e^{At} = \begin{bmatrix} 0.0000 & 0.4000 & 0.6000 \\ 0.0000 & 0.4000 & 0.6000 \\ 0.0000 & 0.4000 & 0.6000 \end{bmatrix} +$$

(3.6)

$$\begin{bmatrix} 0.5981 & 0.3731 & -0.9712 \\ 0.2355 & 0.1469 & -0.3824 \\ -0.1570 & -0.0979 & 0.2549 \end{bmatrix} e^{-0.9604t} +$$

$$\begin{bmatrix} 0.4019 & -0.7731 & 0.3712 \\ -0.2355 & 0.4531 & -0.2175 \\ 0.1570 & -0.3021 & 0.1450 \end{bmatrix} e^{-5.2603t}$$

Before evaluating the final solution, the initial condition vector must be given. For convenience, we will assume homogeneous initial conditions, which means that the initial condition vector is assumed to be the zero vector $Q_0 = [0]$. In physical terms, this means that the given system is completely at rest or quiescent. For this particular example, it means that there are no circulating charges at the time $t = 0$.

Substituting equation (3.6), the initial condition vector and performing the integrations indicated in equation (3.4) leads to the final result.

$$Q = \{1 - \cos(2t)\}\, E_1$$

$$\{0.7580\, e^{-0.9604t} + 0.8408 \sin(2t - 64.35°)\}\, E_2 +$$

$$\{-0.2441\, e^{-5.2603t} - 0.6869 \sin(2t - 20.82°)\}\, E_3$$

This expression has been simplified by the introduction of the phase angles in the harmonic terms, and the eigenvectors of A are given above.

EXAMPLE 5

Consider the circuit given in figure 8.8 below. Determine the two circulating charges subject to the given homogeneous initial conditions

$$q_1(0) = q_2(0) = 0, \; q_1'(0) = q_2'(0) = 0.$$

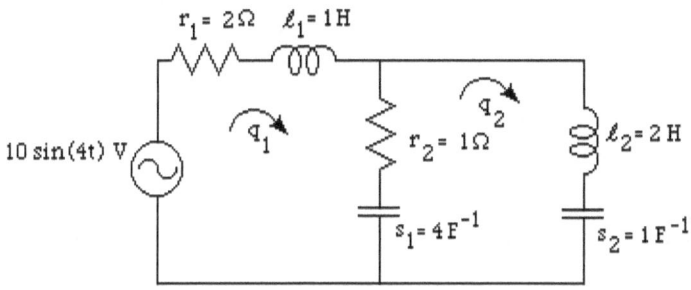

Fig. 8.8

SOLUTION

Following the formation rules for the coefficient matrices stated above, equation (3.1) takes the following specific form for this circuit.

$$\begin{bmatrix} 1 & 0 \\ 0 & 2 \end{bmatrix} Q'' + \begin{bmatrix} 3 & -1 \\ -1 & 1 \end{bmatrix} Q' + \begin{bmatrix} 4 & -4 \\ -4 & 5 \end{bmatrix} Q = 10 \sin(4t) \begin{bmatrix} 1 \\ 0 \end{bmatrix}$$

Converting this second-order equation to the first-order form of (3.5) yields gives the following results for the matrices A and F.

$$A = \begin{bmatrix} -3 & 1 & -4 & 4 \\ \frac{1}{2} & -\frac{1}{2} & 2 & -\frac{5}{2} \\ 1 & 0 & 0 & 0 \\ 0 & 1 & 0 & 0 \end{bmatrix}, \; F = 10 \sin(4t) \begin{bmatrix} 1 \\ 0 \\ 0 \\ 0 \end{bmatrix}$$

For this coefficient matrix, the eigenvalues and eigenvectors are

$$\mu_1 = -1.2771 + 1.7386j \,, \ E_1 = \begin{bmatrix} 0.0747 \\ -0.3490 \\ 0.2875 \\ 0.0485 \end{bmatrix} + j \begin{bmatrix} 0.8245 \\ -0.1264 \\ -0.2541 \\ 0.1651 \end{bmatrix}$$

$$\mu_3 = -0.4729 + 0.4539j \,, \ E_3 = \begin{bmatrix} -0.2836 \\ -0.2696 \\ 0.6578 \\ 0.5089 \end{bmatrix} + j \begin{bmatrix} 0.3272 \\ 0.2008 \\ -0.0605 \\ 0.0638 \end{bmatrix}$$

$$\mu_2 = \overline{\mu_1} \,, \quad E_2 = \overline{E_1} \,, \quad \mu_4 = \overline{\mu_3} \,, \quad E_4 = \overline{E_3} \quad .$$

The exponential function $\exp(At)$ will be evaluated by use of the similarity transformation discussed in chapter VI, section 13. This procedure gives the following result.

(3.7)
$$e^{At} = \{C \cos(1.7386t) - D \sin(1.7386t)\} \, e^{-1.2771t}$$
$$+ \{M \cos(0.4539t) - N \sin(0.4539t)\} e^{-0.4729t},$$

where the matrices C, D, M, and N are defined as

$$C = \begin{bmatrix} 0.1845 & -1.0846 & -1.3115 & 1.2233 \\ -0.5423 & 0.3382 & -0.8210 & 0.9381 \\ 0.4160 & 0.1762 & 1.3444 & -1.4125 \\ 0.0881 & -0.2343 & -0.1609 & 0.1329 \end{bmatrix}$$

$$D = \begin{bmatrix} 1.2489 & -0.3250 & 2.6353 & -2.8825 \\ -0.1625 & -0.3787 & -1.0336 & 1.0448 \\ -0.4116 & 0.4944 & -0.2331 & 0.3339 \\ 0.2472 & -0.0224 & 0.5903 & -0.6371 \end{bmatrix}$$

$$M = \begin{bmatrix} 0.8154 & 1.0847 & 1.3114 & -1.2233 \\ 0.5423 & 0.6618 & 0.8209 & -0.9381 \\ -0.4160 & -0.1762 & -0.3444 & 1.4125 \\ -0.0882 & 0.2343 & 0.1609 & 0.8670 \end{bmatrix}$$

$$N = \begin{bmatrix} 0.4558 & 0.9633 & 1.0404 & 0.0625 \\ 0.4816 & 0.9113 & 1.0077 & -0.1566 \\ -1.3631 & -2.2060 & -2.5304 & 1.2234 \\ -1.1030 & -1.7022 & -1.9762 & 1.1633 \end{bmatrix}$$

Since the initial condition vector is zero, the solution to this example problem as given by the general solution in (3.4) is just the second term.

$$Q = e^{At} \int_0^t e^{-A\eta} F \, d\eta$$

Substituting the above matrices and evaluating the integral leads to the following expression.

$$\int_0^t e^{-A\eta} F \, d\eta = \begin{bmatrix} 16.9458 \\ 0.5483 \\ -4.5748 \\ -0.2090 \end{bmatrix} +$$

$$\left\{ \begin{bmatrix} -8.2765 \\ 5.4451 \\ -1.2906 \\ -2.0727 \end{bmatrix} \cos(1.7386\,t) + \begin{bmatrix} -9.5332 \\ -1.5490 \\ 5.7077 \\ -1.6092 \end{bmatrix} \sin(1.7386\,t) \right\} e^{-2.7229\,t}$$

$$\left\{ \begin{bmatrix} -8.2765 \\ 5.4451 \\ -1.2906 \\ -2.0727 \end{bmatrix} \cos(1.7386\,t) + \begin{bmatrix} -9.5332 \\ -1.5490 \\ 5.7077 \\ -1.6092 \end{bmatrix} \sin(1.7386\,t) \right\} e^{-2.7229\,t}$$

The complete solution is obtained by multiplying this result by the value of the exponential function given in (3.7).

EXAMPLE 6

Find the solution for the circulating currents for the simple electrical circuit shown in figure 8.9.

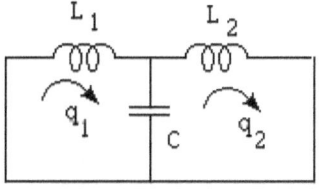

Fig. 8.9

SOLUTION

This circuit is composed of two inductances L_1 and L_2 and the capacitance c. Each mesh has been assigned a circulating charge q_1 and q_2. Again, following the coefficient matrix formation procedure leads to the following matrix differential equation for the circulating charges.

(3.8) $$LQ'' + CQ = [0],$$

where L and C are the inductance and capacitance coefficient matrices

$$L = \begin{bmatrix} L_1 & 0 \\ 0 & L_2 \end{bmatrix}, \quad S = \begin{bmatrix} \frac{1}{C} & \frac{-1}{C} \\ \frac{-1}{C} & \frac{1}{C} \end{bmatrix}.$$

Following the method presented above, we now introduce a new fourth-order partitioned vector X as defined in (3.2) so that (3.8) may be written as

$$X' = \begin{bmatrix} [0] & -L^{-1}C \\ I & [0] \end{bmatrix} X = AX.$$

For convenience, we shall assume that $1/l_1c = 1$ and $1/l_2c = 3$. Evaluating $L^{-1}C$ and employing this assumption gives the matrix A as

$$A = \begin{bmatrix} 0 & 0 & -1 & 1 \\ 0 & 0 & 3 & -3 \\ 1 & 0 & 0 & 0 \\ 0 & 1 & 0 & 0 \end{bmatrix}.$$

If we use Gaussian elimination or many of the standard matrix calculation packages, the eigenvalues and eigenvectors may be found to be

$$\mu_1 = \mu_2 = 0, \ E_2 = \begin{bmatrix} 0 \\ 0 \\ 1 \\ 1 \end{bmatrix}, \ \mu_3 = 2j, \ E_3 = \begin{bmatrix} 2 \\ -6 \\ -j \\ 3j \end{bmatrix}$$

$$\mu_4 = -2j, \ E_4 = \begin{bmatrix} 2 \\ -6 \\ j \\ -3j \end{bmatrix}.$$

The zero eigenvalue is repeated, and the standard approach for finding eigenvectors yields only the single eigenvector E_2.

To find the remaining eigenvector, the method of section 16, chapter VI will be employed. As shown in that section, the columns of the adjoint of the characteristic matrix of A will contain the desired eigenvectors. Equation (16.4) is applied to evaluate the adjoint of A for the zero eigenvalue.

$$\text{Adj } K(0) = K(0)K(2j)K(-2j),$$

where $K(\mu) = [\mu I - A]$, the characteristic matrix. Performing these computations gives the following result.

$$\text{Adj } K(0) = \begin{bmatrix} 0 & 0 & 0 & 0 \\ 0 & 0 & 0 & 0 \\ -3 & -1 & 0 & 0 \\ -3 & -1 & 0 & 0 \end{bmatrix}$$

This matrix contains only the eigenvector E_2, which is already known. Following the method of section 16 of chapter VI, the first derivative of the adjoint of the characteristic matrix must be evaluated. This means that the following product will give the desired result.

$$\text{Adj}^{(1)} K(0) = K(2j)K(-2j)$$

Computing this product gives

$$\text{Adj}^{(1)} K(0) = \begin{bmatrix} 3 & 1 & 0 & 0 \\ 3 & 1 & 0 & 0 \\ 0 & 0 & 3 & 1 \\ 0 & 0 & 3 & 1 \end{bmatrix}.$$

In this matrix, we see the eigenvector E_2 again but another independent column as well. This other column is the desired vector to be associated with the multiple eigenvalue. If tested, it can be shown that it is not an eigenvector but is the appropriate vector with which the solution may be completed. Since this vector is not an eigenvector, it will be designated as G_1.

$$G_1 = \begin{bmatrix} 1 \\ 1 \\ 0 \\ 0 \end{bmatrix}$$

A quick check will show that all four vectors comprise a linearly independent set or basis. The general solution to the original problem may now be written as

$$(3.9) \qquad X = c_1G_1 + c_2E_2t + c_3E_3e^{2jt} + c_4E_4e^{-2jt}.$$

The arbitrary constants may now be evaluated by applying known initial conditions at $t = 0$. Since we have not specified any initial conditions as yet, let us assume no initial circulating charges and equal but opposite initial circulating currents of unit magnitude. These give the initial condition vector as

$$X_0 = \begin{bmatrix} 1 \\ -1 \\ 0 \\ 0 \end{bmatrix}.$$

Imposing this initial condition on (3.9) allows us to solve for the arbitrary constants.

$$\begin{bmatrix} c_1 \\ c_2 \\ c_3 \\ c_4 \end{bmatrix} = \begin{bmatrix} 1 & 0 & 2 & 2 \\ 1 & 0 & -6 & -6 \\ 0 & 1 & -j & j \\ 0 & 1 & 3j & -3j \end{bmatrix}^{-1} \begin{bmatrix} 1 \\ -1 \\ 0 \\ 0 \end{bmatrix} = \begin{bmatrix} 1/2 \\ 0 \\ 1/8 \\ 1/8 \end{bmatrix}$$

Substituting these results into (3.9) gives the desired particular solution.

$$(3.10) \qquad X = \frac{1}{2}\begin{bmatrix} 1 \\ 1 \\ 0 \\ 0 \end{bmatrix} + \frac{1}{8}\begin{bmatrix} 2 \\ -6 \\ -j \\ 3j \end{bmatrix}e^{2jt} + \frac{1}{8}\begin{bmatrix} 2 \\ -6 \\ j \\ -3j \end{bmatrix}e^{-2jt}$$

The imaginary terms may be removed from this expression by utilizing the well-known Euler formulas.

$$e^{ajt} = \cos(at) + j\sin(at) \ , \ e^{-ajt} = \cos(at) - j\sin(at)$$

Applying these expressions to the two exponential terms in (3.10) and rearranging gives

$$X = \frac{1}{2}\begin{bmatrix} 1 \\ 1 \\ 0 \\ 0 \end{bmatrix} + \frac{1}{2}\begin{bmatrix} 1 \\ -3 \\ 0 \\ 0 \end{bmatrix}\cos(2t) + \frac{1}{4}\begin{bmatrix} 0 \\ 0 \\ 1 \\ -3 \end{bmatrix}\sin(2t)$$

It should be recalled that the upper half of this solution vector is the circulating current vector I. From the above, we may therefore find the circulating currents to be given by

$$I = \frac{1}{2}\begin{bmatrix} 1 \\ 1 \end{bmatrix} + \frac{1}{2}\begin{bmatrix} 1 \\ -3 \end{bmatrix} \cos(2t) .$$

EXERCISES

1.   Consider the following circuit.

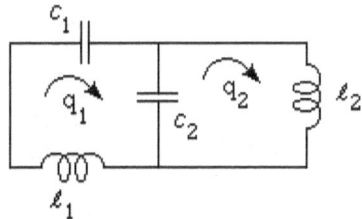

Write the loop equation for the circulating charge Q, and reduce it to a single first-order matrix differential equation. Express the general solution in terms of the matrix exponential function.

2.   For the electric circuit below, write the loop equations for the circulating charges q^1 and q^2. Transform these equations to a single first-order matrix differential equation. Assuming the initial conditions are all zero and the switch is closed at $t = 0$, express the answer in general terms without evaluating the matrix exponential functions.

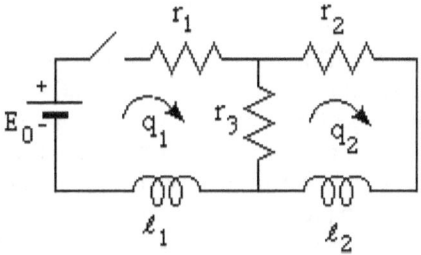

3.   Evaluate the solution to exercise 8 above if $E_0 = 10\ V$, $r_1 = r_y = 10\ \Omega$, $r_y = 20\ \Omega$, and $L_1 = L_2 = 1H$.

4.   For the circuit of exercise 7, evaluate the eigenvalues and eigenvectors if $L_1 = L_2 = 10^{-3}$ H, $C_1 = 10^{-3}$ F, $C_2 = 2 \times 10^{-3}$ F. Evaluate the matrix exponential function, and write out the general solution for the circulating charges.

5.   For the circuit given below, and assuming that all initial conditions are zero, determine the transient and steady state solutions if the switch is closed at $t = 0$.

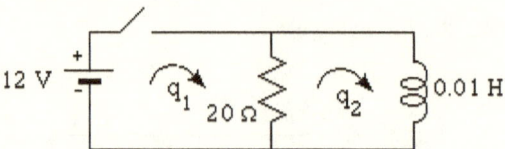

6.   Show that the vector G_1 found in the last example problem is not an eigenvector of the redefined coefficient matrix A.

7.   Show that the set of vectors G_1, E_2, E_3, and E_4 found in the last example problem form a linear independent set.

8.   Using the four vectors referred to in the previous exercise, use them as columns to form a fourth-order matrix P and evaluate the product $P^{-1}AP$. Discuss your findings.

***

4. THE   ANALYSIS   OF   TRANSLATIONAL   MECHANICAL SYSTEMS

As previously pointed out in this text, the analysis of all types of physical systems leads to a mathematical model or description in terms of first- or second-order differential equations with constant coefficients. In this section, this point will be illustrated through the consideration of translational mechanical systems. In these cases, the equations of motion are derived by the application of Newton's laws of motion. However, at the conclusion of the section, a method similar to the one for electrical circuits will be presented, which is based on examination of the physical geometry of a given problem.

As an illustration, let us consider the following system of coupled, springs, and masses. The object is to study the vertical oscillations of the two masses shown in figure 8.10 below about their equilibrium positions. The vertical displacements of the lower and upper masses from their rest or equilibrium positions are designated by the variables x_1 and x_2. The system may be started into motion by displacing one or both masses from equilibrium and releasing them at the instant $t = 0$.

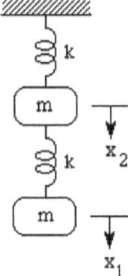

Fig. 8.10

By application of Newton's laws, the simultaneous differential equations describing the motion of the two masses may be written in matrix form as

(4.1)   MX" +KX = [0] ,

where M is the system mass matrix, K the spring stiffness matrix, and X the displacement vector. In terms of elements, these matrices are

$$M = \begin{bmatrix} m & 0 \\ 0 & m \end{bmatrix} = ml, \quad K = k\begin{bmatrix} 1 & -1 \\ -1 & 2 \end{bmatrix}, \quad X = \begin{bmatrix} x_1 \\ x_2 \end{bmatrix}.$$

In this system, the masses are m and the spring constants are k.

In order to introduce a matrix solution to (4.1), it is convenient to premultiply by the inverse of the mass matrix and introduce a new matrix W, which is called the inverse dynamical matrix[63] defined by $W = M^{-1}K$. The result is

(4.2) $X" + WX = [0]$.

[63.]   One may wonder why this matrix is called the inverse dynamical matrix. The reason is that historically dynamic equations were analyzed by premultiplying through by the inverse of the stiffness matrix. The resulting product $K^{-1}M$ was defined as the dynamical matrix of the system. In studying dynamic systems, an analyst is usually most interested in the lowest frequency of oscillation. Since the frequencies are inversely related to the square roots of the eigenvalues of the dynamical matrix, it is the largest eigenvalue one wishes to find. In section 4 of this chapter, it was shown how an iteration scheme would converge to the largest eigenvalue. When working by hand or with a rudimentary calculator, it is easy to understand why the dynamical matrix was formed and iteration used to find the lowest frequency.

To develop the solution to this problem, the method of section 12 of chapter VII will be employed. It has been left as an exercise to obtain the solution by reduction to a first-order system.

As already discussed, equation (4.1) or (4.2) is a general form that occurs when analyzing dynamic systems in which energy-dissipating elements do not occur. In other words, equation (4.1) represents the motion of any conservative or dissipationless system be it electrical, mechanical, acoustical, etc. In the absence of energy dissipation, once the motion of the system is initiated by means of some initial conditions, it will continue to oscillate forever, exhibiting what is termed harmonic motion. Using the approach already cited, the solution to (4.2) is be assumed to be

(4.3) $X = \sin(\sqrt{W}\ t)A + \cos(\sqrt{W}\ t)B,$

where A and B are constant column vectors to be evaluated. This assumed solution may be verified as correct by substitution into (4.3).

To complete a solution, some initial conditions must be specified. For convenience, let us assume that the lower mass is given a unit displacement downward while the upper mass is held fixed. Because the system equation is second order, the initial velocities of the masses must be specified in addition to the initial displacements. For convenience, assume that the masses have no initial velocities when they are released. These statements define the initial condition vectors at $t = 0$ as being

(4.4) $X_0 = \begin{bmatrix} 1 \\ 0 \end{bmatrix}, \ X'_0 = \begin{bmatrix} 0 \\ 0 \end{bmatrix}.$

These conditions are imposed on the assumed solution to evaluate the constant vectors A and B. Imposing the initial displacement condition on (4.3) gives $B = X_0$. To impose the initial velocity condition requires taking the derivative of (4.3). Employing the information from section 13 of chapter VI, the time derivative of (4.3) is

$$X' = \sqrt{W}\ \cos(\sqrt{W}\ t)A - \sqrt{W}\ \sin(\sqrt{W}\ t)B$$

The fact that the initial velocity vector is zero at $t = 0$ requires $A = [0]$.

We now have the particular solution that matches the given initial conditions as

$$X = \cos(\sqrt{W}\ t)X_0,$$

with X_0 given in (4.4).

To continue the solution, we now need to evaluate the matrix function $\cos(\sqrt{W}\, t)$, which is easiest to accomplish by means of the similarity transformation of chapter VI, section 13. To do this, the eigenvalues and eigenvectors of the W matrix are required. From the definitions above, we have

$$W = M^{-1}K = \frac{k}{m}\begin{bmatrix} 1 & -1 \\ -1 & 2 \end{bmatrix}.$$

If we ignore the factor k/m for a moment, the eigenvalues and eigenvectors of the constant matrix above are

$$\mu_1 = 0.3820,\ E_1 = \begin{bmatrix} 1.0000 \\ 0.6180 \end{bmatrix},\ \mu_2 = 2.6180,\ E_2 = \begin{bmatrix} 1.0000 \\ -1.6180 \end{bmatrix}.$$

Since eigenvectors can only be determined up to an arbitrary constant multiplier, these eigenvectors are also the eigenvectors of W. Since the term k/m is a constant multiplying factor, the eigenvalues of W are $\mu_1 = 0.3820k/m$ and $\mu_2 = 2.6180k/m$ respectively.

Knowing the eigenvalues of W allows us to write the following expression for the evaluation of the cosine function.

$$\cos(\sqrt{W}\, t) = E \begin{bmatrix} \cos\left(0.6181\sqrt{\frac{k}{m}}\, t\right) & 0 \\ 0 & \sin\left(1.6180\sqrt{\frac{k}{m}}\, t\right) \end{bmatrix} E^{-1}$$

In this expression, E is the modal matrix of W. Substituting the modal matrix and carrying out the multiplications gives the following result.

$$\cos(\sqrt{W}\, t) = \begin{bmatrix} 0.7236 & 0.4472 \\ 0.4472 & 0.2764 \end{bmatrix} \cos\left(0.6180\sqrt{\frac{k}{m}}\, t\right) + \begin{bmatrix} 0.2764 & -0.4472 \\ -0.4472 & 0.7236 \end{bmatrix} \cos\left(1.6180\sqrt{\frac{k}{m}}\, t\right)$$

Introducing this expression and the initial displacement vector given in (4.4), the desired solution is obtained.

(4.5)
$$X = \begin{bmatrix} 0.7236 \\ 0.4472 \end{bmatrix} \cos\left(0.6180\sqrt{\frac{k}{m}}\, t\right) + \begin{bmatrix} 0.2760 \\ -0.4472 \end{bmatrix} \cos\left(1.618\sqrt{\frac{k}{m}}\, t\right)$$

At this point, it is of interest to note the physical meaning of the eigenvalues and eigenvectors of the matrix W and their relationship to this solution. Since the differential equation is second order, the frequencies of oscillation are the square roots of the eigenvalues as can be seen in (4.5). The lowest frequency is called the fundamental frequency. The amplitude vector of the fundamental frequency term in the above solution can be seen to be a scalar multiple of the eigenvector associated with the fundamental frequency E_1. This eigenvector is termed the fundamental mode, and it indicates that both masses are moving together or in synchronization with the amplitude of the upper mass being about 62% of the amplitude of the lower mass. In engineering terms, this means that the masses are moving in phase at the fundamental frequency.

The square root of the second eigenvalue is the first overtone or second harmonic frequency. In the solution above, this is the second term, and its amplitude vector can be seen to be proportional to the second eigenvector of the W matrix. The second eigenvector is called the second mode, and it indicates that the masses are moving in opposite directions at the second harmonic frequency. In this mode, the upper mass has an amplitude of about 162% of that of the lower mass. In technical terms, this motion is described as completely out of phase.

An important implication of the above analysis is that any possible motion will always be expressible in terms of a linear combination of the eigenvectors of the coefficient matrix as amplitudes that are multiplied by harmonic functions whose frequencies are given by the eigenvalues. What this means is that rather than assuming a solution to (4.2) in the form of (4.3) based on matrix functions, one could equally as well assume a solution in the general form

$$X = c_1 E_1 \sin(\sqrt{\mu_1}\ t) + c_2 E_2 \sin(\sqrt{\mu_2}\ t) +$$

(4.6)

$$c_3 E_1 \cos(\sqrt{\mu_1}\ t) + c_4 E_2 \cos(\sqrt{\mu_2}\ t).$$

The four arbitrary constants c_1 through c_4 would be evaluated by matching prescribed initial conditions. This form for writing the solution vector of a differential equation is called an eigenvector expansion as previously discussed.

EXAMPLE 7

Consider the problem of an idealized train of three identical cars of equal mass m attached by identical coupling springs with spring constant k. This system is depicted in figure 8.11 below.

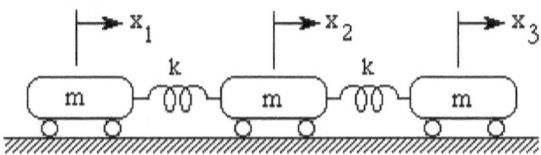

Fig. 8.11

The absolute displacements of each car are given by the coordinates x_1, x_2, and x_3. Analyze the general motion of the train in terms of these coordinates.

SOLUTION

Using these coordinates as elements of a general displacement vector X, Newton's laws of motion will yield the following matrix equation of motion for the train.

(4.7) $$MX'' + KX = [0],$$

where the mass and spring stiffness matrices are

(4.15)

$$M = \begin{bmatrix} m & 0 & 0 \\ 0 & m & 0 \\ 0 & 0 & m \end{bmatrix} = mI,$$

$$K = \begin{bmatrix} k & -k & 0 \\ -k & 2k & -k \\ 0 & -k & k \end{bmatrix} = k \begin{bmatrix} 1 & -1 & 0 \\ -1 & 2 & -1 \\ 0 & -1 & 1 \end{bmatrix}.$$

As we did in the last example of the previous section, we premultiply (4.7) by the inverse of the mass matrix and introduce the inverse dynamical matrix W.

(4.8) $$X'' + WX = [0],$$

where W is

$$W = \frac{k}{m} \begin{bmatrix} 1 & -1 & 0 \\ -1 & 2 & -1 \\ 0 & -1 & 1 \end{bmatrix}.$$

In computing the eigenvalues and eigenvectors of the matrix W, we find

$$\mu_1 = 0, \; E_1 = \begin{bmatrix} 1 \\ 1 \\ 1 \end{bmatrix}, \; \mu_2 = \frac{k}{m}, \; E_2 = \begin{bmatrix} 1 \\ 0 \\ -1 \end{bmatrix}, \; \mu_3 = 3\frac{k}{m}, \; E_3 = \begin{bmatrix} 1 \\ -2 \\ 1 \end{bmatrix}.$$

The occurrence of the zero eigenvalue is a result of the fact that the K matrix is singular. In this problem, the presence of the zero eigenvalue denotes a rigid body motion of the entire system of three cars uniformly to the left or right without the presence of any relative oscillatory motion. The first eigenvector clearly underscores this uniform motion with all the displacements being equal. The second mode shows the central mass remaining motionless with the outer masses moving with the same frequency but out of phase with one another. This means that they are moving equally away from or toward the central mass. The third mode has the outer masses moving in phase with one another back and forth but out of phase with the central mass, which is moving in opposition with twice the amplitude.

***

There is a very important aspect of the previous discussion that should be noted by the reader. It should be clear that just by obtaining the eigenvalues and eigenvectors of a problem, a great deal of the system's behavior is known without even completing the solution.

Since this problem involves a second-order differential equation, an eigenvector expansion of the form of equation (4.6) will be assumed as the solution to (4.8).

(4.9)
$$X = c_1 E_1 \sin(\sqrt{\mu_1}\, t) + c_2 E_2 \sin(\sqrt{\mu_2}\, t) + c_3 E_3 \sin(\sqrt{\mu_3}\, t) +$$
$$c_4 E_1 \cos(\sqrt{\mu_1}\, t) + c_5 E_2 \cos(\sqrt{\mu_2}\, t) + c_6 E_3 \cos(\sqrt{\mu_3}\, t) \quad ,$$

where the six arbitrary constants must be evaluated by applying given initial conditions.

The general form of the solution given in (4.9) is appropriate in all cases except when a zero eigenvector occurs. In these situations, the assumed general solution must be modified. To see how this should be done, we must reflect on the physical meaning of a zero eigenvalue.

As we have already pointed out, the zero eigenvalue implies a rigid body translation of the system. Since no external forces are applied, this means that at most, the masses may have any combination of an equal constant displacement and an equal constant velocity. This means that the existence of the zero eigenvalue requires the inclusion of its associated eigenvector multiplied by a scalar function such as $(a + bt)$, where a and b represent arbitrary constants.

Making these modifications for (4.9) as well as substituting the specific eigenvalues and eigenvectors for this problem, we have.

(4.10)

$$
X = c_1 \begin{bmatrix} 1 \\ 1 \\ 1 \end{bmatrix} + c_2 \begin{bmatrix} 1 \\ 0 \\ -1 \end{bmatrix} \sin(\sqrt{\tfrac{k}{m}}\, t) + c_3 \begin{bmatrix} 1 \\ -2 \\ 1 \end{bmatrix} \sin(\sqrt{\tfrac{3k}{m}}\, t) +
$$

$$
c_4 \begin{bmatrix} 1 \\ 1 \\ 1 \end{bmatrix} t + c_5 \begin{bmatrix} 1 \\ 0 \\ -1 \end{bmatrix} \cos(\sqrt{\tfrac{k}{m}}\, t) + c_6 \begin{bmatrix} 1 \\ -2 \\ 1 \end{bmatrix} \cos(\sqrt{\tfrac{3k}{m}}\, t) .
$$

EXAMPLE 8

Consider the three-car train of example 7 above. As a simple initial condition, assume that all the masses are at rest and the right-hand mass is given a unit velocity in the positive direction. Determine the subsequent motion of the train.

SOLUTION

In terms of vectors, the initial conditions become for $t = 0$.

(4.11)

$$
X_0 = \begin{bmatrix} 0 \\ 0 \\ 0 \end{bmatrix}, \quad X'_0 = \begin{bmatrix} 0 \\ 0 \\ 1 \end{bmatrix}
$$

Imposing the first condition to (4.10) gives

$$
\begin{bmatrix} 0 \\ 0 \\ 0 \end{bmatrix} = c_1 \begin{bmatrix} 1 \\ 1 \\ 1 \end{bmatrix} + c_5 \begin{bmatrix} 1 \\ 0 \\ -1 \end{bmatrix} + c_6 \begin{bmatrix} 1 \\ -2 \\ 1 \end{bmatrix}
$$

$$
= \begin{bmatrix} 1 & 1 & 1 \\ 1 & 0 & -2 \\ 1 & -1 & 1 \end{bmatrix} \begin{bmatrix} c_1 \\ c_5 \\ c_6 \end{bmatrix}.
$$

Since the eigenvectors are linearly independent (in fact, they are all mutually orthogonal), the coefficient matrix is not singular and has an inverse. Thus, the only solution is $c_1 = c_5 = c_6 = 0$.

Now taking the derivative of (4.10) and applying the initial velocity condition, the result is after collecting terms

$$
\begin{bmatrix} c_4 \\ \sqrt{\dfrac{k}{m}}\, c_2 \\ \sqrt{\dfrac{3k}{m}}\, c_3 \end{bmatrix} = \begin{bmatrix} 1 & 1 & 1 \\ 1 & 0 & -2 \\ 1 & -1 & 1 \end{bmatrix}^{-1} \begin{bmatrix} 0 \\ 0 \\ 1 \end{bmatrix}.
$$

Solving[64] this equation gives $c_4 = 0.3333$, $c_2 = -0.5000\sqrt{\dfrac{m}{k}}$, $c_3 = 0.1667\sqrt{\dfrac{m}{3k}}$.

Placing these results into (4.10) gives the final solution.

(4.12)
$$
X = 0.3333 \begin{bmatrix} 1 \\ 1 \\ 1 \end{bmatrix} t - 0.5000\sqrt{\dfrac{m}{k}} \begin{bmatrix} 1 \\ 0 \\ -1 \end{bmatrix} \sin(\sqrt{\dfrac{k}{m}}\, t)
$$
$$
+ 0.1667\sqrt{\dfrac{m}{3k}} \begin{bmatrix} 1 \\ -2 \\ 1 \end{bmatrix} \sin(\sqrt{\dfrac{3k}{m}}\, t)
$$

It has been left as an exercise for the reader to verify that this solution satisfies the differential equation of the system for all time as well as the stated initial conditions.

***

[64] The MATLAB system previously mentioned was used to solve this system.

Geometrical Construction of Mechanical System Equations

Let us now show how system equations may be written in matrix form directly by observation of the geometrical structure. Before doing this, it is important to review the linear behavior of basic mechanical elements. A mechanical mass and spring have already been employed in the previous two examples. If the variable x represents the absolute displacement of a mass m, then it is related to a force f acting on it by Newton's law, which states that $f = mx''$. In the SI system of units, the appropriate units of force, displacement, time, and mass are newtons, meters, seconds, and kilograms respectively. A linear spring is characterized by the relation $f = kx$ where k is the spring constant whose magnitude would be measured in newtons per meter. For the spring, the displacement x must be the relative displacement of the two ends. The force is the force required to extend or compress the spring. The symbols for the mass and spring have already been illustrated in the figures with the two example problems above.

A third basic mechanical translational element that has not been introduced thus far is the dashpot or mechanical damper. The function of a dashpot is like that of an air damper on a screen door. If f is the force applied to extend or retract the dashpot and x is the relative displacement of its two ends, then the linear behavior is given by the relation $f = bx'$. Since the time derivative of displacement is velocity, this relation indicates that the force acting on a dashpot is linearly related to the velocity by which the ends are pulled apart or pushed together. The dashpot constant is b, and it must have units of Newton seconds per meter.

In general, a mechanical system will be described by the following linear constant coefficient differential equation.

$$MX'' + BX' + KX = F$$

In this equation, M is the system mass matrix, B the damping matrix, K the stiffness matrix, and F the forcing function vector. Each mass is considered an inertial node and has a displacement x associated with it. All displacements are represented by the displacement vector X. In some complicated mechanical systems, additional nodes not associated with masses may have to be introduced for a complete description of the system.

By examining an actual system, each of the matrices M, B, and K may be developed directly from the structure in the same manner as previously described for electrical systems. Exactly as expressed in for electrical circuits, the rules for forming the coefficient matrices in mechanical systems are expressed as follows.

$$m_{ij} = \begin{cases} \text{mass of the } i^{th} \text{ node when } i = j \\ m_{ij} \text{ is identically 0 for all cases when } i \neq j \end{cases}$$

$$b_{ij} = \begin{cases} \text{sum of all dashpot coefficients attached} \\ \quad \text{to the } i^{th} \text{ node for } i = j \\ -\text{sum of all dashpot coefficients connecting} \\ \quad \text{the } i^{th} \text{ and } j^{th} \text{ nodes when } i \neq j \end{cases}$$

$$k_{ij} = \begin{cases} \text{sum of all spring constants attached} \\ \quad \text{to the } i^{th} \text{ node for } i = j \\ -\text{sum of all spring constants connecting} \\ \quad \text{the } i^{th} \text{ and } j^{th} \text{ nodes when } i \neq j \end{cases}$$

$$f_i = \text{sum of all forces applied to } i^{th} \text{ node}$$

EXAMPLE 9

Apply the general method described above to construct the matrix equations of motion for the system pictured in the following mechanical circuit diagram.

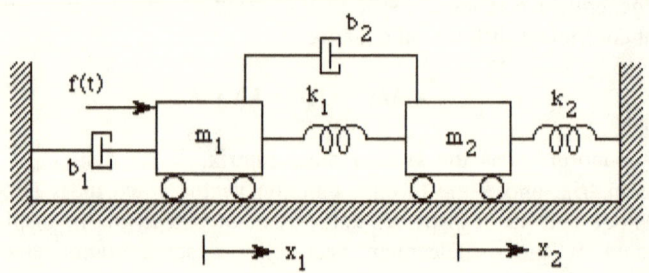

Fig. 8.12

SOLUTION

Following the rules described above, the equations of motion for this system are given as follows.

(4.13)
$$\begin{bmatrix} m_1 & 0 \\ 0 & m_2 \end{bmatrix} X'' + \begin{bmatrix} b_1 + b_2 & -b_2 \\ -b_2 & b_2 \end{bmatrix} X' + \begin{bmatrix} k_1 & -k_1 \\ -k_1 & k_1 + k_2 \end{bmatrix} X = \begin{bmatrix} f(t) \\ 0 \end{bmatrix} = F$$

***

We may now introduce the reader to an interesting and sometimes very useful concept. Assume that in consistent units, the elements in the above system are given the following values:

$$m_1 = 1, \; m_2 = 2, \; b_1 = 2, \; b_2 = 2,$$
$$k_1 = 4, \; k_2 = 1 \; \text{and} \; f(t) = 10 \sin(4t)$$

With these value,s (4.27) becomes

$$\begin{bmatrix} 1 & 0 \\ 0 & 2 \end{bmatrix} X'' + \begin{bmatrix} 3 & -1 \\ -1 & 1 \end{bmatrix} X' + \begin{bmatrix} 4 & -4 \\ -4 & 5 \end{bmatrix} X = 10 \sin(4t) \begin{bmatrix} 1 \\ 0 \end{bmatrix}.$$

If we now look back to equation (3.7) in the previous section, it is easy to see that the equation given there is exactly the same as this one. This means that the electrical system of figure 8.12 and the mechanical system of figure 8.12 are described by precisely the same differential equation. The only differences are the units and the physical meaning of the unknown vectors X and Q. Clearly, the solution developed in section 3 also applies here. There are, in fact, other systems such as mechanical rotational, hydraulic, and acoustic, which also have the identical differential equation.

Systems that have the same differential equation and hence same solution are called either analogs or dualogs (duals). The two systems contrasted here are duals rather than analogs because the variable in the electrical problem is a flow variable (charge), whereas the variable in the mechanical problem is a potential variable (displacement). These distinctions are somewhat subtle and, as a subject, are developed fully in courses in circuits or systems. They are mentioned here only to inform the reader that although systems may appear significantly different in their physical aspects, when expressed in mathematical terms, they are all the same. The only real differences between electrical, mechanical, and other systems are the units and the magnitudes of the quantities. Understanding these similarities and being able to derive and solve the mathematical formulations or models then allows you to be a true generalist.

EXERCISES

1.   Solve the problem given in figure 8.10 if the upper mass has a magnitude of $2m$ instead of m and the upper spring constant is $2k$ rather than k.

2.   Solve the problem given in figure 8.10 if the magnitude of the lower mass is $2m$ rather than m and the spring constant of the upper spring is $2k$ rather than k.

3.   Consider the mechanical translational system shown below.

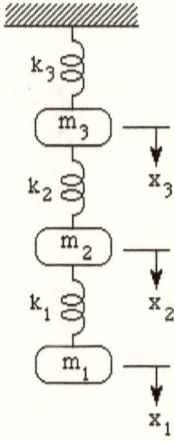

a.   Given $k_1 = k_2 = k_3 = k$, $m_1 = m_2 = m_3 = m$ write the matrix equations of motion.

b.   Determine the particular solution for the initial condition $x_1 = 1$, $x_2 = x_3 = 0$ at $t = 0$.

4.   Repeat exercise 3 except let $k_3 = 3k$, $k_2 = 2k$, $k_1 = k$ and $m_3 = 3m$, $m_2 = 2m$, $m_1 = m$.

5.   In the general solution to exercise 2, discuss the physical meaning of the eigenvalues and the role of the eigenvectors using the idea of vibrational modes. Based on the concepts of linear vector spaces, why are the eigenvectors important for describing an arbitrary motion? If the system were to oscillate with the form of either eigenvector only (one mode or

the other), describe how the masses would be moving and with what frequency. Note that this problem deals with the concept of normal modes as frequently used in the study of mechanical vibrations.

6   Verify that the solution given by equation (4.12) satisfies (4.7) and the initial conditions (4.11).

5. MECHANICAL ROTATIONAL SYSTEMS

As discussed in the previous section, there really is no difference between different linear physical systems when they are examined at the level of their mathematical system equations. This point will be illustrated further in this brief section as we examine mechanical rotational systems.

Physically, there are two distinct types of rotational systems that will be discussed in this section. One concerns rotation about an axis that relates to the twisting of crankshafts, etc. The other is rotation in a plane such as the swinging of a pendulum. This latter problem is really nonlinear but may be reduced to linear form if the angle of swing of the pendulum is restricted to small angles as will be illustrated later in this section.

As an illustration of the first type of rotational system, consider the system shown in figure 8.13.

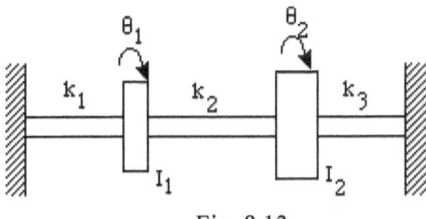

Fig. 8.13

In this system, two flywheels of moment of inertia I are connected by three axial shafts with rotary spring coefficients of k. The outer ends of the system are mounted rigidly into supports, which do not alloy any rotation. The system has two degrees of freedom, which are the angular rotations of each wheel. Applying Newton's law of motion in rotational form gives the following equation of motion for the system.

(5.1)
$$\begin{bmatrix} I_1 & 0 \\ 0 & I_2 \end{bmatrix} \Theta'' + \begin{bmatrix} k_1 + k_2 & -k_2 \\ -k_2 & k_2 + k_3 \end{bmatrix} \Theta = [0]$$
,

where the state vector is

(5.2)
$$\Theta = \begin{bmatrix} \theta_1 \\ \theta_2 \end{bmatrix} .$$

For simplicity, any driving torques or rotary dashpots for energy dissipation have not been included. If we premultiply (5.1) through by the inverse of the rotary inertia matrix, the system equation can be placed into the following standard form.

(5.3)
$$\Theta'' + W\Theta = [0]$$

The coefficient matrix W is

(5.4)
$$W = \begin{bmatrix} \dfrac{k_1 + k_2}{I_1} & \dfrac{-k_2}{I_1} \\ \dfrac{-k_2}{I_2} & \dfrac{k_2 + k_3}{I_2} \end{bmatrix} .$$

Since (5.3) is of the same form as (4.3) in the previous section, then the solution must be of the same form. Following equation (4.4), we may write

(5.5)
$$\Theta = \sin(\sqrt{W}\, t)A + \cos(\sqrt{W}\, t)B.$$

EXAMPLE 10

Determine the solution for the above system assuming it is completely symmetrical with $k_1 = k_2 = k_3 = k$ and $I_1 = I_2 = I$.

SOLUTION

Substituting these values into equation (5.4), a common parameter may be factored out of the coefficient matrix W leaving the following result.

(5.6)
$$W = \frac{k}{I} \begin{bmatrix} 2 & -1 \\ -1 & 2 \end{bmatrix}$$

The eigenvalues and associated eigenvectors of this matrix are

(5.7)
$$\mu_1 = \frac{k}{I}, \quad E_1 = \begin{bmatrix} 1 \\ 1 \end{bmatrix}, \quad \mu_2 = \frac{3k}{I}, \quad E_2 = \begin{bmatrix} 1 \\ -1 \end{bmatrix} .$$

Applying a similarity transformation to evaluate the matrix trigonometric functions in (5.5) yields the following result for the general solution.

(5.8)

$$\Theta = \frac{1}{2}\left\{ \sin\left(\sqrt{\frac{k}{I}}\,t\right)\begin{bmatrix} 1 & 1 \\ 1 & 1 \end{bmatrix} + \sin\left(\sqrt{\frac{3k}{I}}\,t\right)\begin{bmatrix} 1 & -1 \\ -1 & 1 \end{bmatrix} \right\} A +$$

$$\frac{1}{2}\left\{ \cos\left(\sqrt{\frac{k}{I}}\,t\right)\begin{bmatrix} 1 & 1 \\ 1 & 1 \end{bmatrix} + \cos\left(\sqrt{\frac{3k}{I}}\,t\right)\begin{bmatrix} 1 & -1 \\ -1 & 1 \end{bmatrix} \right\} B$$

EXAMPLE 11

Assume the following initial conditions, and determine the particular solution to the previous example. The left disk is given a single unit of rotational displacement in the positive direction but with no initial angular velocity. The right disk is held at zero rotation but given a unit of rotational velocity in the positive direction. These conditions are to be imposed at $t = 0$.

SOLUTION

Based on the stated initial conditions, the following initial vectors are found.

(5.9)
$$\Theta_0 = \begin{bmatrix} 1 \\ 0 \end{bmatrix}, \quad \Theta'_0 = \begin{bmatrix} 0 \\ 1 \end{bmatrix}$$

Imposing the first of these conditions on the general solution given in (5.8) yields

(5.10)
$$B = \begin{bmatrix} 1 \\ 0 \end{bmatrix} = \Theta_0 .$$

Taking the derivative of (5.8) and applying the second initial condition from (5.9) leads to the following result for the constant vector A.

(5.11)
$$A = \frac{1}{2}\sqrt{\frac{I}{3k}}\begin{bmatrix} -1 + \sqrt{3} \\ 1 + \sqrt{3} \end{bmatrix}$$

Given these values for the constant vectors and carrying out the matrix multiplications in (5.8) gives the final particular solution.

(5.12)

$$\Theta = \frac{1}{2}\sqrt{\frac{I}{3k}}\left\{\sqrt{3}\sin\left(\sqrt{\frac{k}{I}}t\right)\begin{bmatrix} 1 \\ 1 \end{bmatrix} - \sin\left(\sqrt{\frac{k}{I}}t\right)\begin{bmatrix} 1 \\ -1 \end{bmatrix}\right\} +$$
$$\frac{1}{2}\left\{\cos\left(\sqrt{\frac{k}{I}}t\right)\begin{bmatrix} 1 \\ 1 \end{bmatrix} + \cos\left(\sqrt{\frac{k}{I}}t\right)\begin{bmatrix} 1 \\ -1 \end{bmatrix}\right\}$$

EXAMPLE 12

Study the oscillatory behavior for the case of three pendulums coupled by springs as shown in figure 8.14.

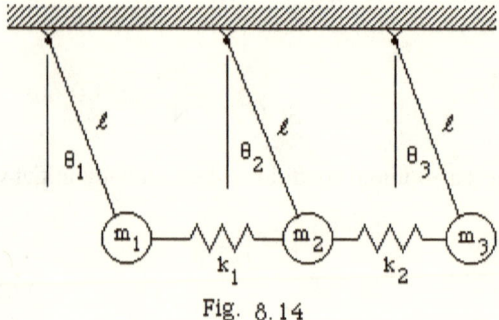

Fig. 8.14

SOLUTION

In this system, the three pendulums of equal length but different masses are coupled by two springs. It is assumed that the free lengths of the springs are such that when there is no motion, the pendulums will hang perfectly vertically downward. The variables of the system are the three displacement angles shown in the figure. To avoid having the equation of motion for the system be nonlinear, the angular displacements must be assumed small so that the following approximations will hold.

(5.13) $\sin(\theta_i) \approx \theta_i$, $\cos(\theta_i) \approx 1 - \frac{1}{2}\theta_i^2$, i = 1,2,3

By application of Newton's laws or Lagrange's equations,[65] the matrix equations of motion are as follows.

(5.14) $M\Theta'' + K\Theta = [0]$

In this expression, the matrices are

$$M = \begin{bmatrix} I^2 m_1 & 0 & 0 \\ 0 & I^2 m_2 & 0 \\ 0 & 0 & I^2 m_3 \end{bmatrix},$$

(5.15)

$$K = \begin{bmatrix} m_1 g I + k_1 I^2 & -k_1 I^2 & 0 \\ -k_1 I^2 & m_2 g I + (k_1 + k_2) I^2 & -k_2 I^2 \\ 0 & -k_2 I^2 & m_3 g I + k_2 I^2 \end{bmatrix}$$

$$\Theta = \begin{bmatrix} \theta_1 \\ \theta_2 \\ \theta_3 \end{bmatrix}$$

If the assumption is made that $m_1 = m_2 = m_3 = m$ and $k_1 = k_2 = k$, the above M and K matrices reduce to the following simpler form.

$$M = I^2 m \begin{bmatrix} 1 & 0 & 0 \\ 0 & 1 & 0 \\ 0 & 0 & 1 \end{bmatrix},$$

(5.16)

$$K = \begin{bmatrix} mgI + kI^2 & -kI^2 & 0 \\ -kI^2 & mgI + 2kI^2 & -kI^2 \\ 0 & -kI^2 & mgI + kI^2 \end{bmatrix}$$

Equation (5.14) can be further simplified by multiplying through by the inverse of the M.

(5.17) $\Theta'' + W\Theta - [0]$,

[65.] Lagrange's equation provides a relatively direct way of determining the equations of motion of a system other than the geometric way discussed in previous sections. For information on Lagrange's equations, the reader is referred to any text on applied mechanics or applied mathematics such as *Applied Mathematics for Engineers and Physicists* by Pipes and Harvill [1970].

where the coefficient matrix is given as follows.

(5.18)

$$W = \begin{bmatrix} \left(\dfrac{g}{I} + \dfrac{k}{m}\right) & -\dfrac{k}{m} & 0 \\[2ex] -\dfrac{k}{m} & \left(\dfrac{g}{I} + \dfrac{2k}{m}\right) & -\dfrac{k}{m} \\[2ex] 0 & -\dfrac{k}{m} & \left(\dfrac{g}{I} + \dfrac{k}{m}\right) \end{bmatrix}$$

The characteristic equation in factored form for this matrix is

(5.19)

$$p(\mu) = \left(\mu - \frac{g}{I}\right)\left(\mu - \left(\frac{g}{I} + \frac{k}{m}\right)\right)\left(\mu - \left(\frac{g}{I} + \frac{3k}{m}\right)\right) = 0 .$$

The eigenvalues are therefore

(5.20)

$$\mu_1 = \frac{g}{I}, \ \mu_2 = \left(\frac{g}{I} + \frac{k}{m}\right), \ \mu_3 = \left(\frac{g}{I} + \frac{3k}{m}\right).$$

The associated eigenvectors are now found to be as follows.

(5.21)

$$E_1 = \begin{bmatrix} 1 \\ 1 \\ 1 \end{bmatrix}, \ E_2 = \begin{bmatrix} 1 \\ 0 \\ -1 \end{bmatrix}, \ E_3 = \begin{bmatrix} 1 \\ -2 \\ -1 \end{bmatrix}$$

Since the equation of motion is a standard second-order equation with no energy dissipation, the solution will be pure harmonic as seen in the previous example. The oscillatory frequencies are the square roots of each of the respective eigenvalues. At this point, the actual solution will not be developed since several similar examples have already been carried out in detail.

***

It should be of interest to the reader to visualize the physical meaning of the eigenvectors for the above problem. The first eigenvector indicates that all three pendulums are swinging together in unison. In this mode, the springs are not extending and the frequency of oscillation is simply that for a freely swinging pendulum. The second eigenvector is a vibrational mode in which the central pendulum remains perfectly still and the outer two swing in exact opposition to each other at the same frequency. The third eigenvector indicates a vibrational mode in which the outer two pendulums are swinging in synchronization while the central one swings in opposition or completely out of phase. Again, all three pendulums are swinging with the same higher frequency. As the author has indicated previously, any complex motion of any system as initiated by the initial conditions will always be represented by some linear combination of the system eigenvectors or modes.

EXERCISES

1. Determine the general solution for the rotational mechanical system given in figure 8.13 if $k_1 = k_3 = 2k$, $k_2 = k$, $I_1 = I$, $I_2 = 2I$.

2. Write the general solution for equation (5.7).

3. Discuss the physical meaning of the eigenvalues and eigenvectors in exercise 1.

***

6. APPLICATION OF FINITE DIFFERENCES TO THE ANALYSIS OF STEADY STATE FOUR TERMINAL NETWORKS

Another area in which finite difference equations occur is that of the transmission of signals through a series of four terminal networks. The most prevalent type of network is electrical, with specific applications to transmission lines, waveguides, and filter networks. However, at the end of this section, applications to other nonelectrical systems will be presented. To begin with, we shall confine the considerations to electrical networks.

For simplicity, steady state alternating current signals will be assumed. A generalized four-terminal network is shown in figure 8.15 below.

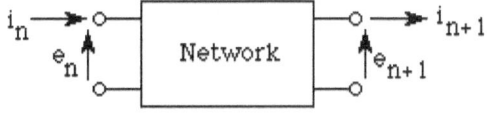

Fig. 8.15

In this figure, e represents a voltage acting across the network terminals and i the current flowing into or out of the network via a terminal. The subscript n denotes the input side of the network and $n+1$ the output side. If the circuit is passive (has no power elements such as a transistor), linear and bilateral (there are no diodes), then the relationship between the input and output variables are described by the following linear difference equations.

(6.1)
$$e_n = t_{11}e_{n+1} + t_{12}i_{n+1}$$

$$i_n = t_{21}e_{n+1} + t_{22}i_{n+1}$$

The coefficients t_{ij} describe the physical characteristics as will be seen shortly. It may appear that the logical order of writing these equations has been reversed; however, the reason for writing the equations in this manner will be explained shortly. The voltage and current variables are combined in a state vector S_n as

$$S_n = \begin{bmatrix} e_n \\ i_n \end{bmatrix}.$$

In matrix notation, the characteristics of the network may be described by a transmission matrix T whose elements are the t_{ij} from equation (6.1) above. The input state vector defines the output state vector in terms of the following difference equation.

(6.2) $S_n = TS_{n+1}$

The transmission matrix T has the following simple form:

$$T = \begin{bmatrix} t_{11} & t_{12} \\ t_{21} & t_{22} \end{bmatrix}.$$

From basic circuit principles, it may be shown that the determinant of T is unity, that is, $t_{11}t_{22} - t_{12}t_{21} = 1$. The values of the elements depend upon the actual internal structure of the network. Two simple configurations for a series impedance and a parallel admittance are given in figure 8.20.

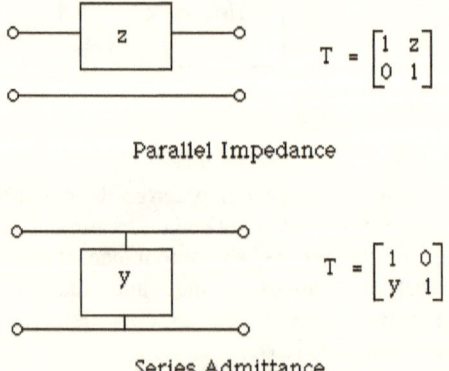

Parallel Impedance

Series Admittance

Fig. 8.16

It is of value to note that more complicated networks may be formed from these two basic forms directly by matrix multiplication. This process is illustrated in the following example.

EXAMPLE 13

Determine the overall transmission matrix for the network shown in figure 8.17 below.

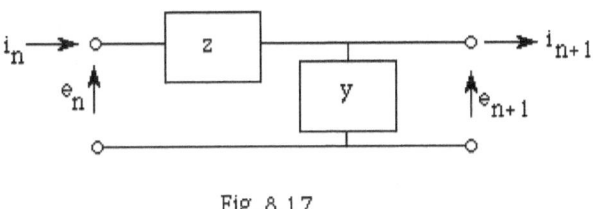

Fig. 8.17

SOLUTION

To calculate the transmission matrix for this network, we need to visualize how it may be assembled from the two basic networks above. Figure 8.18 below has been constructed to aid in this visualization.

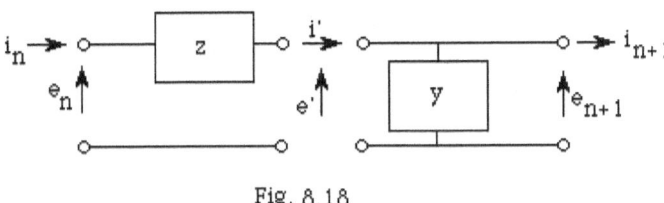

Fig. 8.18

In this figure, we have temporarily broken the two elements apart and introduced a voltage e' and current I' at the interface. Where the state vector is designated by S'. From figure 8.20 and equation (6.2), the following two matrix equations may be written for the two subsections of figure 8.18 noting that S' is the output of the left section and the input to the right.

$$S_n = \begin{bmatrix} 1 & z \\ 0 & 1 \end{bmatrix} S' \quad , \quad S' = \begin{bmatrix} 1 & 0 \\ y & 1 \end{bmatrix} S_{n+1}$$

Substituting the first equation giving S' into the second gives the overall transmission matrix for the network of figure 8.15.

$$S_n = \begin{bmatrix} 1 & z \\ 0 & 1 \end{bmatrix} \begin{bmatrix} 1 & 0 \\ y & 1 \end{bmatrix} S_{n+1} = \begin{bmatrix} 1+yz & z \\ y & 1 \end{bmatrix} S_{n+1}$$

***

The reason for the order in which we chose to write equations (6.1) is illustrated in this example. The apparent backward order permits the formulation of transmission matrices for more complex circuits by multiplying transmission matrices of the subsections in the same order as they occur geometrically in the network. Had we chosen to write equations (6.1) for the output in terms of the input, then the development of more complex transmission matrices would require that the subcomponent transmission matrices be multiplied in reverse order.

Another point that should be noted is that if it is desired to write S_{n+1} in terms of S_n, the fact that the determinant of T is unity makes its inversion quite simple.

$$S_{n+1} = T^{-1}S_n = \begin{bmatrix} t_{22} & -t_{12} \\ -t_{21} & t_{11} \end{bmatrix} S_n$$

In some applications, an analyst is interested in knowing the input and output currents when the input and output voltages are given. Equation (6.1) can be solved for this situation, and the result is the following modified transmission equation.

(6.3)
$$\begin{bmatrix} i_n \\ i_{n+1} \end{bmatrix} = \frac{1}{t_{12}} \begin{bmatrix} t_{22} & -1 \\ 1 & -t_{11} \end{bmatrix} \begin{bmatrix} e_n \\ e_{n+1} \end{bmatrix}$$

To illustrate the application of this last result, consider the following example.

EXAMPLE 14

Consider the circuit below, which has a voltage source applied to the input and the output has been short-circuited. It is desired to find the input and output currents i_n and i_{n+1}. A short circuit means that the voltage is necessarily zero as a shorting wire is assumed to have no impedance.

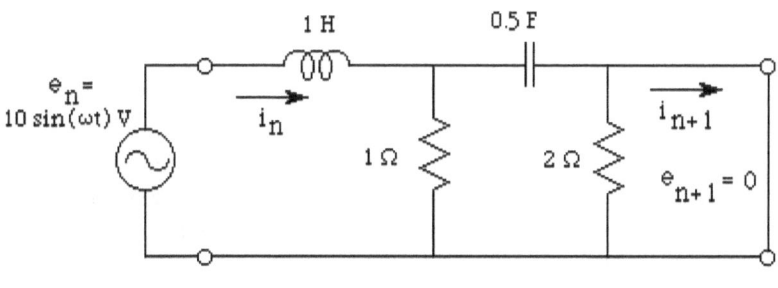

Fig. 8.19

SOLUTION

This circuit can be visualized as a cascade of two circuits of the one given in figure 8.17 with different elements in each circuit. For the left circuit we have $z = j\Omega$ ohm and $y = 1$ mho. An mho is a reciprocal unit of an ohm. For the right circuit $z = 2 / j\Omega$ ohm and $y = 1/2$ mho. As usual, the symbol j is the unit of imaginaries $\sqrt{-1}$. Many students who have studied basic physics or electric circuits will recognize that these circuit elements do not have realistic magnitudes but have been selected simply to make the numbers easier to deal with.

From the previous example, we note that to form the overall transmission matrix of two networks the individual transmission matrices must be multiplied in the same order. This means that the overall transmission matrix for the circuit of this example problem is

$$
T = \begin{bmatrix} 1+j\omega & j\omega \\ 1 & 1 \end{bmatrix} \begin{bmatrix} 1-\dfrac{j}{\omega} & \dfrac{-2j}{\omega} \\ \dfrac{1}{2} & 1 \end{bmatrix}
$$

$$
= \begin{bmatrix} 2+j\left(\dfrac{3\omega}{2}-\dfrac{1}{\omega}\right) & 2+j\left(\omega-\dfrac{2}{\omega}\right) \\ \left(\dfrac{3}{2}-\dfrac{j}{\omega}\right) & 1-\dfrac{2j}{\omega} \end{bmatrix} = \begin{bmatrix} t_{11} & t_{12} \\ t_{21} & t_{22} \end{bmatrix}.
$$

Substituting these values into (6.3) gives the desired transformation relation.

$$(6.4) \quad \begin{bmatrix} i_n \\ i_{n+1} \end{bmatrix} = \frac{1}{\left\{ 2 + j\left(\omega - \dfrac{2}{\omega}\right) \right\}} \begin{bmatrix} 1 - \dfrac{2j}{\omega} & -1 \\ 1 & -2 - j\left(\dfrac{3\omega}{2} - \dfrac{1}{\omega}\right) \end{bmatrix} \begin{bmatrix} e_n \\ e_{n+1} \end{bmatrix}$$

With the output short-circuited, we have $e_{n+1} = 0$. As shown in figure 8.19, the input voltage is $10 \sin(\omega t)$ volts. Because of the presence of the complex numbers in the transfer matrix, it is convenient to utilize a standard circuit-analysis technique based on the classic Euler equation.

$$e^{j\omega t} = \cos(\omega t) + j \sin(\omega t)$$

We assume that the input voltage to the above circuit is $10e^{j\omega t}$, which means that the actual input voltage is the imaginary part of this assumed voltage. The given problem is solved with the complex input voltage for the desired currents. The actual or real currents are the imaginary parts of these answers. It may appear that this method is somewhat awkward, but because it is easier to operate with exponential functions rather than trigonometric ones, the algebraic manipulations required are greatly reduced.

If we make these substitutions into (6.4) and convert the complex numbers from rectangular to polar form, the following relation may be obtained.

$$\begin{bmatrix} i_n \\ i_{n+1} \end{bmatrix} = \frac{10e^{-j\theta}}{\sqrt{4 + \left(\omega - \dfrac{2}{\omega}\right)^2}} \begin{bmatrix} \sqrt{1 + \dfrac{4}{\omega^2}} \; e^{-j\phi} \\ 1 \end{bmatrix} e^{j\omega t} , \text{ Amps}$$

where the two phase angles are defined by

$$\theta = \tan^{-1}\left(\frac{\omega}{2} - \frac{1}{\omega}\right), \quad \phi = \tan^{-1}\left(\frac{2}{\omega}\right) .$$

The actual or real input and output currents can be determined by taking the imaginary parts of the above expression. The results are

$$i_n = \frac{10\sqrt{1 + \dfrac{4}{\omega^2}}}{\sqrt{4 + \left(\omega - \dfrac{2}{\omega}\right)^2}} \sin\{\omega t - (\theta + \phi)\}$$

$$i_{n+1} = \frac{10}{\sqrt{4 + \left(\omega - \dfrac{2}{\omega}\right)^2}} \sin(\omega t - \theta)$$

***

Propagation Along Symmetrical Networks

One particularly interesting type of four-terminal network is one in which the structure is symmetrical. This form is particularly useful for investigating the propagation of electrical waves along transmissions lines or filter circuits. These are also useful for the analysis of heat flow through walls, the propagation of sound waves in tubes, and the propagation of waves along a spring or solid bar, among other similar problems.

For a symmetrical structure, the diagonal elements of the transmission matrix are equal, that is, $t_{11} = t_{22}$. Because powers of the transmission matrix are frequently computed, it is convenient to introduce the following definitions.

$$a = \text{propagation function} = \cosh^{-1}(t_{11})$$

$$z_o = \text{characteristic impedance} = \sqrt{\frac{t_{11}}{t_{21}}}$$

From these relations, the following relations may be derived.

$$t_{12} = z_o \sinh(a), \quad t_{21} = \sinh(a)/z_o$$

It is now a relatively easy task to derive the following result for a general power of the transmission matrix.

$$T^n = \begin{bmatrix} \cosh(na) & -z_o\sinh(na) \\ -\dfrac{\sinh(na)}{z_o} & \cosh(na) \end{bmatrix}$$

Attenuation, Passbands and Stopbands

In physical problems, the propagation function a is generally a complex number that can be written as $a = \alpha + j\beta$. The real part, α, is called the attenuation function as it controls the amplitude of a signal travelling along the network. The imaginary part, β, is the phase function that controls the phase shift of a propagating signal.

If a network or cascade of networks is to pass certain frequencies without any attenuation, then the attenuation function must be zero, which means that the propagation function is pure imaginary. The frequencies for which this condition holds is called a passband, that is, $\alpha = 0$ and $a = j\beta$. We now have

$$\cosh(a) = \cosh(j\beta) = \cos(\beta) = t_{11}.$$

From this relation, it follows that for, a passband the network constant t_{11} must be real and

(6.5) $$|t_{11}| \leq 1$$

since the cosine function must always be less than one.

Frequencies for which a network will not pass a signal or greatly diminish its amplitude are called stopbands. The basic condition for these bands is that t_{11} be real and $|t_{11}| > 1$. To further define this condition, recall that

$$t_{11} = \cosh(a) = \cosh(\alpha + j\beta) = \cosh(\alpha)\cos(\beta) + j\sinh(\alpha)\sin(\beta).$$

To satisfy that t_{11} be real, we must have

(6.6) $$\sinh(\alpha)\sin(\beta) = 0.$$

Under this restriction, we have

$$t_{11} = \cosh(\alpha)\cos(\beta).$$

Given the requirement $|t_{11}| > 1$ we can see that $|\alpha| > 0$. Now returning to (6.6), the only way it can be satisfied is for $\beta = 0, \pm n\pi, n = 1, 2, \ldots$. Thus, the conditions for stopbands are

$$|\cosh(\alpha)| > 1 \text{ or } |\alpha| > 0$$

and

$$\beta = 0 \text{ or } \beta = \pm n\pi, \; n = 1, 2, \ldots$$

EXAMPLE 15

Determine the passbands for the symmetrical structure given in figure 8.20 below.

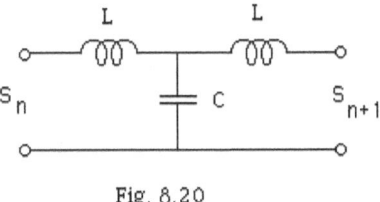

Fig. 8.20

SOLUTION

The transmission matrix for this circuit is

(6.7)
$$T = \begin{bmatrix} 1 + yz & -z(2 + yz) \\ -y & 1 + yz \end{bmatrix},$$

for which $z = j\omega L$ and $y = j\omega C$. To examine the transmission characteristics of this circuit, the above conditions must be applied. For passbands, we have $\cosh(\alpha) = 1 - \omega^2 LC$, which meets the requirement of being real. From the second condition given by (6.5), we must have

(6.8)
$$| 1 - \omega^2 LC | \le 1.$$

Since signal frequencies are real, the above circuit will pass all signals without any attenuation whose frequencies are in the following range.

$$0 \le \omega \le \frac{1}{\sqrt{LC}}$$

The upper frequency that occurs when equation (6.8) is satisfied as an equality is called the critical or cutoff frequency of the filter. All frequencies above the above cutoff frequency are attenuated and hence are in the stopband of the circuit. Because only frequencies between zero and the cutoff frequency are passed without attenuation, the filter circuit of figure 8.20 is called a low pass filter.

***

If the inductors and capacitor are switched in figure 8.20 so that $z = -j/\omega C$ and $y = -j/\omega L$, the cutoff frequency will be the same, but the passband becomes

$$\frac{1}{\sqrt{LC}} \leq \omega \leq \infty \ .$$

This filter is termed a high pass filter.

As stated at the outset of this section, applications of the theory of four-terminal networks to other than electric circuits would be given. The figures in appendix A provide details of how this theory is applicable to the propagation of waves along elastic springs, sound waves through tubes, heat flow through homogeneous media, electrical waves on transmission lines, the torsion of non niform shafts, and the flexure of nonuniform beams. In these last two applications, it must be noted that the transfer matrix is used to generate the $i^{th} + 1$ state vector from the ith as noted in each table.

EXERCISES

1. From the basic transmission matrices given in figure 8.20, derive the transmission matrix given by equation (6.7).

2. If the inductance and capacitance elements in the circuit of figure 8.20 are interchanged, how are the passbands and stopbands affected?

3. Derive overall transmission matrices for the following four pole networks.

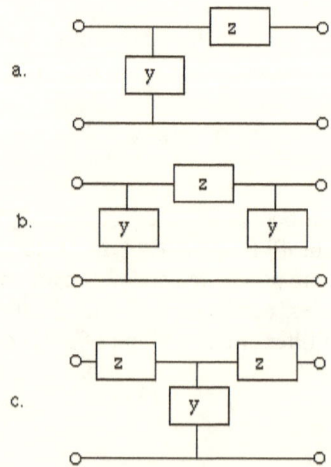

4.   For the circuit of exercise $3b$ above, let z be an inductor of value $0.01H$ ($z = 0.01j\omega$) and the y's be capacitors of $0.01\mu F$ ($y = 10^{-8}j\omega$). Find the passbands and stopbands.

ADDITIONAL EXERCISES

1.*   For the flow system shown below, all tanks have 100 gal capacity. Determine the solution for the salt concentrations as a function of time if the initial concentration of the first tank is 1 lb salt/gal and the others are filled with freshwater. The incoming water is also fresh.

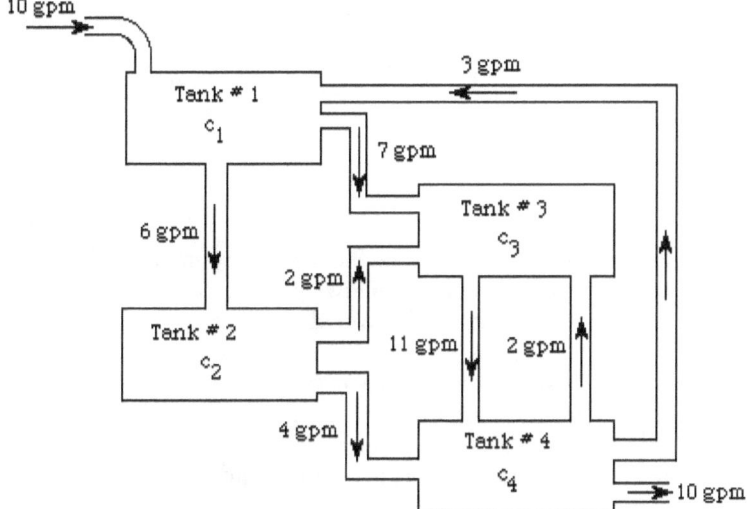

2.*   Solve exercise 1 if all the tanks have fresh water initially except for the second tank, to which salt is initially added to start its concentration at 1 lb salt/gal.

3.*   Consider the two-tank system of example 3. The second tank is 3.5 m tall with an open top. Adjust the output pipe resistance so that after draining is initiated, liquid will not quite overflow out of the second tank.

4.*   Establish the matrix differential equation for the liquid levels in the three-tank system shown below.

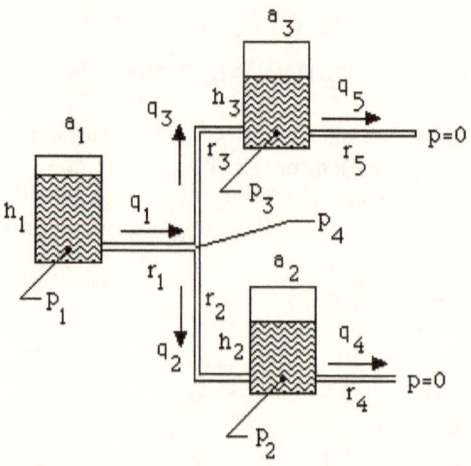

5.*   For problem 4 above, determine the time solution for the liquid levels in all three tanks given the following values and assuming that all three tanks have an initial liquid level of 5 m.

weight density = γ = 10,000 N/m^3
area of first and third tanks = $a_1 = a_3 = 10$ m^2
area of second tank = $a_2 = 6$m^2
resistance of first pipe = $r_1 = 3 \times 10^5$ Ns/m^5
resistance of second and third pipes = $r_2 = r_3 = 2 \times 10^5$ Ns/m^5
resistance of fourth pipe = $r_1 = 1.5 \times 10^5$ Ns/m^5
resistance of fifth pipe = $r_1 = 2 \times 10^5$ Ns/m^5

6.   For the circuit given below, and assuming that all initial conditions are zero, determine the transient and steady state solutions if the switch is closed at $t = 0$.

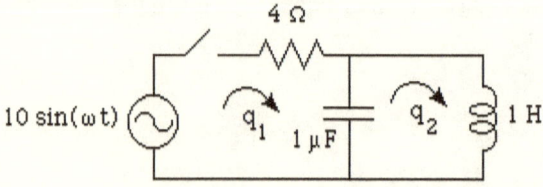

7. Consider the example problem given in figure 8.19. If the initial velocity vector is zero, what initial displacement vector would produce the following motion?

    a. Vibration in the first mode only?

    b. Vibration in the second mode only?

    c. Could the vibration in part *a* be achieved by some initial velocity vector if the initial displacement vector were zero? If so, what would it have to be?

8. For the mechanical system given below, derive the equations of motion for the displacement of the two masses, and write them in matrix form. If $m_1 = m_2 = m$ and $k_1 = k_2 = k_3 = k$, find the eigenvalues and eigenvectors.

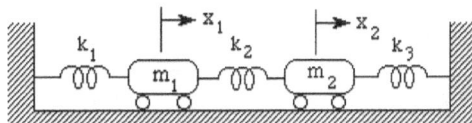

    Hint: Absorb the constants m and k into the definition of the eigenvalue as done in the examples. Neglect friction.

9. Consider the same problem as given in exercise 8 above except include the effect of friction (coefficient *b*) acting on the wheels of the masses. Let $m_1 = m_2 = m$ and $k_1 = k_2 = k_3 = k$. Parts *b* and *c* below require the use of a matrix computational package such as MAX or MATLAB.

    a. Derive the matrix equations of motion.

    b. Set $k/m = 1$ and find the eigenvalues and eigenvectors for each of the following values: $b/m = 1, 2, 3, 3.5$. Compare and discuss the meaning of these results in terms of the motion of the system.

    c. Experiment with the value of b/m to find its value to two decimal places for which oscillatory motion just ceases to exist.

10. In exercise 8 above, write out the general solution, and develop a particular solution that would match a given set of initial conditions at $t = 0$ of $X = X_0$, $X' = V_0$.

11.   Using the general construction method, establish the equations of motion for the following mechanical translational system.

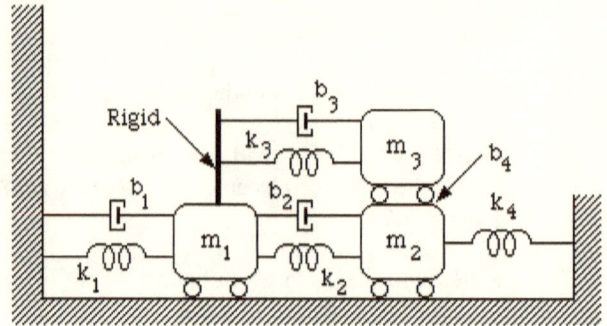

12.*   For the circuit of figure 8.17, let z be a series resistance ($z = R$) and y be a shunt capacitance C ($y = j\omega C$). What are the passbands and stopbands for this circuit?

14.*   A simple ecosystem contains three species of organisms. These are a food specie F and two herbivores H_1 and H_2, which feed on F. The food specie is self-generating and does not require the herbivores to survive. Both herbivores, however, must feed on F in order to survive. They also prey on each other only to protect their territory but not as a source of food. The dynamics of this system may be described by the following set of differential equations.

$$F' = k_{11}F - k_{12}H_1 - k_{13}H_2$$
$$H'_1 = k_{21}F - k_{23}H_2$$
$$H'_2 = k_{31}F - k_{32}H_1$$

If P is the population state vector, this system is characterized by

$$P' = KP, \quad P = \begin{bmatrix} F \\ H_1 \\ H_2 \end{bmatrix}.$$

If K and an initial condition vector are given as follows:

$$P_0 = \begin{bmatrix} 1{,}000 \\ 10 \\ 50 \end{bmatrix}, \quad K = \begin{bmatrix} 0.10 & -0.07 & -0.03 \\ 0.01 & 0 & -0.01 \\ 0.02 & -0.04 & 0 \end{bmatrix}$$

a. Evaluate and discuss the behavior of the system.

b. Change the self-growth coefficient of F (k_{11}) to 0.05 and answer part *a* again.

c. Change the self-growth coefficient of F (k_{11}) to 0.01 and answer part *a* again.

IX

NUMERICAL METHODS

1. INTRODUCTION

One of the more important areas in applied mathematics and computer science is that of numerical methods. Its significance has increased greatly in recent years after the development of digital computers and especially with the rapid evolution of the personal computer. Within the field of numerical analysis, matrix algebra has played an important role since it is the most common framework for analysis.

This chapter will attempt to serve as an introduction and brief overview of this important subject. In addition to introducing several of the current important methods of choice, some of the earlier approaches will be presented as a means of developing the mathematical ideas and approaches behind the methods. However, before proceeding, the author would like to remind the reader that in all situations, numerical solutions are always approximate and thus have some amount of error. These errors typically arise from two sources. First, errors arise from the numerical algorithm itself in that the mathematical formulation involves a certain degree of approximation. The second source of errors arises from the computer itself due to the fact that all numbers can not be represented exactly by means of a finite number of binary digits. As in any situation, it is always good practice to spot-check answers by either back substitution or determining reasonable bounds by means of simplified mathematical models or physical reasoning. In all cases, it is preferable to obtain an analytical solution if at all possible.

2. GAUSSIAN ELIMINATION REVISITED

In section 4 of chapter II, the Gaussian elimination method was introduced as a much more efficient way of solving system of linear algebraic equations. With only slight modifications, this method continues to be the most widely used method for solving systems of any size. It is the fundamental algorithm employed in nearly every application package that involves the solution of algebraic equations. Because of its importance, it will be reviewed briefly here at the outset of this chapter to establish the basis for further refinements and the development of other important numerical procedures.

Basically, the problem is one of determining the solution of a nonhomogeneous system of n linear algebraic equations in n unknowns of the general form

(2.1) $AX = B.$

Schematically, this system is solved by premultiplying the equation through by the inverse of the coefficient matrix A. However, the construction of the inverse is no easy task, and hence, the method of Gaussian elimination provides a very general, efficient procedure of accomplishing the same thing. Fast, efficient algorithms are important as it is not uncommon to encounter systems whose order is 10,000 or greater in many modern applications.

As presented in chapter II, the objective of the Gaussian elimination method is to utilize elementary operations on the given system (2.1) to reduce it to upper triangular form.

(2.2) $UX = D,$

where U is an upper triangular matrix and D is a modified version of the constant vector B. Once in this form, the unknowns are obtained by back substitution.

The augmented matrix of (2.1) is

$$\text{Aug } A = \begin{vmatrix} a_{11} & a_{12} & \cdots & a_{1n} & b_1 \\ a_{21} & a_{22} & \cdots & a_{2n} & b_2 \\ \cdot & \cdot & \cdots & \cdot & \cdot \\ a_{n1} & a_{n1} & \cdots & a_{nn} & b_n \end{vmatrix}.$$

Rather than follow the explicit steps developed in section 4 of chapter II, an alternative approach will be presented. Consider the following elementary matrix E_1.

$$E_1 = \begin{bmatrix} 1 & 0 & 0 & \cdots & 0 \\ -\dfrac{a_{21}}{a_{11}} & 1 & 0 & \cdots & 0 \\ -\dfrac{a_{31}}{a_{11}} & 0 & 1 & \cdots & 0 \\ \vdots & \vdots & \vdots & \ddots & \vdots \\ -\dfrac{a_{n1}}{a_{11}} & 0 & 0 & \cdots & 1 \end{bmatrix}$$

If equation (2.1) is now premultiplied by this matrix, the result is a new augmented matrix in which all the elements of the first column below the first element are zero. If this matrix is called Aug A^1, it will have the following form.

$$\text{Aug } A_1 = E_1 \text{ Aug } A = \begin{bmatrix} a_{11} & a_{12} & \cdots & a_{1n} & b_1 \\ 0 & a_{22}^1 & \cdots & a_{2n}^1 & b_2^1 \\ \cdot & \cdot & \cdots & \cdot & \cdot \\ 0 & a_{n1}^1 & \cdots & a_{nn}^1 & b_n^1 \end{bmatrix}$$

It should be clear that this operation will fail only if the first element were zero. It also can present numerical problems if the first element is not the largest in the first column. If either of these situations occur, then the first row only needs to be exchanged with whatever row has the largest element in the first column. Such an exchange is accomplished by premultiplying Aug A by an appropriate permutation matrix. It should be noted that exchanging any rows in (2.1) does not alter the order of unknowns in the X vector.

Consider a case in which it is desired to exchange the first row with the jth row. The required permutation matrix is simply an identity matrix of proper order in which either the first and j^{th} rows or columns have been exchanged. As presented in section 5 of chapter II, it is always desirable to have the pivot element be the largest in a given column because it minimizes the buildup of computational round-off errors in large or ill-conditioned systems.

Following the Gaussian elimination method the next required elementary matrix would be

$$E_2 = \begin{bmatrix} 1 & 0 & 0 & \cdots & 0 \\ 0 & 1 & 0 & \cdots & 0 \\ 0 & -\dfrac{a_{32}^1}{a_{22}^1} & 1 & \cdots & 0 \\ \vdots & \vdots & \vdots & \ddots & \vdots \\ 0 & -\dfrac{a_{n2}^1}{a_{22}^1} & 0 & \cdots & 1 \end{bmatrix}$$

This matrix is used to premultiply the modified augmented matrix Aug A_1. Repeating this process $n - 1$ times leads to the following desired result.

$$\text{Aug } A_{n-1} = E_{n-1} \cdots E_2 E_1 \text{ Aug } A =$$

$$\begin{bmatrix} a_{11} & a_{12} & a_{13} & \cdots & a_{1n} & b_1 \\ 0 & a_{22}^1 & a_{23}^1 & \cdots & a_{2n}^1 & b_2^1 \\ 0 & 0 & a_{33}^2 & \cdots & a_{3n}^2 & b_3^2 \\ \cdot & \cdot & \cdot & \cdots & \cdot & \cdot \\ 0 & 0 & 0 & \cdots & a_{nn}^{n-1} & b_n^{n-1} \end{bmatrix}$$

This matrix is the augmented matrix for the system (2.2). In other words, we have the following relations.

$$U = E_{n-1} \ldots E_2 E_1 A, \, D = E_{n-1} \ldots E_2 E_1 B$$

In an actual algorithm, the elementary matrices indicated above and any necessary permutation matrices for improvement by column pivoting are not calculated and the matrix multiplications carried out. The numerical operations are performed directly on the original coefficient matrix and constant vector in place in computer memory, which conserves both computational time and memory space. The above notation is simply useful for understanding the operations and clarifying the procedure.

As already mentioned, after a system is reduced to the form of equation (2.2), it is simply solved by the process of back substitution.

EXAMPLE 1

Employ the method to determine the solution for a third-order system whose augmented matrix is the following.

(2.3)
$$\text{Aug } A = \begin{bmatrix} 2 & -2 & 3 & 1 \\ 10 & -4 & 5 & 2 \\ 5 & -4 & 6 & -1 \end{bmatrix}$$

SOLUTION

Since the second element in the first column is the largest value, the first and second rows should be exchanged. Using a permutation matrix to accomplish this, it would be

$$P_1 = \begin{bmatrix} 0 & 1 & 0 \\ 1 & 0 & 0 \\ 0 & 0 & 1 \end{bmatrix}.$$

Premultiplying (2.3) by P_1 gives

$$\text{Aug } A_1 = \begin{bmatrix} 10 & -4 & 5 & 2 \\ 2 & -2 & 3 & 1 \\ 5 & -4 & 6 & -1 \end{bmatrix}.$$

Now the desired first elementary matrix is

$$E_1 = \begin{bmatrix} 1 & 0 & 0 \\ -0.2 & 1 & 0 \\ -0.5 & 0 & 1 \end{bmatrix}.$$

Premultiplying (2.3) by this result gives the following modified augmented matrix.

$$\text{Aug } A_2 = \begin{bmatrix} 10 & -4 & 5 & 2 \\ 0 & -1.2 & 2.0 & 0.6 \\ 0 & -2.0 & 3.5 & -2.0 \end{bmatrix}$$

Again, a permutation matrix is needed to exchange the second and third rows. The required permutation matrix is

$$P_2 = \begin{bmatrix} 1 & 0 & 0 \\ 0 & 0 & 1 \\ 0 & 1 & 0 \end{bmatrix}.$$

The result of premultiplying (2.4) by P_2 is

$$\text{Aug } A_3 = \begin{bmatrix} 10 & -4 & 5 & 2 \\ 0 & -2.0 & 3.5 & -2.0 \\ 0 & -1.2 & 2.0 & 0.6 \end{bmatrix}$$

The next elementary matrix is given by

$$E_2 = \begin{bmatrix} 1 & 0 & 0 \\ 0 & 1 & 0 \\ 0 & -0.6 & 1 \end{bmatrix}.$$

The fourth and final augmented matrix of the system is obtained by premultiplying Aug A_3 by E_2.

$$\text{Aug } A_4 = \begin{bmatrix} 10 & -4 & 5 & 2 \\ 0 & -2.0 & 3.5 & -2.0 \\ 0 & 0 & -0.1 & 1.8 \end{bmatrix}$$

Employing back substitution yields the final result for the unknown vector.

$$X = \begin{bmatrix} -3 \\ -30.5 \\ -18 \end{bmatrix}$$

It is left to the reader as an exercise to verify that the products $E_2 P_2 E_1 P_1 A$ and $E_2 P_2 E_1 P_1 B$ combine to yield the fourth augmented matrix and that the solution vector is indeed correct. This example was selected because it illustrates the Gaussian elimination method including the process of column pivoting.

EXERCISES

Use Gaussian elimination to solve the following equations.

1*. $\begin{bmatrix} 1 & -1 \\ 2 & 1 \end{bmatrix} X = \begin{bmatrix} 2 \\ 3 \end{bmatrix}$

2.* $\begin{bmatrix} 1 & 2 & -1 \\ 2 & -1 & 3 \\ 3 & -2 & 3 \end{bmatrix} X = \begin{bmatrix} 3 \\ -1 \\ 2 \end{bmatrix}$

3.* $\begin{bmatrix} 2 & -1 & 1 \\ 2 & 1 & 0 \\ -2 & 0 & 9 \end{bmatrix} X = \begin{bmatrix} 1 \\ -1 \\ 6 \end{bmatrix}$

$$4.*^{66} \quad \begin{bmatrix} 1 & 2 & 1 & 0 \\ 2 & 0 & -2 & 1 \\ 0 & 4 & 3 & 2 \\ -1 & 6 & -1 & -1 \end{bmatrix} X = \begin{bmatrix} 2 \\ 6 \\ -1 \\ 2 \end{bmatrix}$$

***

3. LU FACTORIZATION

A number of factorization or decomposition methods have been developed as a means of solving linear algebraic systems more efficiently. These methods are based on the fact that matrices may be expressed as a product of two triangular matrices or factors, a lower triangular matrix L times an upper triangular matrix U. Hence, the name LU factorization is obvious. Of the various forms, two prominent ones are those of Crout and Choleski. We will present the Crout factorization in this section and Choleski in the next.

In other words, if A is an arbitrary matrix, then

(3.1) $A = LU,$

where L is a lower triangular matrix and U is an upper triangular matrix. It is a relatively easy matter to show that this factorization is not unique. For example, the factorization process could be carried out in such a manner as to produce ones along either the main diagonal of L or the main diagonal of U. A triangular matrix with ones along the main diagonal is called a unit triangular matrix. This gives two possibilities for factors, and there are an infinity of others.

Crout factorization leads to U being a unit upper triangular matrix. Most current textbooks present this reduction with L being a unit lower triangular matrix. This latter method is sometimes referred to as the Doolittle factorization. If extended or double precision is employed to reduce round-off errors, then the Crout method has the advantage of requiring less computer memory.[67] It should also be noted that Gaussian elimination is also a triangular factorization approach to solving linear algebraic systems. LU factorization has no real advantage over Gaussian elimination if one is attempting to solve a single set of linear algebraic equations. The advantage occurs when a given system must be solved for two or more constant vectors B. In such cases, the L and U factors are saved once and the additional systems solved directly by forward and backward substitution. Gaussian elimination would require complete factorization of the new augmented matrix for every separate equation.

---

[66]   Campbell, [1980], p79.
[67]   See Kraus [1987], p. 117.

To illustrate the use of LU, factorization consider the general linear algebraic system discussed in the previous section.

(3.2) $$AX = B$$

Let the coefficient matrix be factorable, and substitute (3.1) into (3.2) to give

(3.3) $$AX = LUX = B.$$

We now introduce a new intermediate unknown vector Z defined by

(3.4) $$Z = UX$$

so that (3.3) becomes

$$LZ = B.$$

Because L is a lower triangular matrix, this previous equation represents the following system of scalar equations.

$$l_{11}z_1 = b_1$$

$$l_{21}z_1 + l_{22}z_2 = b_2$$

$$\ldots$$

$$l_{n1}z_1 + l_{n2}z_2 + \ldots + l_{nn}z_n = b_n$$

This system is solved directly by forward substitution, that is,

$$z_1 = b_1/l_{11}, \ z_2 = (b_2 - l_{21}z_1)/l_{22}, \text{ etc.,}$$

which leads to the general result for the kth unknown z_k.

$$z_k = \frac{1}{l_{kk}}\left(b_k - \sum_{j=1}^{k-1} l_{kj}z_j \right), \quad k = 1, 2, \ldots, n$$

Now that Z is known, the original unknown X may be solved for from equation (3.4) by the process of back substitution as employed in the Gaussian elimination method. From (3.4), we have

$$x_k = \frac{1}{u_{kk}}\left(z_k - \sum_{j=k+1}^{n} u_{kj}x_j \right), \quad k = n, n-1, \ldots, 1$$

This process is very direct but requires that one knows the factors L and U.

As already noted, most of the common methods differ in how the factors L and U are obtained from A. In this section, the method ascribed to Crout will be presented. If we expand equation (3.1) to determine the general expression for an arbitrary element of A, we find the following result.

(3.5)
$$a_{ij} = \sum_{k=1}^{\min\,(i,j)} l_{ik} u_{kj}$$

In a general triangular matrix, there are $n(n+1)/2$ elements. Equation (3.5) represents n^2 equations for $n^2 + n$ unknowns, the elements of L and U. To obtain a unique solution, the Crout method set $u_{ii} = 1$ for all i so that the number of unknowns is reduced to n^2. With this restriction on the diagonal elements of U, it is called a unit upper triangular matrix.

To illustrate how (3.5) is solved for, the elements of L and U begin with the expression for a_{11}. From (3.5) $a_{11} = l_{11}u_{11}$ and since $u_{11} = 1$, we have $l_{11} = a_{11}$. Now for a_{12} and a_{21}, we have

$$a_{12} = l_{11}u_{12}, \; a_{21} = l_{21},$$

from which $u_{12} = a_{12}/l_{11} = a_{12}/a_{11}$ and $l_{21} = a_{21}$. Continuing, we have

$$a_{22} = l_{21}u_{12} + l_{22}.$$

From this expression, l_{22} may be found to be as follows.

$$l_{22} = a_{22} - l_{21}u_{12} = a_{22} - a_{21}a_{12}/a_{11}$$

At this point, consider the equation for a_{13} and a_{31}. They are $a_{13} = l_{11}u_{13}$ and $a_{31} = l_{31}$, which are easily solved for u_{13} and l_{31}. By continuing the expansion along the first row and column of A leads to the following general result.

$$l_{k1} = a_{k1} \text{ and } u_{1k} = a_{1k}/a_{11}, \, k = 1, 2, \ldots, n$$

After examining the results of a few more expansions, it becomes apparent that the procedure that will yield a single equation with a single unknown at each step is to solve for the column elements of L and the row elements of U in an alternating sequence. Noting this, a general expression for the ith element of the sth column of L may be obtained from (3.5) by expanding with $j = s$ and $i \geq s$.

$$a_{is} = l_{i1}u_{1s} + l_{i2}u_{2s} + \ldots + l_{is}$$

From this we obtain

$$(3.6) \qquad l_{is} = a_{is} - \sum_{k=1}^{s-1} l_{ik}u_{ks} \, , \quad i = s, s+1, \cdots, n$$

Similarly, an expression for the jth element of the sth row of U is given by (3.5) setting $i = s$ and $j \leq s$.

$$a_{sj} = l_{s1}u_{1j} + l_{s2}u_{2j} + \ldots + l_{ss}u_{sj}$$

Solving for u_{sj} gives the desired general expression.

$$(3.7) \qquad u_{sj} = \frac{1}{l_{ss}}\left(a_{sj} - \sum_{k=1}^{s-1} l_{sk}u_{kj}\right), \quad j = s+1, s+2, \cdots, n$$

Equations (3.6) and (3.7) provide the algorithms for sequentially solving for the elements of L and U. With the factors known, the original system is solved by the forward and backward substitution process previously presented.

EXAMPLE 2

For the same example problem used in the previous section, decompose the coefficient matrix into its L, U components by the method presented here.

SOLUTION

From the augnmented of the linear algebraic system given in example 1, we identify the matrices of equation (3.2) as

$$A = \begin{bmatrix} 2 & -2 & 3 \\ 10 & -4 & 5 \\ 5 & -4 & 6 \end{bmatrix}, \quad B = \begin{bmatrix} 1 \\ 2 \\ -1 \end{bmatrix}.$$

The U matrix is formed directly from the A matrix following the Gaussian elimination procedure with some modifications, and the L matrix is constructed from the multipliers employed during the process.

To begin the a_{11} element of A is reduced to unity by dividing the first row by a_{11}. For the a matrix above, the factor is 2. This scale factor becomes the first diagonal element of L. We now have for the modified A matrix and the beginning construction of L

$$A \rightarrow \begin{bmatrix} 1 & -1 & \frac{3}{2} \\ 10 & -4 & 5 \\ 5 & -4 & 6 \end{bmatrix}, \quad L = \begin{bmatrix} 2 & 0 & 0 \\ * & * & 0 \\ * & * & * \end{bmatrix}.$$

The asterisks in L designate the elements that are not determined as yet.

Following the Gaussian elimination procedure, we take -10 times the first row and add it to the second. Also, we take -5 times the first row and add it to the third. In forming L, the element l_{21} is the negative of the multiplier used on the first row to add to the second. This means that l_{21} has the value 10. Similarly, l_{31} must be 5. We have now arrived at

$$A \rightarrow \begin{bmatrix} 1 & -1 & \frac{3}{2} \\ 0 & 6 & -10 \\ 0 & 1 & -\frac{3}{2} \end{bmatrix}, \quad L = \begin{bmatrix} 2 & 0 & 0 \\ 10 & * & 0 \\ 5 & * & * \end{bmatrix}.$$

The a_{22} element is now reduced to unity by dividing the second row by 6 and saving the factor as l_{22}. The result is

$$A \rightarrow \begin{bmatrix} 1 & -1 & \frac{3}{2} \\ 0 & 1 & -\frac{5}{3} \\ 0 & 1 & -\frac{3}{2} \end{bmatrix}, \quad L = \begin{bmatrix} 2 & 0 & 0 \\ 10 & 6 & 0 \\ 5 & * & * \end{bmatrix}.$$

To reduce the last element in the second column to zero, we multiply the second row by -1 and add the result to the third row. The negative of the factor is saved as l_{32}. We now have

$$A \rightarrow \begin{bmatrix} 1 & -1 & \frac{3}{2} \\ 0 & 1 & -\frac{5}{3} \\ 0 & 0 & \frac{1}{6} \end{bmatrix}, \quad L = \begin{bmatrix} 2 & 0 & 0 \\ 10 & 6 & 0 \\ 5 & 1 & * \end{bmatrix}.$$

To finally obtain the U matrix, the last row is divided by 1/6 and the factor retained as l_{33}, which completely defines the L matrix as well. The result is

$$U = \begin{bmatrix} 1 & -1 & \frac{3}{2} \\ 0 & 1 & -\frac{5}{3} \\ 0 & 0 & 1 \end{bmatrix}, \quad L = \begin{bmatrix} 2 & 0 & 0 \\ 10 & 6 & 0 \\ 5 & 1 & \frac{1}{6} \end{bmatrix}.$$

It is left to the reader to verify that the product LU produces the given original matrix A.

EXAMPLE 3

Determine the solution for the system of examples 1 and 2 by the method described above.

SOLUTION

Following the stated method, we solve the system $LZ = B$ by forward substitution. This process is actually recorded in the L matrix so that one only needs to apply the record retained in L to the B vector. For the above system, the given B vector is

$$B = \begin{bmatrix} 1 \\ 2 \\ -1 \end{bmatrix}.$$

To transform B, all the operations in L must be applied in the same order as they were generated. To do this, the first column of L, that is, [2, 10, 5], dictates that the first element of B must be divided by 2 then -10 times this result added to the second element of B, and finally, -5 times the first element of B is added to the last. This leads to the following modified B vector.

$$B \rightarrow \begin{bmatrix} \frac{1}{2} \\ -3 \\ -\frac{7}{2} \end{bmatrix}$$

Now the operations in the second column of L, namely [0, 6, 1], are applied to the modified B vector. These mean that the second element is divided by 6, and -1 times the result is added to the last element. These operations bring us to

$$B \rightarrow \begin{bmatrix} \frac{1}{2} \\ -\frac{1}{2} \\ -3 \end{bmatrix}$$

The last column of L simply indicates that the last element of the modified B vector is divided by 1/6. The result is the desired vector Z.

$$Z = \begin{bmatrix} \frac{1}{2} \\ -\frac{1}{2} \\ -18 \end{bmatrix}$$

Now employing backward substitution to solve $UX = Z$ brings us to the final solution, which agrees with the result obtained in section 2 above.

$$X = \begin{bmatrix} -3 \\ -30\frac{1}{2} \\ -18 \end{bmatrix}$$

***

Although it has already been noted, it is important to emphasize that this method requires essentially the same number of operations as for Gaussian elimination. It has the advantage of providing greater accuracy if the computer system employed would allow the evaluation of (3.6) and (3.7) using extended or double precision arithmetic while the matrix elements are maintained in regular precision. The use of extended or double precision throughout both methods would produce equivalent accuracy with similar increases in memory storage requirements. Most authors and almost all matrix-solving software packages favor the LU factorization approach because it provides increased computation time savings when the same system is to be solved for more than one constant vector B.

In a number of applications such as the solution of partial differential equations, the coefficient matrix is tridiagonal. This means that the matrix A is filled with zeros except for the main diagonal and the diagonals just above and below the main diagonal. In such cases, the Crout procedure is easily reduced to handle only the nonzero elements yielding significant savings in computational time. Also, some algorithms achieve great savings in memory requirements by storing only the nonzero diagonals as vectors. The reader should be aware that there are many subtle modifications made in the algorithms for these procedures that are mostly specific to nuances of the particular programming language used to implement the particular algorithm. This text will not delve into such sophistications as many are well beyond the introductory level.

In following the development of this method, you may have wondered if pivoting is necessary or not. Indeed, it is as can be seen in equation (3.7) if l_{ss} is relatively small in magnitude or zero. Clearly, the method fails in the latter case, and loss of accuracy is likely to occur in the former. Also, as in the Gaussian elimination method, column scaling is important to maintain high accuracy. As a result, an effective Crout algorithm should include column pivoting and scaling as presented in the Gaussian elimination method.

These two aspects are included in the Crout factorization method as follows. To begin, it is convenient to introduce the augmented matrix of the system as is the usual case. For (3.2), we have

$$\text{Aug } A = [A, B].$$

As a means of conserving computer memory, it is the usual procedure to overwrite the coefficient matrix with the elements of L and U as they are computed. The problem of overlapping along the diagonal is avoided by placing the diagonal elements of L there since it is understood that the diagonal elements of U are all unity. As discussed in section 5 of chapter II, column scaling requires the identification of the largest element in each row of A. This is achieved by constructing the column vector M whose elements are given by the following expression.

$$m_i = \max \left(|a_{ij}| \right), j = 1, 2, \ldots, n$$

In other words, M is a vector whose elements are the elements of greatest magnitude of each respective row.

From this point, the process of constructing the L and U matrices is given by (3.6) and (3.7). The only variation is to implement column pivoting and scaling when necessary. To present this procedure, it is convenient to assume that the process has continued without any interchanges being necessary through the calculation of the kth column of L. The elements of this column are given by (3.6) with $s = k$. As previously mentioned, it is desirable to calculate these elements in extended or double precision and store them in rounded single precision if possible. The decision to exchange rows is now based on evaluating the index p from the following relation.

$$\frac{I_{pk}}{m_p} = \max_{k \le i \le n} \left| \frac{I_{jk}}{m_i} \right|$$

If it turns out that $p = k$, then a row exchange is not necessary, and the elements u_{kj}, $k+1 < j < n$ are computed by use of equation (3.7). Should $p > k$, then the pth and k^{th} rows of Aug A and M are exchanged and the procedure continued. Most modern algorithms do not actually perform the row exchanges. An index vector is introduced in which the order of row operations is recorded, which further reduces computational time.

EXERCISES

Find the LU factors for the following matrices, and verify the results.

1*.
$$\begin{bmatrix} -4 & 2 & -2 \\ 2 & 1 & 0 \\ -2 & 0 & 9 \end{bmatrix}$$

2*.
$$\begin{bmatrix} 6 & 6 & 1 \\ -9 & -6 & 6 \\ 4 & -2 & 9 \end{bmatrix}$$

3*.
$$\begin{bmatrix} -9 & 6 & -4 & 7 \\ 6 & 4 & -7 & 3 \\ -4 & -7 & 6 & -4 \\ 7 & 3 & -4 & 8 \end{bmatrix}$$

4*.
$$\begin{bmatrix} -8 & 0 & 4 & 3 \\ -1 & 4 & -4 & 0 \\ 3 & -6 & 5 & -2 \\ -1 & 2 & 0 & 1 \end{bmatrix}$$

5. Show that the product of the U and L matrices calculated for the example in this section reproduces the given A matrix.

Using LU factorization solve, the following systems.

6*. $\begin{bmatrix} 8 & 0 & -6 \\ 0 & -3 & 1 \\ -6 & 1 & 8 \end{bmatrix} X = \begin{bmatrix} -1 \\ 2 \\ 1 \end{bmatrix}$

7*. $\begin{bmatrix} 6 & 6 & 1 \\ -9 & -6 & 6 \\ 4 & -2 & 9 \end{bmatrix} X = \begin{bmatrix} 3 \\ 1 \\ 7 \end{bmatrix}$

8*. $\begin{bmatrix} -9 & 6 & -4 & 7 \\ 6 & 4 & -7 & 3 \\ -4 & -7 & 6 & -4 \\ 7 & 3 & -4 & 8 \end{bmatrix} X = \begin{bmatrix} 1 \\ 0 \\ -1 \\ 2 \end{bmatrix} , \begin{bmatrix} 2 \\ -1 \\ 1 \\ 0 \end{bmatrix}$ and $\begin{bmatrix} -3 \\ 1 \\ 5 \\ -2 \end{bmatrix}$

***

4. CHOLESKY'S METHOD FOR POSITIVE DEFINITE MATRICES

In some applications, the coefficient matrix A in a general system of linear algebraic equations is positive definite.[68] Any symmetric matrix with positive diagonal elements meets this condition and is typical of a large number of applied problems. The positive definite requirement implies the following relation.

$$|a_{ii}| \geq \max_{1 \leq j \leq n} |a_{ij}| , \quad i = 1, 2, \cdots, n$$

With this being true, the result is that column pivoting and scaling are not required in the LU factorization process. In fact, it turns out that the matrix A may be factored in the following special way.

$$A = LL^T,$$

where L is a general lower triangular matrix. Once L is found, the solution to a linear algebraic system such as equation (3.2) of the previous section is obtained by solving the following two successive equations by forward and backward substitution respectively.

$$LZ = B \text{ and } L^T X = Z$$

The advantage of this message is that it requires approximately one-half the number of operations a Crout factorization or Gaussian elimination.

---

68. For a definition of *positive definite*, see exercises 17 and 18 in chapter V.

Following the same sequence of operations as in the previous section to find l_{ij} and u_{ij}, we find that

(4.1) $\qquad\qquad l_{11} = \sqrt{a_{11}}\,,\, l_{j1} = c_{j1}/l_{11}, j = 2, 3, \ldots, n,$

and in general for $i = 2, 3, \ldots, n.$

(4.2) $\qquad\qquad\qquad\qquad l_{ii} = \left(a_{ii} - \sum_{k=1}^{i-1} l_{ik}^2 \right)^{1/2}$

(4.3) $\qquad\qquad\qquad\qquad l_{ji} = \dfrac{1}{l_{ii}} \left(a_{ji} - \sum_{k=1}^{i-1} l_{jk} l_{ik} \right)$

EXAMPLE 4

Apply the method of this section to solve a linear algebraic system that has the coefficient matrix and constant vector given below.

$$A = \begin{bmatrix} 3 & 1 & -1 \\ 1 & 2 & 1 \\ -1 & 1 & 2 \end{bmatrix},\; B = \begin{bmatrix} -1 \\ 2 \\ 1 \end{bmatrix}$$

SOLUTION

From equation (4.1), we have the following:

$$l_{11} = 1.732,\, l_{21} = 0.577,\, l_{31} = -0.577$$

Setting $i = 2$ in (4.2) and (4.3) yields the next two terms.

$$l_{22} = \left(a_{22} - l_{21}^2 \right)^{1/2} = 1.291$$

$$l_{32} = \frac{1}{l_{22}} \left(a_{32} - l_{31} l_{21} \right) = 1.032$$

Now with $i = 3$ in (4.2), the final term is obtained.

$$l_{33} = \left(a_{33} - l_{31}^2 - l_{32}^2\right)^{1/2} = 0.776$$

The desired lower triangular matrix is

$$L = \begin{bmatrix} 1.732 & 0 & 0 \\ 0.577 & 1.291 & 0 \\ -0.577 & 1.032 & 0.776 \end{bmatrix}.$$

The reader may confirm that $A = LL^T$. As already noted, these elements are usually stored in the lower triangular portion of the original A matrix. The original coefficient matrix is rarely needed, and if it is, it can be reconstructed from L.

Now that the L matrix has been obtained, the next step is to solve for the intermediate Z vector by forward substitution. This method proceeds as follows.

$$z_1 = b_1/l_{11} = -0.577$$

$$z_2 = (b_2 - l_{21}z_1)/l_{22} = 1.807$$

$$z_3 = (b_3 - l_{31}z_1 - l_{32}z_2)/l_{33} = -1.543$$

The reader should verify that these operations are identical to the process employed in the example of the previous section.

The final phase is the back substitution process to evaluate the desired unknown vector.

$$x_3 = z_3/l_{33} = -1.989$$

$$x_2 = (z_2 - l_{32}x_3)/l_{22} = 2.990$$

$$x_1 = (z_1 - l_{21}x_2 - l_{31}x_3)/l_{11} = -1.992$$

or

$$X = \begin{bmatrix} -1.992 \\ 2.990 \\ -1.989 \end{bmatrix}.$$

Again it is left for the reader to verify that this vector does satisfy the given equation. The exact solution is

$$X = \begin{bmatrix} -2 \\ 3 \\ -2 \end{bmatrix}.$$

***

The discrepancy between the calculated answer and the exact solution is due to the fact that the calculations were carried out only to three decimal places (rounded).

EXERCISES

Factor the following matrices by Choleski's method.

1*. $\begin{bmatrix} 4 & 2 & -2 \\ 2 & 3 & 1 \\ -2 & 1 & 3 \end{bmatrix}$

2*. $\begin{bmatrix} 8 & 0 & -6 \\ 0 & 3 & 1 \\ -6 & 1 & 8 \end{bmatrix}$

3*. $\begin{bmatrix} 9 & -6 & -4 & 5 \\ -6 & 7 & 4 & -3 \\ -4 & 4 & 5 & -2 \\ 5 & -3 & -2 & 6 \end{bmatrix}$

4*. $\begin{bmatrix} 3 & 1 & 0 & -1 \\ 1 & 2 & -1 & 0 \\ 0 & -1 & 3 & 2 \\ -1 & 0 & 2 & 4 \end{bmatrix}$

Employ Choleski's method to solve the following systems.

5*.
$$\begin{bmatrix} 4 & 2 & -2 \\ 2 & 3 & 1 \\ -2 & 1 & 3 \end{bmatrix} X = \begin{bmatrix} -2 \\ 1 \\ 3 \end{bmatrix}$$

6*.
$$\begin{bmatrix} 9 & -6 & -4 & 5 \\ -6 & 7 & 4 & -3 \\ -4 & 4 & 5 & -2 \\ 5 & -3 & -2 & 6 \end{bmatrix} X = \begin{bmatrix} 1 \\ -1 \\ 2 \\ -1 \end{bmatrix}$$

7.   Carefully examine the solution to exercise 6 above, and discuss your observations.

***

5. JACOBI ITERATION

In section 6 of chapter III, Jacobi iteration was introduced as a means of solving simultaneous linear equations that occur in potential problems. Because of its importance, it will be reviewed again in terms of solving a general system of equations. The advantage of this method as with most iteration techniques is that a nearly constant level of accuracy is maintained. Also, any errors introduced during the iteration sequence die out due to the fact that it is a converging process, only the time to converge is lengthened. Another feature is that in most situations, column pivoting and scaling are not necessary.

If the convergence is rapid, iteration methods such as Jacobi iteration usually require less computation time than the direct methods previously discussed. However, if convergence is slow, these advantages are quickly lost. Given a specific problem, the problem of determining how fast it will converge is not an easy matter to evaluate. In a later section this matter will be examined further.

To introduce the Jacobi iteration procedure, let us consider again a general system of linear algebraic equations.

(5.1) $AX = B$

Now let the coefficient matrix be represented by the sum of another matrix and a diagonal matrix that contains the diagonal elements of A, that is $D = \text{Diag} [a_{ii}]$.

(5.2) $A = A_1 + D$

Clearly, A_1 is just the A matrix with zeros as diagonal elements. Substituting (5.2) into (5.1) and after some manipulations, the following relation may be obtained.

$$X = D^{-1}B - D^{-1}A_1X$$

If we now define two new matrices to be $D_1 = D^{-1}B$ and $A_2 = D^{-1}A_1$, this equation becomes

(5.3) $$X = D_1 - A_2X.$$

Since D is a diagonal matrix of the diagonal elements of A, its inverse is a diagonal matrix whose elements are the inverses of the diagonal elements of A respectively. This means that D_1 is simply the B vector with each element divided by the respective diagonal element of the A matrix.

$$D_1 = \begin{bmatrix} \dfrac{b_1}{a_{11}} \\ \dfrac{b_2}{a_{22}} \\ \vdots \\ \dfrac{b_n}{a_{nn}} \end{bmatrix}$$

Similarly, A_2 will be

$$A_2 = \begin{bmatrix} 0 & \dfrac{a_{12}}{a_{11}} & \cdots & \dfrac{a_{1n}}{a_{11}} \\ \dfrac{a_{21}}{a_{22}} & 0 & \cdots & \dfrac{a_{2n}}{a_{22}} \\ \vdots & \vdots & \ddots & \vdots \\ \dfrac{a_{n1}}{a_{nn}} & \dfrac{a_{n2}}{a_{nn}} & \cdots & 0 \end{bmatrix}$$

If any of the diagonal elements of the original coefficient matrix are zero, then rows must be exchanged so that no zeros appear on the main diagonal.

   The iteration procedure is to make some initial guess of the solution vector X_0, substitute it into the right-hand side of (5.3) to calculate the next iterate. This process is continued until the solution vector has converged to meet some desired degree of accuracy. To better indicate the iteration sequence, equation (5.3) may be written as follows.

$$X_{n+1} = D_1 - A_2X_n$$

The initial vector can be anything, but the most common approach is to choose the B vector as X_0. Also, the evaluation of convergence can be done in a variety of ways. One fairly easy computational way is to check the relative change of the L_1 norms of two successive iterates until the desired accuracy level is achieved. This approach is expressed in the following relation.

(5.4) $$\frac{|\,|X_{n+1}\,|\,|_1 - |\,|X_n\,|\,|_1}{|\,|X_n\,|\,|_1} \leq \text{specified tolerance}$$

EXAMPLE 5

Solve example 4 of the previous section by the method presented in this section.

$$\begin{bmatrix} 3 & 1 & -1 \\ 1 & 2 & 1 \\ -1 & 1 & 2 \end{bmatrix} X = \begin{bmatrix} -1 \\ 2 \\ 1 \end{bmatrix}$$

SOLUTION

For this problem the D_1 and A_2 matrices are the following.

$$D_1 = \begin{bmatrix} -0.3333 \\ 1.0000 \\ 0.5000 \end{bmatrix}, \quad A_2 = \begin{bmatrix} 0 & 0.3333 & -0.3333 \\ 0.5000 & 0 & 0.5000 \\ -0.5000 & 0.5000 & 0 \end{bmatrix}$$

The first trial solution vector is the B vector or

$$X_0 = \begin{bmatrix} -1 \\ 2 \\ 1 \end{bmatrix}.$$

After 41 iterations, the solution vector is

(5.6) $$X_{41} = \begin{bmatrix} -1.992 \\ 2.990 \\ 1.990 \end{bmatrix}.$$

The tolerance as measured by equation (5.4) for this number of iterations is 0.056 % and the exact answer is

$$X = \begin{bmatrix} -2 \\ 4 \\ -2 \end{bmatrix}.$$

***

For interest, a plot of the logarithm of the error versus the number of iterations is given in figure 9.1 below.

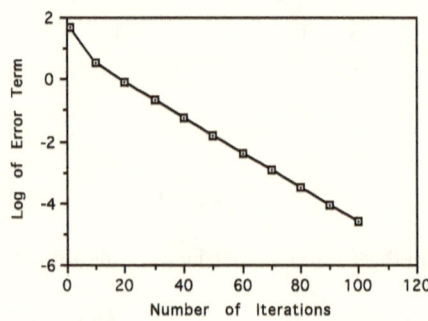

Fig. 9.1

From this plot, it appears that the accuracy improves by an order of magnitude for about every 20 iterations.

EXERCISES

Use Jacobi iteration to solve the following systems.

1*.

$$\begin{bmatrix} 2 & -1 \\ 4 & 3 \end{bmatrix} X = \begin{bmatrix} 5 \\ -1 \end{bmatrix}$$

2*.

$$\begin{bmatrix} 1 & 1 & 1 \\ 2 & -1 & -1 \\ 1 & 2 & -1 \end{bmatrix} X = \begin{bmatrix} 2 \\ 1 \\ 3 \end{bmatrix}$$

3*.
$$\begin{bmatrix} 1 & 1 & 0 \\ 0 & 1 & 6 \\ 2 & 2 & 1 \end{bmatrix} X = \begin{bmatrix} 3 \\ 1 \\ 6 \end{bmatrix}$$

4*.
$$\begin{bmatrix} 3 & -1 & 1 \\ 1 & -1 & 0 \\ 2 & 0 & 1 \end{bmatrix} X = \begin{bmatrix} 4 \\ 0 \\ 4 \end{bmatrix}$$

5*.
$$\begin{bmatrix} 11 & 1 & -3 \\ 2 & 0 & 1 \\ 1 & 3 & -10 \end{bmatrix} X = \begin{bmatrix} 2 \\ 0 \\ 6 \end{bmatrix}$$

6*.
$$\begin{bmatrix} 0 & 0.20 & 0.40 \\ -0.75 & -0.05 & 1.15 \\ -0.25 & -0.15 & 0.45 \end{bmatrix} X = \begin{bmatrix} 0.50 \\ -1.50 \\ 1.00 \end{bmatrix}$$

7*.
$$\begin{bmatrix} 1 & 1 & -1 \\ 1 & -1 & -1 \\ -1 & -1 & -1 \end{bmatrix} X = \begin{bmatrix} 1 \\ 2 \\ 3 \end{bmatrix}$$

***

6. GAUSS-SEIDEL ITERATION

As the example in the previous section illustrated, the Jacobi iteration scheme converges fairly rapidly. However, the rate of convergence can be improved by a fairly direct approach, which is known as the Gauss-Seidel method. To present this method, consider a linear algebraic system that has been rewritten for iterative solution as in equation (5.3) above.

(6.1) $$X = D_1 - A_2X,$$

where, as we previously defined,

$$
D_1 = \begin{bmatrix} \dfrac{b_1}{a_{11}} \\ \dfrac{b_2}{a_{22}} \\ \vdots \\ \dfrac{b_n}{a_{nn}} \end{bmatrix}, \quad A_2 = \begin{bmatrix} 0 & \dfrac{a_{12}}{a_{11}} & \cdots & \dfrac{a_{1n}}{a_{11}} \\ \dfrac{a_{21}}{a_{22}} & 0 & \cdots & \dfrac{a_{2n}}{a_{22}} \\ \vdots & \vdots & \ddots & \vdots \\ \dfrac{a_{n1}}{a_{nn}} & \dfrac{a_{n2}}{a_{nn}} & \cdots & 0 \end{bmatrix}.
$$

In the Jacobi iteration method, a trial solution vector X is substituted into the right-hand side of equation (6.1) to generate an improved estimate. However, it can be seen that once the trial vector is assumed, the first equation in the system (6.1) gives the new value for the first element, x_1, of the solution vector. Using this value in the second equation along with the remaining values from the first trial vector will produce a better estimate of the second element, x_2. These two new estimates along with the initial values of x_3 through x_n may now be used to generate the new value of x_3. If we use superscripts i and $i+1$ to denote the ith and i^{th} + first iterates, (6.1) may be used to express this operation in individual equation form as follows.

$$
x_1^{i+1} = \frac{b_1}{a_{11}} - \frac{1}{a_{11}}\left(a_{12}x_2^i + a_{13}x_3^i + \cdots + a_{1n}x_n^i\right)
$$

$$
x_2^{i+1} = \frac{b_2}{a_{22}} - \frac{1}{a_{22}}\left(a_{21}x_1^{i+1} + a_{23}x_3^i + \cdots + a_{2n}x_n^i\right)
$$

$$
\cdots
$$

$$
x_n^{i+1} = \frac{b_n}{a_{nn}} - \frac{1}{a_{nn}}\left(a_{n1}x_1^{i+1} + a_{n2}x_2^{i+1} + \cdots + a_{n,n-1}x_{n-1}^{i+1}\right)
$$

In matrix form, these equations may be written as the following.

$$
X_{i+1} = D_1 - LX_{i+1} - UX_i,
$$

where L and U are the lower and upper triangular portions of the matrix A_2. By rearranging terms, this last equation may be rewritten as

$$
(I + L)X_{i+1} = D_1 - UX_i = D_2.
$$

With the calculation of a given solution vector X_i, the matrix D_2 is calculated, and then the above equation is used to calculate the next estimation by forward substitution.

EXAMPLE 6

Consider the same example problem used in the two previous sections, and apply the method introduced in this section to determine its solution.

$$\begin{bmatrix} 3 & 1 & -1 \\ 1 & 2 & 1 \\ -1 & 1 & 2 \end{bmatrix} X = \begin{bmatrix} -1 \\ 2 \\ 1 \end{bmatrix}$$

SOLUTION

For this problem, the matrices D_1 and A_2 are given in equation (5.5) of the previous section. From A_2, the matrices $I + L$ and U are easily identified to be

$$I + L = \begin{bmatrix} 1 & 0 & 0 \\ 0.5000 & 1 & 0 \\ -0.5000 & 0.5000 & 1 \end{bmatrix},$$

$$U = \begin{bmatrix} 0 & 0.3333 & -0.3333 \\ 0 & 0 & 0.5000 \\ 0 & 0 & 0 \end{bmatrix}.$$

After just 25 iterations, the accuracy as determined by equation (5.4) slightly exceeds that for (5.6), and the results are the following:

$$X_{25} = \begin{bmatrix} -1.996 \\ 2.995 \\ -1.995 \end{bmatrix} \text{ with an accuracy of } 0.049\%$$

For 41 iterations that were used for the Jacobi iteration example, the results become the following:

$$X_{41} = \begin{bmatrix} -1.999 \\ 2.999 \\ -1.999 \end{bmatrix} \text{ with an accuracy of } 0.00082\%$$

The accuracy for the Jacobi method was just 0.056% for the same number of iterations.

***

As we have seen with this example, the Gauss-Seidel method converges more rapidly than the Jacobi method. In practically all situations, this is true; however, there are cases for which the Jacobi method will converge and the Gauss-Seidel will not. Such situations are fairly unusual, and practically all authors recommend the Gauss-Seidel method.

Another area of concern is that of convergence. The least restrictive case for which convergence can be proven is for the coefficient matrix A to be a positive definite matrix.[69] A more useable but more restrictive sufficient condition is for A to be strictly diagonally dominant. This means that the elements of A must satisfy the following relation.

$$|a_{ii}| > \sum_{j=1 \neq i}^{n} |c_{ij}| \ , \ i = 1, 2, \cdots, n$$

In fact, it has been pointed out by Dorn and McCracken (1972) that a sufficient but not necessary condition is for the above relation to hold for at least one equation of the entire system as long as the greater than (>) is replaced by greater than or equal to (≥) in all the others. In practice, what this means is that, if necessary, the equations should be rearranged so that the diagonal elements are those of largest magnitude. This operation is automatically contained in almost all modern computational algorithms. It is also true that in many physical problems, the matrix elements naturally occur in this pattern.

Perhaps the most important question for an analyst to consider is how to decide when to use an iteration method rather than a direct method such as Gaussian elimination. To this end, Dorn and McCracken (1972) have made the following comparisons. In terms of total algebraic operations, Gaussian elimination requires approximately $2n^3/3$ operations for a system of order n. For large n, Gauss-Seidel requires $2n^2$ per operation. This means that as long as the number of iterations required to reach a desired level of accuracy is $< n/3$, the Gauss-Seidel method is more efficient operation-wise. However, since iteration methods contain only the round-off errors occurring in the last step, iteration methods are recommended even for cases involving more computations. On the other hand, iteration methods will not converge in the general case when the coefficient matrix A fails to meet the conditions discussed above.

There are alternative approaches termed relaxation methods that are quite useful for accelerating the rate of convergence. They will not be presented here but have been introduced in an exercise for the reader.

---

[69].   For example, see the reference by Ralston and Rabinowitz [1978].

EXERCISES

Employ Gauss-Seidel iteration of to solve the same exercise problems as given for section 5 above.

1*.
$$\begin{bmatrix} 2 & -1 \\ 4 & 3 \end{bmatrix} X = \begin{bmatrix} 5 \\ -1 \end{bmatrix}$$

2*.
$$\begin{bmatrix} 1 & 1 & 1 \\ 2 & -1 & -1 \\ 1 & 2 & -1 \end{bmatrix} X = \begin{bmatrix} 2 \\ 1 \\ 3 \end{bmatrix}$$

3*.
$$\begin{bmatrix} 1 & 1 & 0 \\ 0 & 1 & 6 \\ 2 & 2 & 1 \end{bmatrix} X = \begin{bmatrix} 3 \\ 1 \\ 6 \end{bmatrix}$$

4*.
$$\begin{bmatrix} 3 & -1 & 1 \\ 1 & -1 & 0 \\ 2 & 0 & 1 \end{bmatrix} X = \begin{bmatrix} 4 \\ 0 \\ 4 \end{bmatrix}$$

5*.
$$\begin{bmatrix} 11 & 1 & -3 \\ 2 & 0 & 1 \\ 1 & 3 & -10 \end{bmatrix} X = \begin{bmatrix} 2 \\ 0 \\ 6 \end{bmatrix}$$

6*.
$$\begin{bmatrix} 0 & 0.20 & 0.40 \\ -0.75 & -0.05 & 1.15 \\ -0.25 & -0.15 & 0.45 \end{bmatrix} X = \begin{bmatrix} 0.50 \\ -1.50 \\ 1.00 \end{bmatrix}$$

7*.
$$\begin{bmatrix} 1 & 1 & -1 \\ 1 & -1 & -1 \\ -1 & -1 & -1 \end{bmatrix} X = \begin{bmatrix} 1 \\ 2 \\ 3 \end{bmatrix}$$

***

7. ILL-CONDITIONED PROBLEMS

In section 12 of chapter II, a brief mention of ill-conditioned systems was made. This term is used to refer to systems of algebraic equations for which the uncertainty or accuracy of the solution is much greater than the uncertainty of the given coefficient matrix and constant vector. In this section, ill-conditioned systems will be investigated in more detail.

When a solution to a system of linear algebraic equations such as (7.1) below has been found by any method, its accuracy may be tested by substitution.

$$(7.1) \qquad\qquad AX = B.$$

This test is performed by computing the residual vector R once a solution vector X has been determined.

$$(7.2) \qquad\qquad R = B - AX$$

Clearly, if the exact solution has been found, then the residual vector will be the null vector. However, the residual vector usually does not vanish due to the accumulation of computational errors in determining X as well as any uncertainties in the given coefficient and constant matrices. The accuracy of the solution may be measured by the norm of the residual, $\| R \|$, which should be small, or more particularly the ratio $\| R \| / \| X \|$. If this ratio is small compared to one, then the computed solution can be said to be close to the true or exact solution.

To discuss this situation in more detail, let X be the true solution and X_c be a computed solution. In reality, the vector X is usually never be known, and we hope that X_c is quite close to it. An important thing to note is that even though $\| R \| \ll 1$, we may not conclude that $\| X - X_c \| \ll 1$. To be more precise, equation (7.2) should be written as

$$R = B - AX_c.$$

Replacing B by the left-hand side of equation (7.1) and solving for the term $X - X_c$ gives

$$X - X_c = A^{-1}R.$$

Taking the norm of this expression and applying the general property that $\| AB \| \le \| A \| \| B \|$ results in

$$(7.3) \qquad\qquad \| X - X_c \| = \| A^{-1} \| \| R \|.$$

From equation (7.1), we also have $\| B \| \le \| A \| \| X \|$. Dividing (7.3) by $\| X \|$ and applying the previous result gives the following result.

$$\frac{\| X - X_c \|}{\| X \|} \le \| A \| \| A^{-1} \| \frac{\| R \|}{\| B \|}$$

At this point, the systems condition number is defined as follows.

(7.4) $\text{Cond } A = \| A \| \| A^{-1} \|$

Introducing the condition number into the previous result yields

$$\frac{\| X - X_c \|}{\| X \|} \le (\text{Cond } A) \frac{\| R \|}{\| B \|}.$$

The left side of this expression is the relative error in X_c, and the ratio $\| R \| / \| B \|$ is termed the relative residual.

In examining the condition number defined in (7.4), it is easy to conclude that since $AA^{-1} = I$ for any nonsingular matrix, then the value of Cond A should be unity. Because of computational inaccuracies, this is not the usual case. Situations in which Cond $A \le 10$ are termed well-conditioned problems, and one may conclude that a small relative residual implies that the computed solution is close to the true solution.

On the other hand, cases for which Cond $A \ge 100$ are called ill conditioned, and one may not conclude that the computed solution is close to the true solution. Characteristically, an ill-conditioned system occurs when the determinant of the coefficient matrix is small in comparison with its elements.

Matrix Norms

Before continuing, it is useful to introduce the concept of the norm of a matrix similar to the norm of a vector, which was introduced in section 9 of chapter II. As already noted, there are numerous ways by which norms of vectors and matrices can be defined. In section 9 of chapter II, the L_1, L_2, and L_∞ vector norms were defined. In a similar manner, corresponding norms for an nth-order square matrix are defined as follows.

$$\| A \|_1 = \max_{X \neq 0} \frac{\| AX \|_1}{\| X \|_1}$$

$$\| A \|_2 = \max_{X \neq 0} \frac{\| AX \|_2}{\| X \|_2}$$

$$\| A \|_\infty = \max_{X \neq 0} \frac{\| AX \|_\infty}{\| X \|_\infty}$$

In examining these definitions, the reader might assume that it is fairly involved computationally to evaluate these norms since the vector X is a general vector. This is not the case for the L_1 and L_∞ norms since the following relationships may be proved.

$$\| A \|_1 = \max \left\{ \sum_{i=1}^{n} | a_{ij} | \ , \ j = 1, 2, \cdots, n \right\}$$

$$\| A \|_\infty = \max \left\{ \sum_{j=1}^{n} | a_{ij} | \ , \ i = 1, 2, \cdots, n \right\}$$

The first norm is the maximum absolute column sum, and the second is the maximum absolute row sum. Either of these norms is easy to evaluate as may be seen in the following example.

EXAMPLE 7

Determine the L_1 and L_∞ norms for the following matrix

$$A = \begin{bmatrix} 7 & -2 & 3 \\ -4 & 5 & 1 \\ -5 & 0 & -2 \end{bmatrix}$$

SOLUTION

Using the above definitions, it can be seen that

$$\| A \|_1 = \max (16, 7, 6) = 16$$
$$\| A \|_\infty = \max (12, 10, 7) = 12$$

***

To return to our discussion of ill-conditioned systems, we now introduce the following alternative measure of the condition number, which is somewhat easier to apply.

$$(7.5) \qquad \text{Cond } A \sim \frac{\| A \|}{\text{Det } A}$$

As an illustration of this material consider the following example problem.

$$\begin{bmatrix} 1 & 1 \\ -1 & 2 \end{bmatrix} X = \begin{bmatrix} 2 \\ 1 \end{bmatrix}$$

For this system, Det $A = 3$ and $\| A \|_1 = 3$, and using equation (7.5), Cond $A = 1$, which means that the system is well conditioned. For reference, the true solution is

$$(7.6) \qquad X = \begin{bmatrix} 1 \\ 1 \end{bmatrix}.$$

Now consider another second-order system given below.

$$(7.7) \qquad \begin{bmatrix} 1.000 & 1.000 \\ 1.001 & 0.999 \end{bmatrix} X = \begin{bmatrix} 2 \\ 2 \end{bmatrix}$$

It is easy to verify that the true solution to this system is also the vector given by (7.6). For this system, Det $A = -0.002$ and $\| A \|_1 = 2.001$ so that Cond $A = 1000$. This indicates that this system is ill conditioned. It is also of interest to note that the inverse of the coefficient matrix is

$$A^{-1} = \begin{bmatrix} -499.5 & 500.0 \\ 500.5 & -500.0 \end{bmatrix}.$$

For this inverse $\| A^{-1} \|_1 = 1,000$. Applying equation (7.4) to evaluate the formal condition number gives Cond $A = 2,001$, which indicates that the approximate condition number given by (7.5) is low by a factor of two in this case but still indicates an ill-conditioned system.

To complete the illustration, assume that by some means, the following approximate solution to equation (7.7) has been arrived at.

$$X_c = \begin{bmatrix} 2.000 \\ 0.000 \end{bmatrix}$$

For this solution, the residual of (7.7) is

$$R = \begin{bmatrix} 2 \\ 2 \end{bmatrix} - \begin{bmatrix} 1.000 & 1.000 \\ 1.001 & 0.999 \end{bmatrix} \begin{bmatrix} 2.000 \\ 0.000 \end{bmatrix} = \begin{bmatrix} 0.000 \\ -0.002 \end{bmatrix}$$

Using L_1 norms, the relative residual is 0.0005, which is small and might mislead one to assume that the approximate answer is fairly accurate. However, the relative error for the solution is the following:

$$\frac{\| X - X_c \|_1}{\| X \|_1} = \frac{\left\| \begin{bmatrix} 1 \\ 1 \end{bmatrix} - \begin{bmatrix} 2 \\ 0 \end{bmatrix} \right\|_1}{\left\| \begin{bmatrix} 1 \\ 1 \end{bmatrix} \right\|_1} = 1$$

This indicates that the approximate solution differs from the true answer with an error of 100%.

In addition to detecting ill-conditioned systems by the method shown in this section, there are a number of others that may be found in several of the references to this text. One common approach is to actually compute the inverse of the coefficient matrix and evaluate the product AA^{-1} to see how close it comes to being an identity matrix.

EXERCISES

Evaluate the L_1 and L_∞ norms for each of the following matrices.

1. $\begin{bmatrix} 4 & 2 & -2 \\ 2 & 3 & 1 \\ -2 & 1 & 3 \end{bmatrix}$

2. $\begin{bmatrix} 8 & 0 & -6 \\ 0 & 3 & 1 \\ -6 & 1 & 8 \end{bmatrix}$

3. $\begin{bmatrix} 9 & -6 & -4 & 5 \\ -6 & 7 & 4 & -3 \\ -4 & 4 & 5 & -2 \\ 5 & -3 & -2 & 6 \end{bmatrix}$

4. $\begin{bmatrix} 3 & 1 & 0 & -1 \\ 1 & 2 & -1 & 0 \\ 0 & -1 & 3 & 2 \\ -1 & 0 & 2 & 4 \end{bmatrix}$

5. The symmetric real matrix of order n whose elements are defined as $h_{ij} = 1/(i+j-1)$ is called a Hilbert matrix. For example

$$H_3 = \begin{bmatrix} 1 & \frac{1}{2} & \frac{1}{3} \\ \frac{1}{2} & \frac{1}{3} & \frac{1}{4} \\ \frac{1}{3} & \frac{1}{4} & \frac{1}{5} \end{bmatrix}$$

Calculate Cond H_3. Use three-decimal arithmetic to solve the following equation by the Gaussian elimination method, and compare your result to the exact answer $X = [9, -36, 30]$.

$$H_3X = \begin{bmatrix} 1 \\ 0 \\ 0 \end{bmatrix}$$

***

8. ITERATIVE IMPROVEMENT OF SOLUTIONS

When an ill-conditioned system is encountered or one wishes to improve the accuracy of a computed solution to a well-conditioned system, the method of iterative improvement is the recommended approach. The general topic of improving the accuracy of solutions to systems of equations in one of continued interest and study.

The method of iterative improvement proceeds in the following way. Consider that a solution to the linear system below has been obtained by some direct method and is denoted as X_1.

(8.1) $\qquad AX = B$

The residual for this solution is

(8.2) $\qquad R_1 = B - AX_1$

The same method used to compute X_1 is now employed to find the solution Y_1 to the following equation.

(8.3) $\qquad AY_1 = R_1$

To see what the solution vector Y_1 represents Equation (8.3) is solved schematically and (8.1) and (8.2) are substituted to obtain

$$Y_1 = A^{-1}R_1 = A^{-1}B - X_1 = X - X_1$$

Considering that X represents the true solution of (8.1) then Y_1 is the difference between the true solution and the first computed solution X_1. Solving for X gives the result

(8.4) $$X = X_2 = X_1 + Y_1$$

Due to computational inaccuracies we cannot expect X to be the true solution and hence the notation X_2 has been used to indicate that it is an improved solution.

This procedure can be continued by calculating the next residual R_2 from (8.2) then finding Y_2 from (8.3) and constructing X_3 from (8.4). In general we may write the following iteration scheme.

$$X_i = X_{i-1} + Y_{i-1}$$

In ill-conditioned problems this procedure may be continued until $\| Y_{i-1} \| / \| X_i \|$ reaches some predetermined level of accuracy. It must be noted that any calculation of iterative improvement must be carried out in extended or double precision arithmetic or else any theoretical gain in accuracy will be lost in the actual computations. In practically all applications iterative improvement is applied only once as the gain, if any, in further iterations is small. The exceptions would be in dealing with strongly ill-conditioned problems.

As an illustration consider the second example problem of the previous section given by Equation (7.7). Assume also that we employed a very faulty method and came up with the following vector as the first solution which we have already seen is in error by 100%.

$$X_1 = \begin{bmatrix} 2.000 \\ 0.000 \end{bmatrix}$$

We are also assuming that the calculator we are using has only three decimal digits of accuracy in normal precision. In double precision the first residual is

$$R_1 = \begin{bmatrix} 2.000 \\ 2.000 \end{bmatrix} - \begin{bmatrix} 1.001 & 0.999 \\ 1.000 & 1.000 \end{bmatrix} \begin{bmatrix} 2.000 \\ 0.000 \end{bmatrix} = \begin{bmatrix} -0.002000 \\ 0.000000 \end{bmatrix}$$

In setting up this equation we have interchanged the first two rows of (7.7) to reflect column pivoting to improve accuracy. It should be recalled that interchanging rows in linear equations does not affect the order of unknowns in the solution vector. The equation for the Y_1 vector is now

$$\begin{bmatrix} 1.001000 & 0.999000 \\ 1.000000 & 1.000000 \end{bmatrix} Y_1 = \begin{bmatrix} -0.002000 \\ 0.000000 \end{bmatrix}$$

Solving this system by Gaussian elimination gives the solution for Y_1.

$$Y_1 = \begin{bmatrix} -1.000000 \\ 1.000000 \end{bmatrix}$$

From Equation (8.4) the improved solution is found.

$$X_2 = \begin{bmatrix} 1.000000 \\ 1.000000 \end{bmatrix}$$

This illustration shows how by iterative improvement a solution with three decimal digits of accuracy and 100% error is improved to a solution correct to six decimal digits in a single iteration.

EXERCISES

1.*   Consider that an approximate solution to the equation below is $X = [10.0, -8.1, 5.2]$. Use the method of iterative improvement to find the solution to at least four decimal places.

$$\begin{bmatrix} 0 & 0.20 & 0.40 \\ -0.75 & -0.05 & 1.15 \\ -0.25 & -0.15 & 0.45 \end{bmatrix} X = \begin{bmatrix} 0.50 \\ -1.50 \\ 1.00 \end{bmatrix}$$

2.*   Consider that an approximate solution to the equation[70] below is $X = [2.31, -0.42, -0.51, -0.53]$. Use the method of iterative improvement to find the solution to at least four decimal places.

$$\begin{bmatrix} 1 & 1 & 1 & 1 \\ 1 & 1 & 1 & -1 \\ 1 & 1 & -1 & -1 \\ 1 & -1 & -1 & -1 \end{bmatrix} X = \begin{bmatrix} 1 \\ 2 \\ 3 \\ 4 \end{bmatrix}$$

***

---

[70]   Hohn, [1958], p120.

9. MATRIX INVERSION BY GAUSS-JORDAN METHOD

In most applied problems, the actual inverse of a matrix is not needed. However, there are a few exceptions such as calculating the produce AA^{-1} as a means of investigating the ill condition of a system.

Of several methods available for computing the inverse of a matrix, the method of Gauss-Jordan is perhaps the most widely known and used. This method is primarily an alternative procedure for solving systems of linear algebraic equations. However, it is not as efficient as the Gaussian elimination method and hence has not been presented earlier in this text.

To invert a given matrix A, the procedure is to construct a special augmented matrix comprising the matrix A and an identity matrix I of the same order as shown below.

$$[A, I]$$

The first step is to apply Gaussian elimination to the rows of the matrix reducing A to upper triangular form with the added feature that the diagonal elements must also be reduced to ones. Schematically, the result of this operation would have the following appearance.

$$[U, M]$$

Here, U is an upper triangular matrix with ones on its main diagonal, and M represents the matrix resulting from the row operations on I. At this point, a back substitution process is begun, operating on the rows to reduce all the upper triangular elements to zeros, leaving only ones on the main diagonal. After this process, the previous matrix will have been converted to the following form.

$$[I, N]$$

What the row operation process has accomplished is that the identity matrix has effectively been moved from the right-hand side to the left-hand side of the n by $2n$ array. It may be shown now that the remaining matrix N on the right-hand side is actually the desired inverse of the original A matrix, that is, $N = A^{-1}$.

EXAMPLE 8

Apply the above procedure to determine the inverse of the following matrix.

$$A = \begin{bmatrix} 1 & 2 & -1 \\ -3 & 3 & 2 \\ -2 & 1 & 4 \end{bmatrix}$$

SOLUTION

The special augmented matrix will be the following.

$$\begin{bmatrix} 1 & 2 & -1 & 1 & 0 & 0 \\ -3 & 3 & 2 & 0 & 1 & 0 \\ -2 & 1 & 4 & 0 & 0 & 1 \end{bmatrix}$$

Performing Gaussian elimination on the first column and dividing the second row by the 2, 2 element to reduce it to unity leads to

$$\begin{bmatrix} 1 & 2 & -1 & 1 & 0 & 0 \\ 0 & 1 & -1/9 & 1/3 & 1/9 & 0 \\ 0 & 5 & 2 & 2 & 0 & 1 \end{bmatrix}$$

Performing the same procedure on the second column gives the following result.

$$\begin{bmatrix} 1 & 2 & -1 & 1 & 0 & 0 \\ 0 & 1 & -1/9 & 1/3 & 1/9 & 0 \\ 0 & 0 & 1 & 3/23 & -5/23 & 9/23 \end{bmatrix}$$

We now continue with the Gaussian elimination method, reducing the super diagonal elements of the third column to zero.

$$\begin{bmatrix} 1 & 2 & 0 & 3/23 & -5/23 & 9/23 \\ 0 & 1 & 0 & 8/23 & 2/23 & 1/23 \\ 0 & 0 & 1 & 3/23 & -5/23 & 9/23 \end{bmatrix}$$

The process is concluded by reducing the element above the main diagonal of the second column to zero. The final result is

$$\begin{bmatrix} 1 & 0 & 0 & 10/23 & -9/23 & 7/23 \\ 0 & 1 & 0 & 8/23 & 2/23 & 1/23 \\ 0 & 0 & 1 & 3/23 & -5/23 & 9/23 \end{bmatrix}.$$

In this matrix, the inverse of the given matrix is the right-hand third-order matrix or

$$A^{-1} = \begin{bmatrix} 10/23 & -9/23 & 7/23 \\ 8/23 & 2/23 & 1/23 \\ 3/23 & -5/23 & 9/23 \end{bmatrix}.$$

It is left for the reader to verify that this result is the inverse of the given matrix A.

EXERCISES

Evaluate inverses for the following matrices by the Gauss-Jordan method.

1*.
$$\begin{bmatrix} -4 & 2 & -2 \\ 2 & 1 & 0 \\ -2 & 0 & 9 \end{bmatrix}$$

2*.
$$\begin{bmatrix} 6 & 6 & 1 \\ -9 & -6 & 6 \\ 4 & -2 & 9 \end{bmatrix}$$

3*.
$$\begin{bmatrix} -9 & 6 & -4 & 7 \\ 6 & 4 & -7 & 3 \\ -4 & -7 & 6 & -4 \\ 7 & 3 & -4 & 8 \end{bmatrix}$$

4*.
$$\begin{bmatrix} 1 & -3 & -1 & -1 \\ 2 & 8 & -9 & -4 \\ -2 & 3 & -2 & 4 \\ -1 & 2 & 9 & -5 \end{bmatrix}$$

***

10. HOUSEHOLDER   TRANSFORMATIONS   AND   QR
   FACTORIZATION

A very important tool in matrix algebra is the QR factorization method, which states that any matrix A of order n by m may be expressed as a product of an nth-order orthogonal matrix Q and an upper (right) triangular matrix R of order n by m. The only restriction is that the columns of the given matrix A must be linearly independent. In other words,

(10.1) $A = QR.$

This factorization is of great value in least squares problems and the determination of eigenvalues as will be illustrated in following sections. It is almost the universal method for solving these problems in all software packages available today.

In section 10 of chapter II, the Gram-Schmidt method was introduced as a means of producing a set of orthogonal vectors from a set of linearly independent vectors. In many current textbooks, it is the method presented for constructing the QR factorization. However, the Gram-Schmidt procedure becomes computationally unstable in some situations, and hence, an alternative approach to forming the QR factors will be presented. This approach is by use of Householder transformations based on Householder matrices, which are orthogonal. Because of its improved stability, this approach is employed in practically all matrix-computational software programs available.

If X is any n^{th}-order arbitrary vector other than the zero vector, then the following matrix is the Householder matrix based on X.

(10.2) $H_x = I - \dfrac{2X^T X}{|X|^2}$

In this expression, I is an nth-order identity matrix and $|X|$ represents the L_2 (Euclidian) norm of X. It has been left as an exercise to prove that H_x is symmetric and orthogonal, that is,

$$H_x = H_x^T = H_x^{-1}$$

Another property of the Householder matrix is that $H_x X = -X$, which demonstrates that one eigenvalue is -1. It turns out that all of the eigenvalues of H_x are ± 1.

Now consider a general n by m matrix A whose first column is A_1. We desire to find a sequence of transformations to apply to A, which will reduce it to upper triangular form. Begin by constructing a vector X from A_1 by the following relation.

(10.3) $X = A_1 - \|A_1\| \, U_1$,

where U_1 is a column vector of proper order with a one as the first element and the remainder zeros. Now use equation (10.2) to generate a Householder matrix from X, which will be called H_1. If we now premultiply A by H_1, the following result occurs.

(10.4) $H_1 A = \begin{bmatrix} \|A_1\| & \vdots & [\,a\,] \\ \cdots\cdots & \vdots & \cdots\cdots \\ [\,0\,] & \vdots & B \end{bmatrix}$

The result has been shown in partitioned form. The norm $\|A_1\|$ is the L_2 norm of the first column of A. The matrix $[\,a\,]$ is a general row vector of order $m-1$, and $[\,0\,]$ is a null column of order $n-1$. The matrix B is a generated matrix of order $n-1$ by $m-1$.

The process is continued by taking the first column of B, B_1, and constructing a vector Y in the same way X was produced.

(10.5) $Y = B_1 - \|B_1\| \, U_1$

The order of Y is now $n-1$. Using Y, a Householder matrix is formed, which is denoted as H_y.

(10.6) $H_y = I - \dfrac{2Y^T Y}{\|Y\|^2}$

An n by m Householder matrix H_2 is now constructed from H_y as follows.

$$H_2 = \begin{bmatrix} 1 & \vdots & [\,0\,] \\ \cdots\cdots & \vdots & \cdots\cdots \\ [\,0\,] & \vdots & H_y \end{bmatrix}$$

The product $H_y B$ has the same effect as $H_1 A$, except the matrix is one smaller in each dimension. From this product, the same procedure is applied to the next smaller matrix in the lower right-hand corner yielding another smaller Householder matrix H_z. From H_z, the next full Householder matrix H_3 is constructed.

$$
H_3 = \left[
\begin{array}{c:c}
I_{(2,2)} & [\ 0\ ]_{(2,m-2)} \\
\hdashline
[\ 0\ ]_{(n-2,2)} & H_y
\end{array}
\right]
$$

The process is continued until the n by m Householder matrix H_{n-1} is formed. The original matrix A may now be transformed to upper triangular form by the following operation.

(10.7) $H_{n-1}H_{n-2} \ldots H_2 H_1 A = R$

EXAMPLE 9

Apply Householder transformations to determine the Q and R factors for the following 4 × 3 matrix.

$$
A = \left[
\begin{array}{rrr}
1 & 2 & 0 \\
2 & -2 & -1 \\
0 & -1 & 3 \\
-1 & 1 & 1
\end{array}
\right]
$$

SOLUTION

For this matrix, A_1 and X are

$$
A_1 = \left[
\begin{array}{r}
1 \\
2 \\
0 \\
-1
\end{array}
\right], \quad
X = A_1 - \|A_1\| U_1 = \left[
\begin{array}{r}
-1.4495 \\
2.0000 \\
0 \\
-1.0000
\end{array}
\right].
$$

From equation (10.2), the first Householder matrix is

$$H_1 = I - \frac{2XX^T}{||X||^2} =$$

$$\begin{bmatrix} 0.4082 & 0.8165 & 0 & -0.4082 \\ 0.8165 & -0.1266 & 0 & 0.5633 \\ 0 & 0 & 1 & 0 \\ -0.4082 & 0.5633 & 0 & 0.7184 \end{bmatrix}.$$

Now taking the product H_1A gives the following result.

$$H_1A = \begin{bmatrix} 2.4495 & -1.2247 & -1.2247 \\ 0.0000 & 2.4495 & 0.6899 \\ 0 & -1.0000 & 3.0000 \\ 0.0000 & -1.2247 & 0.1551 \end{bmatrix}$$

This result corresponds to the general equation (10.4). From the lower right corner, the B matrix is identified as

$$B = \begin{bmatrix} 2.4495 & 0.6899 \\ -1.0000 & 3.0000 \\ -1.2247 & 0.1551 \end{bmatrix}.$$

Now taking the first column again and applying (10.5), the vector Y is found to be

$$Y = \begin{bmatrix} -0.4660 \\ -1.0000 \\ -1.2247 \end{bmatrix}.$$

From (10.6), the Householder matrix H_y is found.

(10.8) $$H_y = \begin{bmatrix} 0.8402 & -0.3430 & -0.4201 \\ -0.3430 & 0.2639 & -0.9015 \\ -0.4201 & -0.9015 & -0.1041 \end{bmatrix}$$

At this point, we actually do not need to construct H_2 to proceed. This can be done later. The process can continue by computing the product H_yB to move down to the next order.

$$H_yB = \begin{bmatrix} 2.9155 & -0.5145 \\ 0.0000 & 0.4152 \\ 0.0000 & -3.0105 \end{bmatrix}$$

The procedure continues by identifying the next column from the lower right-hand matrix to operate on. This column is

$$C_1 = \begin{bmatrix} 0.4152 \\ -3.0105 \end{bmatrix}.$$

The next generating vector would be called Z if we followed the notational scheme above. Actually, the procedure is repetitious, and we are actually determining new X vectors of lower order. Thus, we should have

$$X = C_1 - \|C_1\|U_1 = \begin{bmatrix} -2.6238 \\ -3.0105 \end{bmatrix}$$

For this X, the Householder matrix is

(10.9) $$H_x = \begin{bmatrix} 0.1366 & -0.9906 \\ -0.9906 & -0.1366 \end{bmatrix}.$$

From (10.8) and (10.9), the desired transformation matrices H_2 and H_3 may be constructed.

$$H_2 = \begin{bmatrix} 1 & 0 & 0 & 0 \\ 0 & 0.8402 & -0.3430 & -0.4201 \\ 0 & -0.3430 & 0.2639 & -0.9015 \\ 0 & -0.4201 & -0.9015 & -0.1041 \end{bmatrix}$$

$$H_3 = \begin{bmatrix} 1 & 0 & 0 & 0 \\ 0 & 1 & 0 & 0 \\ 0 & 0 & 0.1366 & -0.9906 \\ 0 & 0 & -0.9906 & -0.1366 \end{bmatrix}$$

Now as indicated in equation (10.7), the original matrix can be transformed to upper triangular form by computing the product $H_3H_2H_1A$. The result of this operation is

$$R = \begin{bmatrix} 2.4495 & -1.2247 & -1.2247 \\ 0.0000 & 2.9155 & -0.5145 \\ 0.0000 & 0.0000 & 3.0390 \\ 0.0000 & 0.0000 & 0.0000 \end{bmatrix}$$

Now let us return to the idea of QR factorization. As stated in equation (10.1), the given matrix A may be expressed as the product of the two factors Q

and R, where Q is an orthogonal matrix and R is upper triangular. From equation (10.7), we may solve for R for which we have the following result.

(10.10) $A = H_1 H_2 H_3 R$

In solving for R, we have made use of the fact that the Householder matrices are orthogonal and symmetric, hence $H_i^{-1} = H_i^T = H_i$. If we now compare equations (10.1) and (10.10), we see that $Q = H_1 H_2 H_3$. In general, we may conclude that

(10.11) $Q = H_1 H_2 \ldots H_n,$

where n is the number of rows in the given matrix A.

Calculating Q for our example problem yields

$$Q = \begin{bmatrix} 0.4082 & 0.8575 & 0.3097 & -0.0461 \\ 0.8165 & -0.3430 & -0.0581 & 0.4608 \\ 0.0000 & -0.3430 & 0.9291 & -0.1383 \\ -0.4082 & 0.1715 & 0.1936 & 0.8755 \end{bmatrix}.$$

***

It should be noticed that Q is an orthogonal matrix in that $Q^{-1} = Q^T$, but it is not symmetrical as the Householder matrices were. It is left as an exercise for the reader to verify this as well as show that the product QR does indeed produce the original A matrix.

EXERCISES

Compute the QR factors for the matrices below, and verify your results.

1*. $\begin{bmatrix} 4 & 2 & -2 \\ 2 & 3 & 1 \end{bmatrix}$

2*. $\begin{bmatrix} 8 & 0 & -6 \\ 0 & 3 & 1 \\ -6 & 1 & 8 \end{bmatrix}$

3*.
$$\begin{bmatrix} 9 & -6 & -4 & 5 \\ -6 & 7 & 4 & -3 \\ -4 & 4 & 5 & -2 \\ 5 & -3 & -2 & 6 \end{bmatrix}$$

4*.
$$\begin{bmatrix} 3 & 1 & 0 \\ 1 & 2 & -1 \\ 0 & -1 & 3 \\ -1 & 0 & 2 \end{bmatrix}$$

5.*   For the example problem, show that Q is orthogonal and verify that QR = A.

***

11. REDUCTION TO TRIDIAGONAL OR HESSENBERG FORM

As we will see very shortly the QR factorization can be used in a powerful algorithm for computing the eigenvalues of matrices. For this algorithm to be efficient in its operation, it is very helpful to have the given matrix for which the eigenvalues are desired, converted to either tridiagonal or Hessenberg form. A tridiagonal matrix is like a diagonal matrix except that the diagonals just above and below the main diagonal are filled with nonzero[71] elements as well as the main diagonal. A matrix for which all the elements lying below the first diagonal below the main diagonal are zero is said to be in the Hessenberg form. From the discussion of the previous section, it should be clear that this is accomplished by means of Householder transformations. If the given matrix A is symmetric, then a tridiagonal matrix is produced. If A is nonsymmetric, then a Hessenberg form is the result.

Since the basic method has already been developed in the previous section, the reduction to tridiagonal form will be illustrated directly by means of two examples. First we will examine the reduction of a symmetrical matrix as they occur more frequently than any other type in applied problems. The second example will be a nonsymmetrical matrix.

A given symmetrical matrix A is reduced to tridiagonal form by means of the general transformation

(11.1) $H_n H_{n-1} \ldots H_2 H_1 A H_1 H_2 \ldots H_n,$

[71.]   Some of these elements may be zero, but not all of them.

where the Householder matrices are found in the same way as presented in the previous section. If we introduce the Q matrix of equation (10.25) above, it may be seen that (11.1) is a similarity transformation given by $Q^T A Q$.

EXAMPLE 10

Apply the method of this section to reduce the following symmetric matrix to tridiagonal form.

$$A = \begin{bmatrix} 4 & 2 & -1 & 1 \\ 2 & 3 & 3 & 0 \\ -1 & 3 & 2 & -2 \\ 1 & 0 & -2 & 3 \end{bmatrix}$$

SOLUTION

Since the object is to reduce the last two elements of the first column and the first row to zero, we begin by selecting our first vector as being

$$A_1 = \begin{bmatrix} 2 \\ -1 \\ 1 \end{bmatrix}.$$

From equation (10.4) of the last section, X is found to be

$$X = A_1 - \|A_1\| \begin{bmatrix} 1 \\ 0 \\ 0 \end{bmatrix} = \begin{bmatrix} -0.4495 \\ -1.0000 \\ 1.0000 \end{bmatrix}.$$

Now using (10.2) of the previous section and this result, H_x may be found.

$$H_x = \begin{bmatrix} 0.8165 & -0.4082 & 0.4082 \\ -0.4082 & 0.0918 & 0.9082 \\ 0.4082 & 0.9082 & 0.0918 \end{bmatrix}$$

Thus, the first desired Householder matrix is

$$H_1 = \begin{bmatrix} 1 & 0 & 0 & 0 \\ 0 & 0.8165 & -0.4082 & 0.4082 \\ 0 & -0.4082 & 0.0918 & 0.9082 \\ 0 & 0.4082 & 0.9082 & 0.0918 \end{bmatrix}$$

To proceed to the next level, the first reduction must be evaluated. This is

(11.2) H_1AH_1 = $\begin{bmatrix} 4.0000 & 2.4495 & 0.0000 & 0.0000 \\ 2.4495 & 1.5000 & 1.4289 & 1.4289 \\ 0.0000 & 1.4289 & 2.4335 & -2.7500 \\ 0.0000 & 1.4289 & -2.7500 & 4.0665 \end{bmatrix}.$

We now select

$$A_1 = \begin{bmatrix} 1.4289 \\ 1.4289 \end{bmatrix}.$$

From this A_1, the next X vector is calculated, which then gives H_x, from which H_2 is determined.

$$H_2 = \begin{bmatrix} 1 & 0 & 0 & 0 \\ 0 & 1 & 0 & 0 \\ 0 & 0 & 0.7071 & 0.7071 \\ 0 & 0 & 0.7071 & -0.7071 \end{bmatrix}$$

Applying this transformation to (11.2) or essentially forming (11.1), the desired reduction to tridiagonal form is achieved.

$$H_2H_1AH_1H_2 = \begin{bmatrix} 4.0000 & 2.4495 & 0.0000 & 0.0000 \\ 2.4495 & 1.5000 & 2.0207 & 0.0000 \\ 0.0000 & 2.0207 & 0.7071 & -0.8165 \\ 0.0000 & 0.0000 & -0.8165 & 6.0000 \end{bmatrix}$$

As may be seen in this expression the resulting matrix is tridiagonal. Because the given matrix was symmetric and fourth order only two Householder transformation matrices were required.

***

When the given matrix is not symmetric, the procedure produces a Hessenberg form as previously stated. This result is illustrated in the next example.

EXAMPLE 11

Reduce the following matrix to Hessenburg form.

(11.3)
$$A = \begin{bmatrix} 3 & -2 & 1 & 0 \\ 1 & 2 & 0 & -3 \\ 2 & 1 & 1 & 0 \\ 0 & 1 & -1 & 2 \end{bmatrix}$$

SOLUTION

As the method describes above, we select as the beginning vector

$$A = \begin{bmatrix} 1 \\ 2 \\ 0 \end{bmatrix}.$$

By following exactly the same steps just completed in the previous example, one finds H_1.

$$H_1 = \begin{bmatrix} 1 & 0 & 0 & 0 \\ 0 & 0.4472 & 0.8944 & 0 \\ 0 & 0.8944 & -0.4472 & 0 \\ 0 & 0 & 0 & 1.0000 \end{bmatrix}$$

Computing the product $H_1 A H_1^T$ gives the following result.

$$H_1 A H_1^T = \begin{bmatrix} 3.0000 & 0.0000 & -2.2361 & 0 \\ 2.2361 & 1.6000 & 1.2000 & -1.3416 \\ -0.0000 & 0.2000 & 1.4000 & -2.6833 \\ 0 & -0.4472 & 1.3416 & 2.0000 \end{bmatrix}$$

The next vector is

$$A_1 = \begin{bmatrix} 0.2000 \\ -0.4472 \end{bmatrix},$$

which leads to H_2 as

$$H_2 = \begin{bmatrix} 1 & 0 & 0 & 0 \\ 0 & 1 & 0 & 0 \\ 0 & 0 & 0.4083 & -0.9129 \\ 0 & 0 & -0.9129 & -0.4083 \end{bmatrix}.$$

The desired reduction of (11.3) may now be computed.

$$H_2 H_1 A H_1^T H_2^T = \begin{bmatrix} 3.0000 & 0.0000 & -0.9129 & 2.0412 \\ 2.2361 & 1.6000 & 1.7146 & -0.5477 \\ 0 & 0.4899 & 2.4000 & 1.7889 \\ 0.0000 & 0.0000 & -2.2361 & 1.0000 \end{bmatrix}$$

The result is the expected Hessenberg form.

EXERCISES

Reduce the following matrices to tridiagonal or Hessenberg form.

1*. $\begin{bmatrix} 4 & 2 & -2 \\ 2 & 3 & 1 \end{bmatrix}$

2*. $\begin{bmatrix} 8 & 0 & -6 \\ 0 & 3 & 1 \\ -6 & 1 & 8 \end{bmatrix}$

3*. $\begin{bmatrix} 9 & -6 & -4 & 5 \\ -6 & 7 & 4 & -3 \\ -4 & 4 & 5 & -2 \\ 5 & -3 & -2 & 6 \end{bmatrix}$

4*. $\begin{bmatrix} 3 & 1 & 0 \\ 1 & 2 & -1 \\ 0 & -1 & 3 \\ -1 & 0 & 2 \end{bmatrix}$

***

12. APPLICATIONS OF QR FACTORIZATION

QR factorization is the most common algorithm used in nearly all software packages available at the present time for evaluating eigenvalues and related applications. As illustrations, we will examine its application to a least squares problem and one of finding eigenvalues of a given matrix.

The method of least squares was presented in section 2 of chapter III. The example problem presented there dealt with finding a linear equation for predicting the weight of a man of average build from his height as represented in the following linear equation.

$$(12.1) \qquad w = a_0 + a_1 h$$

This equation is conceptually written down for each of the nine sets of measurements, and the object is to find the best values for the intercept and slope parameters a_0 and a_1, which give the best fit for all the data in the least squares sense. The height and weight measurements for nine men were given as the data values.

For the previously given data, equation (12.1) represents nine equations for the two unknowns a_0 and a_1. In matrix form, this is

$$(12.2) \qquad HA = W,$$

where

$$H = \begin{bmatrix} 1 & 72 \\ 1 & 68 \\ 1 & 66 \\ 1 & 74 \\ 1 & 62 \\ 1 & 70 \\ 1 & 64 \\ 1 & 76 \\ 1 & 64 \end{bmatrix}, \quad W = \begin{bmatrix} 174 \\ 152 \\ 154 \\ 180 \\ 135 \\ 161 \\ 140 \\ 174 \\ 157 \end{bmatrix}, \quad A = \begin{bmatrix} a_0 \\ a_1 \end{bmatrix}.$$

To apply QR factorization to solving equation (12.2) for the coefficient vector A, the matrix H is factored as

$$H = QR$$

so that (12.2) may be written as

$$QRA = W.$$

Now taking advantage of the fact that Q is orthogonal, this equation may be written as follows.

(12.3) $$RA = Q^T W$$

Since R is upper triangular, the elements of the A vector may be solved for by back substitution. Carrying[72] out the factorization for the example cited produces the following result for (12.3). For convenience, only the first two rows need be retained since there are only two unknowns to solve for.

$$RA = \begin{bmatrix} 3 & 205.3333 \\ 0 & 13.7921 \end{bmatrix} \begin{bmatrix} a_0 \\ a_1 \end{bmatrix},$$

$$Q^T W = \begin{bmatrix} 475.6667 \\ 39.8618 \end{bmatrix}$$

Solving the equations by back substitution gives the same answer as obtained in section 2 of chapter III, that is, $a_0 = -39.2617$, $a_1 = 2.8902$.

Determination of Eigenvalues by QR Factorization

The application of QR factorization to find the eigenvalues of a matrix is sometimes called the QR algorithm. To illustrate the method, consider the standard eigenvalue problem, which is defined as

$$AX = \mu X.$$

The A matrix for which the eigenvalues are desired is factored as follows.

$$A = Q_1 R_1$$

The subscripts are used to denote the first factors since the process is to be continued. A new matrix is formed by multiplying the factors in the reverse order.

(12.4) $$A_1 = R_1 Q_1$$

---

[72] All the calculations for this example were carried out using matrix functions in an EXCEL spreadsheet program. It should be noted that instead of actual zero values occurring in the tridiagonalization process, numbers such as 2.71 $E-16$ occurred. This is due to the fact that the computer carried sixteen-digit accuracy for arithmetic and round off produced the noted errors or differences from exact zero values.

This product is possible since A must be square. This new matrix is now factored to give

$$A_1 = Q_2 R_2$$

These new factors are now multiplied in the reverse order to form another matrix.

$$A_2 = R_2 Q_2$$

For the continued process, the following general relation may be written.

(12.5) $A_i = R_i Q_i = Q_{i+1} R_{i+1}$

It may be shown that as the process continues, the matrices A_i converge to a quasi-upper triangular matrix of the following schematic form.

$$\text{Lim}_{i \to \infty} A_i = \begin{bmatrix} \mathrm{x} & \mathrm{x} & \mathrm{x} & \cdots & \mathrm{x} & \mathrm{x} \\ * & \boxed{\mathrm{x} \ \ \mathrm{x}} & \cdots & \mathrm{x} & \mathrm{x} \\ 0 & * & \mathrm{x} & \cdots & \mathrm{x} & \mathrm{x} \\ 0 & 0 & * & \cdots & \mathrm{x} & \mathrm{x} \\ \vdots & \vdots & \vdots & \ddots & \vdots & \vdots \\ 0 & 0 & 0 & \cdots & * & \mathrm{x} \end{bmatrix}$$

The x's denote elements of the upper triangular portion that have general values. The asterisks may or may not be zeros. The small box is used to denote a two by two array of elements along the main diagonal. Depending on the form of the original A matrix, the above result may have three possible forms.

If the original matrix is real and symmetric, then the QR algorithm produces a diagonal matrix. The diagonal elements are the eigenvalues of A. If A is real and nonsymmetric with real eigenvalues, the matrix above is upper triangular with the eigenvalues again appearing along the main diagonal.

When A is real and nonsymmetric with some eigenvalues being complex conjugate pairs and real arithmetic is used, the above general form results. Each complex conjugate pair of roots results in a two by two block like the one shown schematically above. The eigenvalues of the two by two block are the eigenvalues of the original matrix and may be computed by a simple direct calculation.

If complex arithmetic is used for the computations, then the resulting matrix will be simply upper triangular with all the eigenvalues appearing along the main diagonal. The same result as just described will occur if the original A matrix is complex.

If A is tridiagonal or in Hessenberg form, then the process converges more rapidly. The higher the order of the matrix, the greater the time saving. Thus, most modern algorithms for finding eigenvalues employ the Householder transformation method as a preprocessor to reduce A to tridiagonal or Hessenberg form prior to initiating the QR process.

Another aspect of the QR algorithm is that the eigenvalues that appear along the main diagonal will occur in order of decreasing magnitude.

EXAMPLE 12

Determine the eigenvalues for the following matrix.

$$A = \begin{bmatrix} 4 & -1 & 2 \\ -1 & -2 & 1 \\ 2 & 1 & 3 \end{bmatrix}$$

SOLUTION

For this matrix, the two Householder matrices required to form R_1 and Q_1 are the following:

$$H_1 = \begin{bmatrix} 0.8729 & -0.2182 & 0.4364 \\ -0.2182 & 0.6254 & 0.7491 \\ 0.4364 & 0.7491 & -0.4983 \end{bmatrix}$$

$$H_2 = \begin{bmatrix} 1 & 0 & 0 \\ 0 & -0.1157 & -0.9933 \\ 0 & -0.9933 & 0.1157 \end{bmatrix}$$

From these matrices, we find R_1 and Q_1 to be the following matrices.

$$R_1 = \begin{bmatrix} -4.5826 & 0 & -2.8368 \\ 0 & 2.4495 & -0.4082 \\ 0 & 0 & 2.4054 \end{bmatrix}$$

$$Q_1 = \begin{bmatrix} -0.8729 & -0.4082 & -0.2673 \\ 0.2182 & -0.8165 & 0.5345 \\ -0.4364 & 0.4082 & 0.8018 \end{bmatrix}$$

The first iterated matrix may now be found using equation (12.4).

$$A_1 = \begin{bmatrix} 5.2381 & 0.7127 & -1.0498 \\ 0.7127 & -2.1667 & 0.9820 \\ -1.0498 & 0.9820 & 1.9286 \end{bmatrix}$$

Repeating the process, we find the next iterate to be

$$A_2 = R_2 Q_2 = \begin{bmatrix} 5.5131 & -0.3830 & 0.3944 \\ -0.3830 & -2.3138 & 0.8380 \\ 0.3944 & 0.8380 & 1.8007 \end{bmatrix}.$$

After 38 iterations, the diagonal values converge to four-decimal-place accuracy. The result is

$$A_{38} = \begin{bmatrix} 5.5658 & -0.0000 & -0.0000 \\ -0.0000 & -2.5035 & 0.0000 \\ 0.0000 & 0.0000 & 1.9377 \end{bmatrix}$$

From this matrix, the eigenvalues are seen to be $\mu_1 = 5.5658$, $\mu_2 = -2.5035$, and $\mu_3 = 1.9377$.

EXAMPLE 13

Find the eigenvalues for the following matrix, which is already in Hessenberg form.

$$A = \begin{bmatrix} 3 & -2 & 1 & 0 \\ 1 & 2 & 0 & -3 \\ 0 & 1 & 1 & 0 \\ 0 & 0 & -1 & 2 \end{bmatrix}$$

SOLUTION

After twenty iterations of the QR algorithm, the process converges to four-digit accuracy, giving the following result.

$$A_{20} = \begin{bmatrix} 3.7379 & -1.2500 & -1.1528 & 0.1761 \\ 1.5487 & 2.0033 & 1.2046 & -2.3923 \\ 0.0000 & 0.0000 & 0.3705 & -1.6963 \\ 0 & -0.0000 & 1.7145 & 1.8883 \end{bmatrix}$$

In this result there are two two-by-two blocks indicating that there will be two pairs of complex conjugate eigenvalues. The upper block is

$$B_1 = \begin{bmatrix} 3.7379 & -1.2500 \\ 1.5487 & 2.0033 \end{bmatrix}.$$

The eigenvalues of this submatrix are $2.8706 \pm 1.0880i$. Similarly the eigenvalues of the lower block are $1.1294 \pm 1.5272i$, which completes the set of four eigenvalues of the original matrix.

Shifting and Deflation to Accelerate Convergence

In addition to converting a given matrix to tridiagonal or Hessenberg form before applying the QR algorithm as a means of accelerating convergence, further improvements are available. The first improvement is based on eigenvalue shifting. Given a matrix A and a scalar r, a new matrix may be formed as $B = A - rI$. The eigenvalues of B are the eigenvalues of A reduced by the factor r. The eigenvectors of A and B are the same. If the value of r is selected to be a close estimate of the smallest eigenvalue of A, then the smallest eigenvalue of B is close to zero. The process is applied to the shifted matrix B. Since the rate of convergence of the QR algorithm depends on the ratio of the magnitudes of the smallest to next largest eigenvalues, the rate is increased as a result of this shift. Once it has converged the factor r is added back to obtain the smallest eigenvalue of A. Convergence is deemed to occur when all the elements of the last row of B reach zero to a preselected number of decimal digits except for $B_i(n, n)$, which is converging to the smallest eigenvalue of B. At this point, the smallest eigenvalue of A is given by $r + B_i(n, n)$.

To continue the process, the B matrix is deflated by one order by eliminating the last row and columns. This gives a submatrix $SB1$ whose order is one less than that of the original A matrix. Since the new matrix $SB1$ has been operated on by the QR algorithm, its last element should be an approximate value of its smallest eigenvalue. This means that $SB1$ can be shifted by $SB1(n-1, n-1)$. The QR algorithm is now applied to this new shifted matrix until convergence is again reached. The second smallest eigenvalue of A is now given by $r + B_i(n, n) + SB1_i(n-1, n-1)$. This matrix is deflated again and the procedure repeated until a two by two matrix is obtained from which the eigenvalues can be calculated directly.

Modern algorithms only follow this procedure roughly as they utilize a number of sophisticated and specialized aspects of matrix algebra and programming techniques. The purpose here is to give the reader only a general idea of how the method proceeds as the details of any given algorithm are beyond an introductory level text.

To illustrate the approach just described, consider the following real nonsymmetrical matrix whose eigenvalues are exactly 1, 2, 3, and 4.

$$A = \begin{bmatrix} 7 & -3 & 5 & -6 \\ -1 & 4 & -1 & 1 \\ 1 & 0 & 3 & -1 \\ 5 & -3 & 5 & -4 \end{bmatrix}$$

If the QR algorithm is applied to this matrix directly without any shifting or deflation, it takes 33 factorizations to obtain convergence to four-decimal-place accuracy. To apply the technique of shifting and deflation, a first rough estimate of the smallest eigenvalue is required. This is obtained by applying one QR factorization step to the matrix A. The result is

$$A_1 = \begin{bmatrix} 4.1579 & -0.9453 & 1.1693 & 13.6122 \\ -0.3725 & 3.4519 & -0.1560 & -0.9404 \\ 0.2606 & 0.3652 & 2.1582 & 1.1135 \\ -0.1777 & 0.0317 & -0.0979 & 0.2320 \end{bmatrix}.$$

The first shifting factor is $A_1(4,4) = 0.2320$, and the first shifted matrix is $AS = A_1 - A_1(4, 4)I$. Seven steps of the process indicated in (12.5) are needed to reach convergence with the following result.

$$AS_7 = \begin{bmatrix} 3.7301 & -0.4087 & 0.9229 & -12.8977 \\ -0.0890 & 2.8082 & -0.2028 & -4.9719 \\ 0.0018 & 0.0193 & 1.7657 & 1.3889 \\ 0.0000 & 0.0000 & -0.0000 & 0.7680 \end{bmatrix}$$

Now the smallest eigenvalue of A is given by the sum of the first shifting factor, $A_1(4, 4)$, and the last element in AS_7, $AS_7(4, 4)$, or 1.0000.

This last matrix is deflated to given the following submatrix.

$$SA1 = \begin{bmatrix} 3.7301 & -0.4087 & 0.9229 \\ -0.0890 & 2.8082 & -0.2028 \\ 0.0018 & 0.0193 & 1.7657 \end{bmatrix}$$

From this matrix, the second shifting factor is $SA1(3,3) = 1.7657$. After shifting this matrix, only two QR factorization steps are required for convergence. The result is

$$SA1_2 = \begin{bmatrix} 1.9946 & -0.3251 & 0.9402 \\ -0.0236 & 1.0101 & -0.1585 \\ 0.0000 & 0.0000 & 0.0023 \end{bmatrix}.$$

From this matrix, the second smallest eigenvalue is the sum of the two shifting factors and the last element $SA1_2(3, 3)$ which gives $0.2320 + 1.7657 + 0.0023 = 2.0000$. The last two eigenvalues are calculated directly from the two by two matrix obtained by deflating the above matrix. These final two values are 3.0000 and 4.0000. It should be noted that this process of shifting and deflation required only 10 QR factorization steps to reach four-decimal-place accuracy in comparison to 33 by applying the method directly.

EXERCISES

1.*   Determine the best

    a.   quadratic and
    b.   cubic

least squares that fit the following data.

x	-1	0	1	2	3
y	-22	-1	4	5	14

2.*   A test run has been made of a turbine-powered vehicle on the Bonneville Salt Flats in Utah. The results of six time trials are given below.

Time, t (s)	10.0	20.0	30.0	40.0	50.0	60.0
Speed, v (m/s)	15.1	32.2	63.4	84.5	118.0	139.0

Determine the best least squares linear equation to predict the speed s on the basis of the time t.

Use the QR algorithm to determine the eigenvalues for the matrices below.

3*. $\begin{bmatrix} -1 & -1 & -1 \\ -1 & -1 & 1 \\ -1 & 1 & -1 \end{bmatrix}$

4*. $\begin{bmatrix} 1 & 0 & 1 \\ 0 & 2 & 0 \\ 1 & 0 & 1 \end{bmatrix}$

5*. $\begin{bmatrix} 1 & -1 & 1 \\ -1 & 1 & -1 \\ 1 & -1 & 1 \end{bmatrix}$

6*. $\begin{bmatrix} 2 & -1 & 0 \\ -1 & 2 & -1 \\ 0 & -1 & 2 \end{bmatrix}$

7*. $\begin{bmatrix} 2 & 1 & 0 \\ -1 & 0 & 1 \\ 1 & 3 & 1 \end{bmatrix}$

***

13. EIGENVECTOR DETERMINATION: THE SHIFTED INVERSE POWER METHOD

As we have just discussed, the QR algorithm is the most widely employed procedure for determining eigenvalues. However, once the eigenvalues are known, the problem is to find the associated eigenvectors. In this section, the most common approach to computing the eigenvectors associated with given eigenvalues is presented. This method is known as the shifted inverse power method. Before presenting this method, the case of real symmetric matrices will be discussed since the QR algorithm provides the eigenvectors directly.

As mentioned above and illustrated by the first example problem, the QR algorithm produces a diagonal matrix by means of the process given by equation (12.5). Let us assume that at the ith step, the process has produced a diagonal matrix of eigenvalues, D_i, from an original real symmetric matrix A. From equation (12.5), we may write

(13.1) $A_i = D_i = R_i Q_i.$

From the previous step, we have

$$A_{i-1} = Q_i R_i.$$

Solving for R_i and substituting the result into (13.1) yields

(13.2) $$D_i = Q_i^T A_{i-1} Q_i.$$

But from the previous step, we also have $A_{i-1} = R_{i-1} Q_{i-1}$. Substituting this into (13.2) gives

$$D_i = Q_i^T R_{i-1} Q_{i-1} Q_i.$$

If this process of substituting from previous steps is carried out to the beginning, we will arrive at the following result.

(13.3) $$D_i = Q_i^T Q_{i-1}^T \dots Q_1^T A Q_1 \dots Q_{i-1} Q_i$$

Now let the continued product of orthogonal matrices be denoted by M, which must also be orthogonal.

(13.4) $$M = Q_1 Q_2 \dots Q_{i-1} Q_i$$

Introducing this into (13.3) now gives the result

$$D_i = M^T A M.$$

From our prior discussion of the eigenvalue problem, this result may be seen as a similarity transformation, which means that M is the modal matrix of A and hence contains all the associated eigenvectors of A. In the QR algorithm, it is a straightforward matter to accumulate the continued products of the Q matrices. This means that when applied to a real symmetric matrix to determine its eigenvalues, the QR algorithm also yields the eigenvectors as well.

EXAMPLE 14

Apply the QR method to determine the eigenvalues for the matrix of example 12 above. For that example, the given matrix is

$$A = \begin{bmatrix} 4 & -1 & 2 \\ -1 & -2 & 1 \\ 2 & 1 & 3 \end{bmatrix}.$$

SOLUTION

For this matrix, the modal matrix given by (13.4) is

$$M = \begin{bmatrix} 0.7930 & 0.2223 & -0.5672 \\ -0.0244 & 0.9419 & 0.3351 \\ 0.6087 & -0.2512 & 0.7524 \end{bmatrix}.$$

The columns of this matrix are the eigenvectors of A associated with the eigenvalues in descending order.

***

The Shifted Inverse Power Method

In section 4 of chapter VII, the power method for determining eigenvalues of a given matrix A was presented. As discussed, this method converges to the eigenvalue of greatest magnitude. It is a fairly easy exercise to show that the same procedure applied to A^{-1} will converge to the eigenvalue of smallest magnitude. If μ_1 and μ_2 are the eigenvalues of smallest and next to smallest magnitudes, then the rate of convergence of this process depends upon the ratio $|\mu_1|/|\mu_2|$. The smaller the ratio, the faster the convergence. Interest in determining the smallest eigenvalue in many areas of application comes from the fact that it is related to the lowest frequency or response time of a system. The iteration process of this method is characterized by the following difference equation.

(13.5) $$V_{k+1} = A^{-1}X_k,$$

where X_k is the kth estimate of the eigenvector associated with the smallest eigenvalue. The vector V_{k+1} simply results from the evaluation of the product in (13.5) and is used to generate the next estimate from the relation

(13.6) $$X_{k+1} = \frac{V_{k+1}}{||V_{k+1}||},$$

where typically, either the L_2 or L_∞ norm is used. Instead of computing the inverse of the matrix A, equation (13.5) is usually written in the form

$$AV_{k+1} = X_k.$$

This system is solved by LU factorization since the factors need only be computed once.

The shifted inverse power method for determining the eigenvectors of a matrix A is based on the following difference equation.

(13.7) $$(\mu I - A)V_{k+1} = K(\mu)V_{k+1} = X_k$$

The vector X_k is a normalized estimate of an eigenvector, and the next estimate is computed from the solution to (13.7), V_{k+1}. The next estimate is given by (13.6).

If an exact value of an eigenvalue of A is used for μ in (13.7), the characteristic matrix $K(\mu)$ will be singular and (13.7) could not be solved. However, by using an approximate result for an eigenvalue, $K(\mu)$ will be ill conditioned, but the process will converge to the associated eigenvector. The reason for this is that small discrepancies are minimized or swept out in the iteration process.

As an illustration, consider the matrix employed in the third example of the previous section. For this problem, we considered the following matrix.

$$A = \begin{bmatrix} 7 & -3 & 5 & -6 \\ -1 & 4 & -1 & 1 \\ 1 & 0 & 3 & -1 \\ 5 & -3 & 5 & -4 \end{bmatrix}$$

We know the smallest eigenvalue is exactly 1, and the QR algorithm of the previous section gave the result 1.0000. To ensure we are a little off the true value, let us introduce a 10% deviation and take 1.1 for the first value of μ. For this value, the characteristic matrix of (13.7) is as follows.

$$K(\mu) = \mu I - A = \begin{bmatrix} -5.9 & 3 & -5 & 6 \\ 1 & -2.9 & 1 & -1 \\ -1 & 0 & -1.9 & 1 \\ -5 & 3 & -5 & 5.1 \end{bmatrix}$$

The L and U factors for this matrix are the following:

$$L = \begin{bmatrix} 1 & 0 & 0 & 0 \\ -0.1695 & 1 & 0 & 0 \\ 0.1695 & 0.2126 & 1 & 0 \\ 0.8475 & -0.1914 & 0.6767 & 1 \end{bmatrix}$$

$$U = \begin{bmatrix} -5.9 & 3 & -5 & 6 \\ 0 & -2.3915 & 0.1525 & 0.0169 \\ 0 & 0 & -1.0850 & -0.0206 \\ 0 & 0 & 0 & 0.0324 \end{bmatrix}$$

The reader may verify that $A = LU$.

At this point, the question must be addressed as to what should be used as the initial trial vector. The most common choices are usually $V_0 = [1, 0, 0, 0]^T$ or $V_0 = [1, 1, 1, 1]^T$. As stated above, either the L_2 or L_∞ norm is used for the normalizing process. Using the first suggested form for V_0 and the L_2 norm, the following result is obtained after six iterations to four-decimal-place accuracy for the eigenvector associated with the lowest eigenvalue of 1.0000.

$$E_1 = \begin{bmatrix} 1.0000 \\ 0.0000 \\ -0.0000 \\ 1.0000 \end{bmatrix}$$

From this result, it should be clear that the true eigenvector is $E_1 = [1, 0, 0, 1]^T$.

Using estimates for the second through fourth eigenvalues that are adjusted to be 10% high gives the following associated eigenvectors for the true eigenvalues of 2, 3, and 4.

$$E_2 = \begin{bmatrix} 1.0000 \\ 0.0000 \\ -1.0000 \\ 0.0000 \end{bmatrix} \quad E_3 = \begin{bmatrix} 1.0000 \\ 1.0000 \\ 1.0000 \\ 1.0000 \end{bmatrix} \quad E_4 = \begin{bmatrix} 1.0000 \\ -1.0000 \\ -0.0000 \\ 1.0000 \end{bmatrix}$$

EXERCISES

Apply the shifted inverse power method to determine eigenvectors for the matrices below.

1*. $\begin{bmatrix} -1 & -1 & -1 \\ -1 & -1 & 1 \\ -1 & 1 & -1 \end{bmatrix}$

2*. $\begin{bmatrix} 1 & 0 & 1 \\ 0 & 2 & 0 \\ 1 & 0 & 1 \end{bmatrix}$

3*. $\begin{bmatrix} 1 & -1 & 1 \\ -1 & 1 & -1 \\ 1 & -1 & 1 \end{bmatrix}$

4*. $\begin{bmatrix} 2 & -1 & 0 \\ -1 & 2 & -1 \\ 0 & -1 & 2 \end{bmatrix}$

5*. $\begin{bmatrix} 2 & 1 & 0 \\ -1 & 0 & 1 \\ 1 & 3 & 1 \end{bmatrix}$

ADDITIONAL EXERCISES

1.*   Solve for X in the example problem of section 4 by direct calculation of $X = (L^T)^{-1}L^{-1}B$ and $X = A^{-1}B$, and compare the answers to the one given. Discuss any differences in significant figures.

2.*   Solve the following system by the method of Gauss-Seidel iteration, continuing the iterative process until the maximum difference between successive value of x_1, x_2, or x_3 is less than 0.02. Does this mean that the approximate solution is within 0.02 of the exact solution?

$$10x_1 + 2x_2 + 6x_3 = 28$$
$$x_1 + 10x_2 + 9x_3 = 7$$
$$2x_1 - 7x_2 - 10x_3 = -17$$

3.*   Solve the following system by the method of Gauss-Seidel iteration, continuing the iterative process until the maximum difference between the successive value of x_1, x_2, or x_3 is less than 0.02. Compare the convergence rate with that of exercise 7. Discuss the difference.

$$20x_1 + 2x_2 + 6x_3 = 38$$
$$x_1 + 20x_2 + 9x_3 = -23$$
$$2x_1 - 7x_2 - 20x_3 = -57$$

4.*   Consider a system of linear algebraic equations $AX = B$. The modified Gauss-Seidel method based on relaxation for solving this system proceeds as follows. Let $\overline{X_i}$ be an iterated solution obtained from equation (6.5) and X_{i-1} be the iterated solution of the modified process at the previous step. A differential vector is defined as

$$\Delta X_i = \overline{X_i} - X_{i-1}.$$

The next iterated solution of the modified process is now given by

$$X_i = \overline{\Delta X_i} + rX_{i-1},$$

where r is a scalar called the relaxation parameter. If $r = 1$, then $X_i = \overline{X_i}$. The rate of convergence of the modified process is controlled by the value of r. Typically, the most rapid convergence is produced when it has a value slightly less than one.

Solve the following second-order system to five or six significant figure accuracy by this modified Gauss-Seidel or relaxation process for a range of values of r from 0.4 to 1.2. Make a plot of the number of iterations to reach convergence versus r, and determine the value that produces the most rapid convergence.

$$x + 2y = 3, x - 4y = -3$$

5.   Show that the sufficient but not necessary conditions for convergence of the Gauss-Seidel iteration method are automatically satisfied for any resistive electrical circuit.

6.   Show that the sufficient but not necessary conditions for convergence of the Gauss-Seidel iteration method are automatically satisfied for any general electrical circuit with linear complex impedances in all branches. Note that this result implies that these convergence conditions are met

for *all* physical linear systems such as mechanical, thermal, fluid, acoustic, etc.

7.   Plot the sets of equations given as the two example problems in section 7 on the same graph of x_1 versus x_2, and discuss their implications.

8.   Consider the following two dimensional linear algebraic system.

$$\begin{bmatrix} 1 & 1 \\ 1 & c \end{bmatrix} X = \begin{bmatrix} 4 \\ -3 \end{bmatrix}$$

The constant c is a parameter $\neq 1$. Calculate the condition number of the coefficient matrix, and sketch the lines of the two equations in x_1, x_2 space. Discuss the problem of finding solutions as the value of c approaches one.

9.   If H_x denotes a Householder matrix based on a vector X, prove that $H_x^T = H_x$ and $H_x^{-1} = H_x$. Hint: expand the product $H_x H_x$.

10.* Apply the QR algorithm to determine the eigenvalues of the following matrix.

$$A = \begin{bmatrix} 7 & -3 & -6 & 5 \\ -1 & 4 & 1 & -1 \\ 5 & -3 & -4 & 5 \\ 1 & 0 & -1 & 3 \end{bmatrix}$$

It should take about 33 steps to reach upper triangular form with four-decimal-place accuracy. Note the diagonal. Continue the process for at least 30 more steps and note the diagonal again. Discuss what happens between the 33rd and 63rd steps? If possible, it is convenient to pause after each step so the elements of the reformed A matrix may be observed.

11.   Given the arbitrary second-order matrix below, derive the general expression for its eigenvalues.

$$A = \begin{bmatrix} a_{11} & a_{12} \\ a_{21} & a_{22} \end{bmatrix}$$

What happens to this result if the elements are complex? These results are convenient for use in the method discussed in section 12.

Use the QR algorithm to determine the eigenvalues and eigenvectors for the following matrices.

12*.
$$\begin{bmatrix} 4 & -1 & -2 & 0 \\ -1 & 3 & -1 & 0 \\ -2 & -1 & 5 & -2 \\ 0 & 0 & -2 & 4 \end{bmatrix}$$

13*.
$$\begin{bmatrix} 2 & 3 & -2 & -5 \\ 0 & -3 & 0 & 1 \\ 2 & -5 & 2 & -1 \\ 0 & 1 & 0 & -3 \end{bmatrix}$$

INDEX

A

adjoint matrix, 243
adjoint method, 9, 417-18
analogs, 469
attenuation function, 484

B

back substitution, 31, 93
basis, 126
biological growth problems, 384, 391, 393, 397, 559
block diagonal matrix, 383
Bocher, M., 295
Bocher's formulas, 7, 295
branch current, 164
branch impedances, 163-64, 559

C

capacitance, 164, 447
Cartesian coordinate system, 256, 280, 559
Cayley, A., 7, 9, 13, 84, 222
Cayley-Hamilton theorem, 7, 326
characteristic determinant, 6, 293, 332
characteristic equation, 252
characteristic function, 252
characteristic impedance, 483
characteristic matrix, 252
characteristic polynomial, 252
characteristic roots, 252
circulant matrices, 9, 298
cofactor, 223
column pivoting, 106
complex eigenvalues, 294, 411
condition number, 152, 521, 523
continuum or distributed variable, 177
convergent series, 321, 559
coordinate systems. *See* basis
Cramer's rule, 6, 241
Crout factorization, 498, 505
cutoff frequency, 485-86

D

dashpot, 467, 559
determinants
    definition of, 222
    fundamental properties of, 229
    determination of eigenvalues by QR factorization, 543
difference equations, 188, 384
    with constant coefficients, 384, 386
    of the kth order with constant coefficients, 390
divergent, 393
Doolittle factorization, 498, 559
dualogs, 469, 559

E

eigenvalue problem, 6, 9, 251, 254, 256
eigenvalues, 293
    distinct, 328, 360
    of the modified matrix, 326
    multiple, 355
eigenvector expansion, 7, 9, 392, 407, 413
eigenvectors, 253
electrical circuit, 163
electrostatics, 178, 559
elementary operations, 63
equilibrium of force systems, 197
Euler's equation, 412

F

finite differences, 7, 9, 178, 318, 477, 559
fluid flow, 437
forcing function, 399, 402, 417, 419, 422-23, 425, 430, 448, 467
four-terminal network, 477, 483, 486
free and forced responses, 419
fundamental frequency, 462
fundamental mode, 462

BIBLIOGRAPHY

Aitken, A. C. *Determinants and Matrices*, 9th ed. Edinburgh: Oliver and Boyd, 1956.

Antosik, P. and C. Swartz. *Matrix Methods in Analysis: Lecture Notes in Mathematics*, no. 1113. New York: Springer-Verlag, 1985.

Barnett, S. *Matrix Methods for Scientists and Engineers*. New York: McGraw-Hill Book Company, 1979.

Barnett, S. and C. Storey. *Matrix Methods in Stability Theory*. New York: Barnes & Noble, 1970.

Bellman, R. *Introduction to Matrix Analysis*, 2nd ed. New York: McGraw-Hill Book Co., 1970.

Bodewig, E. *Matrix Calculus*, New York: Interscience Publishers, Inc., 1956.

Bronson, R. *Matrix Methods: An Introduction*, 2nd ed. New York: Academic Press, 1991.

Burden, R. L., J. D. Faires, and A. C. Reynolds. *Numerical Analysis*, 2nd ed. Boston, Massachusetts: Prindle, Webster and Schmidt, 1981.

Burden, R. L. and P. Rabinowitz. *A First Course in Numerical Analysis*, second edition. McGraw-Hill Book C.: New York, 1978.

Bocher, M. *Introduction to Higher Algebra*. New York: The MacMillan Co., 1931.

Campbell, H. G. *Linear Algebra with Applications*, 2nd ed. Englewood Cliffs, New Jersey: Prentice-Hall, 1980.

Chio, F. "M°emoire sur les fonctions connues sous le nom de r°esultantes ou de d°eterminants." Turin: E. Pons, 1853.

Coleman, T. F. and C. Van Loan. *Handbook for Matrix Computations*. Philadelphia: SIAM, 1988.

Courant, R. and D. Hilbert. *Methods of Mathematical Physics*. New York: Interscience, 1953.

Crandall, S. H. *Engineering Analysis*. New York: McGraw-Hill Book Co., 1956.

Cullen, C. G., *Matrices and Linear Transformations*, Reading, Massachusetts: Addison-Wesley, 1966.

Cullen, C. G. *Linear Algebra and Differential Equations*. Boston: Prindle, Webber & Schmidt, 1979.

Derrick, W. R. and S. I. Grossman. *Elementary Differential Equations with Applications*, Reading, Massachusetts: Addison-Wesley Publishing Co, 1976.

Dorf, R. C. *Matrix Algebra: A Programmed Introduction*. New York: John Wiley and Sons, 1969.

Dorn, W. S. and D. D. McCracken. *Numerical Methods with FORTRAN IV Case Studies*. New York: John Wiley & Sons, Inc., 1972.

Erugin, N. P. *Linear Systems of Ordinary Differential Equation*. New York: Academic Press, 1966.

Faddeeva, V. N. *Computational Methods of Linear Algebra*, New York: Dover Publications, Inc. 1959.

Fraleigh, J. B. and R. A. Beauregard. *Linear Algebra*. Reading, Massachusetts: Addison Wesley, 1987.

Franklin, J. N. *Matrix Theory*. Englewood Cliffs, New Jersey: Prentice Hall, Inc., 1968.

Frazer, R. A., W. J. Duncan, and A. R. Collar. *Elementary Matrices*. New York: Cambridge University Press, 1957.

Friedman, B. *Principles and Techniques of Applied Mathematics*. New York: John Wiley and Sons, 1956.

Gantmacher, F. R. *Applications of the Theory of Matrices*. New York: Interscience, 1959.

Gantmacher, F. R. *The Theory of Matrices*, two vols. Rhode Island: AMS Chelsa Publishing Company, 1959.

Gere, J. M. and W. Weaver Jr. *Matrix algebra for Engineers*, 2nd ed. Boston, Massachusetts: PWS Engineering, 1983.

Gohberg, I., P. Lancaster, and L. Rodman. *Matrix Polynomials*. Maryland Heights, Missouri, Academic Press, 1982.

Golub, G. and C. VanLoan. *Matrix Computations*, 2nd ed. Baltimore, Maryland: Johns Hopkins University Press, 1989.

Graybill, F. A. *Matrices with Applications to Statistics*, 2nd ed. Belmont, California,Wadsworth, 1983.

Hildebrand, F. B. *Methods of Applied Mathematics*. Englewood Cliffs, New Jersey: Prentice Hall, 1952.

Hill, D. R. *Experiments in Computational Matrix Algebra*. New York: Random House, 1988.

Hochstadt, H. *Differential Equations*. New York: Holt, Rinehart and Winston, 1964.

Hohn, F. E., *Elementary Matrix Algebra*. New York: The MacMillan Co., 1958.

Horn, R. A. and C. A. Johnson. *Matrix Analysis*. Cambridge: Cambridge University Press, 1985.

Hotelling, H. "Some New Methods in Matrix Calculation." *The Annals of Mathematical Statistics*, vol. XVI, no. 1.   Ohio: The Institute of Mathematical Statistics, March 1943.

Householder, A. S., *The Theory of Matrices in Numerical Analysis*. New York: Blaisdell, 1964.

Kaufmann, A. and M. Denis-Papin. *Cours de Calul Matricel Applique*. Paris, France: Albin Michel, 1957.

Kellogg, R. B. *Topics in Matrix Theory*, lecture notes, report no. 71.04. Chalmers Institute of Technology and the University of Götenberg, 1971.

Kraus, A. D. *Matrices for Engineers*, Washington: Hemisphere Publishing C., 1987.

Lancaster, P. and M. Tismenetsky. *The Theory of Matrices with Applications*, 2nd ed, Maryland Heights, Missouri: Academic Press, 1985.

MacDuffee, C. C. *The Theory of Matrices*. Rhode Island: AMS Chelsa Publishing Company, 1946.

Marcus, M. and H. Minc. *A Survey of Matrix Theory and Matrix Inequalities.* New Jersey: Allyn and Bacon, 1964.

Marcus, M. and H. Minc. *Introduction to Linear Algebra.* New York: The MacMillan Co., 1965.

Michal, A. D., *Matrix and Tensor Calculus.* New York: John Wiley and Sons, 1946.

Muir, T., *The Theory of Determinants in the Historical Order of Development.* London: 1920. Vols. 1-4 reprinted by Dover in 1960.

Nering, E., *Linear Algebra and Matrix Theory*, 2nd ed. New Jersey: John Wiley & Sons, 1963.

Noble, B. and J. W. Daniel. *Applied Linear Algebra*, 3rd ed. Englewood Cliffs, New Jersey: Prentice-Hall, 1988.

Paige, L. J. and O. Taussky, eds. *Simultaneous Linear Equations and the Determination of Eigenvalues*, National Bureau of Standards, Applied Mathematics Series 29. August 1953.

Perlis, S. *Theory of Matrices.* Reading, Massachusetts: Addison Wesley, 1952.

Pipes, L. A. *Matrix Methods for Engineers.* Englewood Cliffs, New Jersey: Prentice-Hall, 1963.

Pipes, L. A. and L. R. Harvill. *Applied Mathematics for Engineers and Physicists*, 3rd edition. New York: McGraw-Hill Book Co., 1970.

Rice, J. R. *Matrix Computations and Mathematical Software.* New York: McGraw-Hill Book Co., 1981.

Rorres, C. and H. Anton. *Applications of Linear Algebra*, 2nd ed. New York: John Wiley & Sons, Inc., 1977.

Starzak, M. E. *Mathematical Methods in Chemistry and Physics.* New York: Plenum Press, 1989.

Stewart, G. W. *Introduction to Matrix Computations.* Maryland Heights, Missouri: Academic Press, 1973.

Strang, G. *Linear Algebra and Its Applications*, 3rd ed. San Diego: Harcourt Brace Jovanovich 1988.

Turnbull, H. W., *The Theory of Determinants, Matrices and Invariants*. Scotland: Blackie, 1950.

Varga, R. S., *Matrix Iterative Analysis*. New Jersey: Prentice Hall, 1962.

Wedderburn, J. H. M. *Lectures on Matrices*. Ann Arbor, Michigan: J. W. Edwards Publisher, Inc., 1949.

Wilkinson, J. H. *The Algebraic Eigenvalue Problem*. South Cotswolds, England: Clarendon Press, 1965.

www.ingramcontent.com/pod-product-compliance
Lightning Source LLC
Chambersburg PA
CBHW031810170526
45157CB00001B/23